Transformers for Electronic Circuits

SECOND EDITION

by NATHAN R. GROSSNER

Edited by ISABEL S. GROSSNER

McGraw-Hill Book Company

New York   St. Louis   San Francisco   Auckland
Bogotá   Hamburg   Johannesburg   London   Madrid
Mexico   Montreal   New Delhi   Panama   Paris
São Paulo   Singapore   Sydney   Tokyo   Toronto

Library of Congress Cataloging in Publication Data

Grossner, Nathan R.
  Transformers for electronic circuits.

  Includes bibliographical references.
  1. Electronic transformers. I. Grossner, Isabel S. II. Title.
TK7872.T7G76 1983   621.31'4   82–13984
ISBN 0–07–024979–2

1 2 3 4 5 6 7 8 9 0 KGPKGP 8 9 8 7 6 5 4 3

ISBN 0-07-024979-2

The editors for this book were Harry Helms and Charles P. Ray;
the designer was Richard Roth, and the production supervisor
was Sara Fliess. It was set in Gael by The Kingsport Press.

Printed and bound by The Kingsport Press.

Transformers for
Electronic Circuits

*The author and editor dedicate
this book to the memory
of GRETCHEN OLDER,
their daughter*

Contents

Part 2. *THE MAGNETIC CIRCUIT*

6. *The Magnetic Circuit: Inductors without DC* 177

List of Illustrations

List of Tables

Preface to the Second Edition

About four years ago, McGraw-Hill Book Company, in the person of Tyler Hicks, invited me to write a second edition of *Transformers for Electronic Circuits.* For two important reasons, I was not thrilled by the prospect.

First, it took me five years to write the first edition, and I had to leave a position (which paid a regular salary) in order to plow my way through what at times appeared to be a labyrinth instead of a straightforward project. I am not one of those legendary authors (although Tyler is) who can write a book on evenings and weekends; I have to give it my full attention.

Second, I was not convinced that there was that much more to say about this device which I had come to love more than any other. I had already designed the high-voltage transformer used in the camera circuit of the Lunar Orbiter, the first satellite to travel to the moon, and years earlier had taken those exciting photographs down from the wall of my office. If even more exciting things had happened in between, at least theoretically more exciting things, the prospect of plowing through the same ground to find out about them did not seem particularly appealing. Certainly, in the years that intervened between the first edition and my publisher's new invitation, I was far from out of touch with my field.

In those years, I had arrived in Chicago, at Tempel Steel Company, where my prime objective was to help customers select optimum core materials and geometry for their transformers and motors. During the many meetings at this or that locale, I was reinforced in my opinion (expressed at length in the first edition) that circuit and components designers need a common vocabulary.

And then—especially as I began to review the recent literature in my own field—the questions that were asked of me during these meetings with specialists from other companies seemed to acquire a more provocative quality, a more challenging aspect. The engineers were inquiring not only into the state of the art, with which I presumed familiarity, but into the limit of the art, with which I was, because of my publisher's invitation, again familiarizing myself. And although I had written several papers during those years, I found myself listening to my peers' presentations at meetings with closer interest and finding new, challenging questions of my own.

It takes no aficionado of mystery stories to find out what happened to me. My own subject regained its allure and, during the final year of writing this second edition, I again left a salaried position to devote my full time to the work which follows this Preface.

Despite ingenious attempts to displace the transformer, it is indeed a classic component of electronic circuits. Furthermore, its use and its versatility have grown. I have, therefore, expanded or placed new emphasis on the following topics:

Type of transformer	Application	Treated in Chap.
Ignition coil	Automobile ignition	2
Ferroresonant transformer	Microwave oven, battery charger, electrostatic precitator, and computer power supplies	5
Converter transformer	Power electronics	5
Energy-storage inductor	Switch-mode power supplies	7
Hybrid transformer	Data modem, RF amplifiers	8
Transmission-line transformer	RF, video, and pulse circuits	9
Horizontal flyback transformer	TV and video display terminal	10

I have also amplified my treatment of the topics of dielectrics, geometry, and optimization. Computerized design, treated only glancingly in the first edition, has merited more extended treatment. Computerized graphics, now part of the design engineer's arsenal of techniques, is introduced.

Many individuals have made this book possible; some have already been referred to.   The professional dedication of members of the Electronic Transformers Committee of the IEEE Magnetic Society has helped to sustain my own interest in and attachment to the field of applied magnetics.   I have also worked with, corresponded with, met with, and exchanged ideas and information with a large number of colleagues, and there are others whom I have not been privileged to meet but whose work I have read: to all of these, I am grateful.

My editor and wife, Isabel S. Grossner, joins me on the title page of this book.   Her personal encouragement has helped me to complete this second edition, as it did the first, and her professional skill has greatly enhanced the prospect that this book will prove to you, the reader, a clear exposition of a complex subject.

Nathan R. Grossner

Preface to the First Edition

This book has been written by a transformer design engineer for the users of electronic transformers—the circuit, systems, and standards engineers, as well as the many other working engineers in the electronics industry.

Although transformers are to be found in virtually all electronic circuits, most practicing engineers no longer design and build their own transformers, as did the engineers of an earlier generation. This book, therefore, is not a design manual. The discussion of design considerations which it contains are intended to enable the user to follow the logic and understand the methods of the transformer specialist. I hope thereby to have made it easier for him to use the electronic transformer to its best advantage.

Each chapter stresses topics and concepts of major importance in modern electronic circuits. These include:

1. The basic, as well as the newer, functions of the transformer
2. The iron-copper area product; the d-c inverter transformer
3. The vulnerable areas of the transformer
4. Techniques of thermal design
5. Miniaturization
6. Complex permeability; the μQf merit factor; bounds on harmonic distortion
7. Gap design for the polarized inductor
8. The band-span parameters
9. Concepts from network synthesis employed for optimum wide-band performance

10. Relations between rise and fall times of rectangular pulses; bandwidth and the pulse-squareness ratio

11. Concepts from network synthesis employed for optimum pulse performance

Certain types of nonlinear transformers are not treated in the book. These include the constant-voltage transformer, the magnetic amplifier, logic transformers, and other magnetic components—topics originally planned for a final chapter. I now believe that these devices are best understood in the larger context of nonlinear static-magnetic devices. However, I have, in the first chapter, attempted to discuss the relationship of the quasi-linear transformer (the subject of this book) to the nonlinear and multiple flux-path devices. Certain specialized considerations have, in addition, had to be omitted because of the limitations of space. These include shock, corona, and thermal aging; transient loading and heating; and precision transformers. I have written elsewhere on these topics, and these articles are referenced in the text for the convenience of the interested reader.

I have been immersed in a sea of details, hardware, and recipes for half my adult years, because these are necessary to the daily practice of design. It has only gradually become clear to me that no person's design, or program for a design, can be better than his or her comprehension of basic principles. I hope that this point of view is reflected in my book and that the simplicity for which I have striven will be found by my readers to be illuminating rather than naïve. To this end, I have omitted large quantities of data (they are available elsewhere, in any event) and have tried to focus on fundamental relationships and design considerations. My feeling for the unity of the subject has led me to stress the analogies among the electric, thermal, and magnetic fields of the transformer. I have also tried to avoid complicated equations, or to cast them in a form which enables the basic relationships to emerge. I have placed the brief discussion of reliability early in the book so that subsequent design concepts can be formulated within a pragmatic context. When we close a switch and excite a transformer circuit, we want it to work and to continue to work.

Many individuals have made this book possible. These include Albert Cezar and David Wildfeuer, who taught me the art of transformer design. I have also benefited from the considerable experience of Walter H. Winchell and Mynor Payne. I am particularly grateful to Professor Lawrence Arguimbau and to Herbert Sullivan, who encouraged me to think about concepts rather than formulas. I have an obvious debt to the creative investigators in the field (the real experts), many of whom are mentioned throughout the book. The help of E.

Stuart Eichart was twofold.   His company, Technitrol, Inc., of Philadelphia, helped me to survive financially during the early phases of
the writing of this book; in addition, I have benefited from his still
unpublished monograph on pulse transformers.

I am fortunate that my wife, Isabel Shoket Grossner, is an editor.
I therefore owe her more than the usual tribute to wifely patience
and encouragement.   As the editor of this book, she has tried to help
me avoid both jargon and muddiness.   The lacks and lapses which
remain are my own doing.

Nathan R. Grossner

CHAPTER 1

A Survey

The subject of our discussion made its first public appearance in 1831 when Michael Faraday demonstrated electromagnetic induction, and with it was born the whole world of electric technology in which we now live. Described as the induction coil in Faraday's time, and still called that in the telephone industry, the device to which we are about to turn our attention is today known as the *static transformer*. Although some of these transformers have movable parts, they are static in the sense that they do not include any parts that are continuously in motion. The transforming characteristic, originally applied to the transformation of magnetic energy into electric energy, is now defined as the alteration of voltage, current, and impedance from one level to another.

The transformer, or induction coil, stirred the imagination of mathematicians, physicists, inventors, and the first electrical engineers. Lord Rayleigh studied the mathematics of hysteresis; Sir John Fleming wrote a book on the transformer in 1890, the first on the subject in the English language; Charles Steinmetz investigated the relation between hysteresis and flux density; Nikola Tesla built large coils and dreamed of vast systems of wireless power transmission. Later and more glamorous components, such as the vacuum tube, the transistor, and the microprocessor, have caught the imagination of engineers and

1

have been supplanted by electronic systems, the names of which become increasingly more exciting. Even though the modern transformer and such other basic components as the rotating generator and the motor have long since lost their fascination, our entire electric technology is based on the successful exploitation of these three devices. The fact is that no space vehicle can yet be designed without transformers nor, for that matter, can a video recorder, a computer, or even a telephone.

As electric circuits and electronic systems have become more complex, the design of the transformer has necessarily become a more difficult and subtle art. The modern transformer designer must possess a wealth of information and mathematical technique. A corollary development is that the systems and circuit engineers who requisition a practical transformer need to know more about the state and the limit of the art of transformer design than they might have deemed necessary.

And yet, the most modern transformer is still basically the device used by Faraday to demonstrate electromagnetic induction. Its primary function remains the transformation of voltage and current from one level to another. This transforming will be a recurrent theme, and it is well to review briefly how it comes about.[1]

1.1. THE TRANSFORMER'S BEHAVIOR

The transformer is perhaps the most basic application of the principle of induction. Essentially, it consists basically of two copper coils (Fig. 1.1a), physically placed so that a magnetic field in one links the other, either through air or through a ferromagnetic core. When one coil, the primary, is connected to a generator source, and the other coil, the secondary, is connected to a load (Fig. 1.1b), a changing current in the primary coil produces a changing magnetic field common to both coils and results in the *transfer of electric energy* from the input circuit to the output.

When the secondary coil does not exist or remains unconnected, the device is called an *inductor*. The inductor is physically similar to the transformer and is governed by the same principles of induction; moreover, the inductance and quality factor (Q) of a winding play an important role in transformer behavior.

Faraday's law states that the voltage E induced in a coil is equal to the change of magnetic flux linkages $N\phi$ with respect to time. Mathematically:

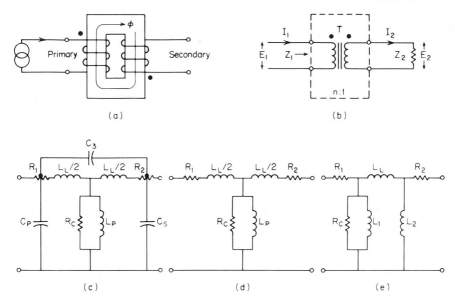

Fig. 1.1   Transformer diagrams. (*a*) Physical schematic. Input current produces core flux ϕ. (*b*) Elementary black-box circuit. (*c*) A complete T equivalent circuit (unity turns ratio, $n = 1$). (*d*) A simplified T circuit neglecting capacitances. (*e*) Alternative π circuit, sometimes preferred when coupling k is low. R_1 = primary resistance; R_2 = secondary resistance; R_c = shunt core-loss resistance; L_1 = primary inductance; $L_P = kL_1$ = primary shunt (mutual) inductance; k = coupling coefficient; L_L = leakage inductance; C_P = primary distributed capacitance; C_D = secondary distributed capacitance; C_3 = direct input-to-output leakage capacitance.

$$E = -\frac{d(N\phi)}{dt} \, 10^{-8} \tag{1.1}$$

where E is in volts, N is the number of turns, and ϕ is the flux in maxwells or lines.*

The voltages induced in the primary and secondary coils (E_1 and E_2 in Fig. 1.1*b*) are proportional to the number of turns (N_1, N_2) in that coil and the turns ratio n:

$$\frac{E_1}{E_2} = \frac{N_1}{N_2} = n \tag{1.2}$$

---

* F. E. Neumann (1845) is credited with the mathematical formulation of Faraday's experiments. The minus sign expresses the observation by Lenz that the direction of the induced voltage (*electromotive force*, emf) is such as to oppose, through its magnetic effects, the change producing the voltage. The factor 10^{-8} results from the *electromagnetic unit* (emu) system of definitions in which 1 abvolt = 10^{-8} volts (V). Later, the *meter-kilogram-second* (mks) system defined the *weber* [weber (Wb) = 10^8 maxwells] as the unit of flux, thus eliminating the factor 10^{-8}.

Current is furnished by the generator to the primary coil as current flows through the secondary coil and into the load. In the *ideal transformer,* where there is perfect magnetic coupling and no loss of energy in the form of heat, the output power equals the input power:

$$E_1 I_1 = E_2 I_2$$

To maintain this relationship, the current ratio $I_2/I_1 = E_1/E_2$ must be inversely proportional to the turns ratio:

$$\frac{I_2}{I_1} = \frac{N_1}{N_2} = n \tag{1.3}$$

Equations (1.2) and (1.3) state that voltage and current are transformed simply by altering the turns ratio n. To obtain a relationship in terms of impedance, we multiply Eqs. (1.2) and (1.3):

$$\frac{E_1/I_1}{E_2/I_2} = \frac{Z_1}{Z_2} = n^2 \tag{1.4}$$

Equation (1.4) states that the primary and secondary impedances Z_1 and Z_2 may also be transformed or matched by altering the turns ratio.

Since the number of turns on both coils may be arbitrarily (and easily) altered, it follows that the number of possible turns ratios is truly astronomical. Thus, the number of possible alterations of voltage, current, and impedance that the transformer can produce is virtually unlimited.

With the advent of alternating current late in the nineteenth century, the profound practical importance of the simple equations which summarize the properties of the transformer became evident. The induction coil was put to use even before alternating current (ac) became available. With a high step-up turns ratio, sparks can be produced at the secondary terminals by interrupting direct current (dc) in the primary. Thus used to obtain electric discharges through gases and to produce electric oscillations, the transformer facilitated the momentous discoveries of Hertzian radio waves, cathode rays, and Roentgen X-rays.

Since current is stepped down when voltage is stepped up, losses of power in copper ($I^2 R$) can be drastically reduced in the transmission lines which furnish electric power between remote locations. This simple observation led to the development of the practical transformer and thus made possible the worldwide use of ac power transmission. By the opening of the twentieth century, the importance of the transformer to the electric utility industry was widely recognized.

Some transformers made then have been in continuous use since their construction, and working transformers made 30 or more years ago are not uncommon. The longevity of the transformer is undoubtedly due to the simplicity of its basic ingredients—copper, iron, and insulation—and also to the fact that it has no continuously moving parts.

1.2. TRANSFORMERS IN THE ELECTRONICS INDUSTRY

The fame of the rectifier circuit developed by Sir John Fleming in 1904 to detect wireless telegraphy signals rests on the introduction of the Fleming valve, later called a vacuum diode. This was followed in 1906 by the triode, in 1928 by the thyratron, in 1948 by the transistor, and since by the newer and still proliferating family of semiconductors. Transformers are now found in circuits built with electron devices such as the Zener diode, varactor diode, *f*ield-*e*ffect *t*ransistor (FET), *s*ilicon *c*ontrolled *r*ectifier (SCR) or thyristor, *tri*ode *ac* switch (triac), and various light-sensitive devices such as the photovoltaic (solar) cell, photodiode, phototransistor, and *l*ight-*e*mitting *d*iode (LED). Control circuits may contain nonlinear components such as the thermistor, varistor, magnetoresistance devices,[2] and Hall effect devices.[3]

The electronics industry has evolved from the successful application of these electron devices in circuits which permit the basic functions of rectification, amplification, oscillation, modulation, and demodulation; thus the term *electronics* has come to stand for the aggregate of the equipment and systems which exploit such devices. The transformer, which was included in Fleming's circuit, remains an indispensable part of the newest electronic circuits. Used in these circuits, it is called the *electronic transformer* and has features which distinguish it from those used in the power utility field.

As a basic component of electronic circuits, the transformer entered the commercial and government markets, whereas the earlier practical transformers, which furnished electric power and aided in telephonic and telegraphic communication, went principally to the industrial market area. Quite apart from technical considerations, each of these market areas selectively stresses *cost, value, reliability,* or *minimum size,* and these requirements have profound repercussions on the design and development of electronic transformers.

The *commercial* sector manufactures expendable consumer products such as radio and television receivers, which are produced in quantity under highly competitive conditions. Thus, the cost of components is a primary consideration. Transformers for this market are engineered for the lowest possible unit price, and increasingly automated mass-production techniques are used.

The *industrial* sector uses electronic equipment in the control of production machinery and the processing of materials. The failure of a transformer means downtime, and the desire for low cost is tempered by the need for reliability. The compromise which results is termed *value,* and it is value—a weighted combination of cost and quality—which influences the specifications for, and determines the design of, components for the industrial market.

The *government* requires compliance with stringent technical specifications which assume the use of high-quality materials and techniques. Failure of a component may mean failure of a military mission or even loss of human life, and so *reliability* is central to a set of military specifications. Because even the highest standards of reliability do not pose a formidable problem in transformer design, requirements for underwater, naval, and ground-based equipment are met with comparative ease. A more difficult requirement has arisen with the development of mobile, airborne, rocket, missile, satellite, and space-vehicle systems, however. The transformer tends to be large and heavy compared with other basic components such as resistors and capacitors, and as a result *miniaturization* (the progressive reduction of size and weight) has also become a dominating motif in the design and construction of transformers for this segment, at least, of the government market.

1.3. QUASI-LINEAR MAGNETICS

The practice of labeling a transformer with the name of its circuit application has created the unfortunate impression that there is a vast variety of transformers. Actually, by making use of a distinction of proven value in circuit theory, we can classify all transformers into two major groups—linear and nonlinear—depending on the application.[4]

Such a distinction, however, obscures the fact that in a fundamental sense *ferromagnetic transformers are all nonlinear.* It has proved convenient, because of tradition and the mathematical simplicity of linear circuit analysis, to treat the iron-core transformer as if it were a linear circuit element, but in actuality only skillful design of the transformer allows this assumption to be made. Over a limited range of voltage, frequency, and temperature, Eqs. (1.2) to (1.4) can be made to hold at least approximately, and it is a mark of good design when the second-order effects, e.g., moderate distortion, produced by the inherent nonlinearities do not disturb a circuit function.

The assumption of a linear relationship between flux ϕ and the magnetizing current i_m which produces it greatly simplifies the circuit analysis of transformers. ϕ and i_m are traditionally expressed in terms of the more general parameters of flux density B and magnetizing force H:

$$B = \phi A \qquad (1.5)$$

$$H = \frac{4\pi}{10} \frac{Ni_m}{l} \qquad (1.6)$$

Flux density is expressed in gauss (G) if ϕ is in lines or maxwells; A is cross-sectional area in square centimeters (sq cm). Magnetizing force is expressed in oersteds (Oe) when i_m is in amperes (A) and when l, the length of the magnetic path, is in centimeters (cm). The relationship between B and H is traditionally given as:

$$\frac{B}{H} = \mu \qquad (1.7)$$

which defines permeability μ as the degree to which the magnetizing force makes the medium more permeable by flux. If the medium is air or a hypothetical linear core material, then μ is a constant and the B-H relationship is graphically expressed by a straight line (as in Fig. 1.2a). This linearity was assumed in deriving Eqs. (1.2) to (1.4).

In actual ferromagnetic core materials, B and H are not linearly related. The graphical plot (Fig. 1.2b) exhibits, in addition to saturation, a hysteresis characteristic which results in a family of concentric loops. At any given value of H, the value of B depends on the prior history of the core material, that is, on the excitations which have preceded the one under consideration. In addition, any variation in ambient temperature or frequency of excitation will affect the area and slope of the entire family of curves. The complicated relationship between B and H requires that a combination of graphical and other mathematical techniques be used to predict the exact behavior.

If the B-H relationship is simplified into a linear one, however, the transformer can be treated as an LCR network susceptible to the mathematics of linear circuit analysis.* This can be accomplished by several methods. When eddy currents are large, the hysteresis loop may be regarded as a tilted ellipse, the slope of whose axis is approximately equal to the average value of permeability (Fig. 1.2c). Another approach, more arbitrary, is to mentally squeeze the loop into a line

---

* Throughout the book, *LC*, *LR*, and *LCR* are used to denote combinations of the inductor L, capacitor C, and resistor R.

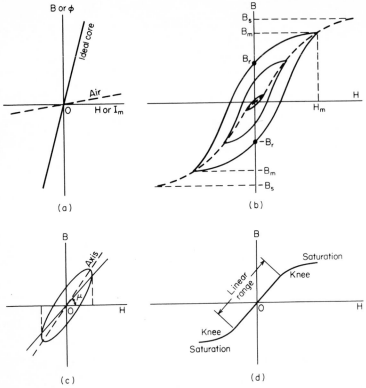

Fig. 1.2 *B-H* curves. (*a*) Ideal linear core with high constant slope and no memory B_r. (*b*) A family of hysteresis curves for soft ferromagnetic core material. B_m is maximum flux density, B_s is saturation flux density. (*c*) Ellipse approximation of a hysteresis curve, when eddy-current losses dominate. (*d*) A single-valued curve with nonlinear knee and saturation regions. Hysteresis (see text) is neglected.

which is approximately straight over a large range (Fig. 1.2*d*) and curved only at a positive and negative knee where saturation begins. If, by constraining the maximum voltage, we bound the range of flux density so that it does not enter the region where saturation occurs, Eqs. (1.2) to (1.4) become valid for a large range of applications, and with a degree of accuracy acceptable for most engineering purposes.

For the universally popular periodic sine wave, the relationship between maximum flux density B_m and *root-mean-square* (rms) voltage E_{rms} is:

$$E_{\text{rms}} = \frac{2\pi}{\sqrt{2}} f N B_m A \, 10^{-8} = 4.44 f N B_m A \, 10^{-8} \qquad (1.8)$$

This equation, derived directly from Faraday's law shown in Eq. (1.1), is of fundamental importance in sine-wave transformers because it establishes the basic interdependency among the five principal parameters. If frequency f, turns, and core area are fixed, and if flux density is not to exceed B_m, voltage must not exceed E_{rms} as determined from Eq. (1.8).

Heat losses in the core material are attributed principally to hysteresis and to eddy currents. They also are a nonlinear function of: (1) the area of the hysteresis loop; (2) the frequency, or pulse repetition rate; (3) the lamination or particle thickness and resistivity; and (4) the temperature. Nevertheless, it is possible, with the aid of empirical data, to reduce such losses to a quasi-linear shunt resistance parameter R_c (Fig. 1.1c) valid for a narrow range of frequency, voltage, and temperature.

The assumption of a linear B-H relationship in the region which precedes core saturation preserves the concept of linear inductance. The definition of inductance bears this out:

$$L_P = \frac{d(N\phi)}{di_m} 10^{-8} \tag{1.9}$$

If ϕ is a linear function of i_m, then L_P equals a constant times μN^2, since $\phi = \mu H A$ and μ is constant.[1]

Thus, when definite constraints are placed on the range of voltage, frequency, and temperature, it becomes possible to regard the transformer and inductor as *quasi-linear*, or approximately linear, components.

1.4. LINEAR APPLICATIONS

So-called linear transformers may be classified in terms of three major characteristics which permit useful circuit functions: (1) the *transformation* of voltage, current, or impedance; (2) behavior as an *LCR network;* and (3) the versatility of *winding connections.* Notably important examples of strictly linear behavior are the *potential*[5] and *current*[6] transformers.

The nonideal linear (i.e., quasi-linear) transformer exhibits the basic characteristics of a network that has discrete values of inductance, capacitance, and resistance and is often used as an *LCR* network. Examples of the constructive use of winding connections are described throughout this book; one which is of particular interest occurs in the hybrid circuit, which is used to combine, split, and isolate signals at different ports of the transformer.

The objective of extreme miniaturization of electronic circuits has resulted in the *i*ntegrated *c*ircuit (IC) and in the *l*arge-*s*cale *i*ntegration (LSI) of an aggregate of complex circuits. The *microprocessor,* a microcomputer the size of a thick postage stamp, is the crowning achievement of LSI. The *mi*cro*p*rocessor *u*nit (MPU) combines the essential functions of the computer: a *p*rogrammable *l*ogic *a*rray (PLA), *r*andom *a*ccess *m*emory (RAM), *r*ead-*o*nly *m*emory (ROM), and a *c*entral *p*rocessing *u*nit (CPU).

A transformer the size of a 1-centimeter (cm) cube looks inordinately large when mounted on a circuit board containing such integrated circuits, and much design talent is devoted to dispensing with the magnetic-core component. Low-power communications systems are traditionally loaded with discrete LC components, but conversion from analog to digital circuits has, in many instances, eliminated the need for bulky inductors. With the advent of thin-film and thick-film technology, the active filter can be substituted for the discrete inductor and capacitor. In still other networks, the operational amplifier, the gyrator, the *n*egative *i*mpedance *c*onverter (NIC), and the mutator[7] are utilized to synthesize the solid-state inductor. Even the isolation property of the transformer can be achieved (albeit with very poor efficiency) with the *o*ptically *c*oupled *i*solator (OCI) and other static devices such as the piezoelectric (ceramic) transformer and the thermomagnetic transformer.

Notwithstanding, the magnetic-core transformer is alive, well, and fertile, judging from its ubiquity in electronic circuits.

1.5. NONLINEAR MAGNETICS

Occasional anomalies in circuit behavior serve as rude reminders that magnetic linearity is a limited as well as arbitrary assumption. Fuses and transistors burn out when the rated bounds of voltage, frequency, or pulse width are exceeded. The resonant frequency of an LC circuit which includes a transformer primary exhibits variations attributable to the nonlinearity of L_P (thus providing a minor illustration of the important phenomenon known as *ferroresonance*). In addition, transformer ratios do not vary linearly with voltage amplitude except within a narrow range and do not remain constant when frequency or temperature changes.

Early in the genesis of audio circuits, telephone engineers became aware of the harmonics and the cross-modulation distortion caused by magnetic cores in audio transformers. They recognized that μ varied not only at the knee of the $B\text{-}H$ curve but also at low flux densities. The idealized Fig. 1.2*d*, which shows a linear range of μ, does not provide a picture sufficiently accurate for estimating harmonic distortion. When a minimum of nonlinear behavior is called for, sophisticated circuit compensation techniques and fastidious design procedures must be employed.

Even before the development of an electronics industry, creative engineers were able to make a virtue of the "evil" of nonlinearity and exploit the nonlinearity of the magnetic medium. The magnetic amplifier, introduced into American radio broadcasting in 1916 by E. F. W. Alexanderson, uses the saturable reactor (or saturable transformer) in a circuit which amplifies power. This was followed by an increasing number of patent applications for devices exploiting nonlinear core behavior. In the mid-1940s, when rectangular-loop magnetic cores became commercially available, applications of nonlinear circuits increased dramatically.

The rectangular or square loop ($B\text{-}H$ characteristic) shown in Fig. 1.3a is only approximately rectangular, and its squareness is defined as the quotient of residual flux density B_r and saturation flux density B_s. In soft (easily magnetized) core materials, a value of B_r/B_s close to 1 indicates high retentivity and an abrupt transition to staturation.

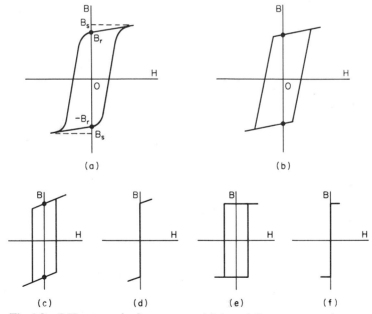

Fig. 1.3 $B\text{-}H$ rectangular-loop curves. (a) Actual characteristic. Approximations: (b) tilted straight lines; (c) perpendicular sides with hysteresis, $B_r \neq B_s$; (d) perpendicular sides, no hysteresis, $B_r \neq B_s$; (e) perfect rectangle, $B_r = B_s$; (f) ideal characteristic with no hysteresis, $B_r = B_s$.

Retentivity B_r ignored in Fig. 1.2d, serves in switching and logic circuits to produce the *memory* feature.   For this reason, simplifications of the hysteresis loop (and thus, of its mathematical analysis) preserve the B_r characteristic shown in Fig. 1.3d and f.

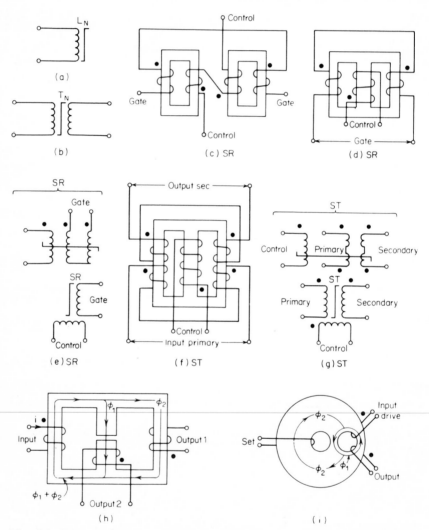

Fig. 1.4 Basic nonlinear magnetic components. (*a*) Nonlinear inductor L_N. (*b*) Nonlinear saturating transformer T_N. (*c*) Two-core saturable reactor *SR*. (*d*) Single-core *SR*. (*e*) Alternative schematics of the *SR* of (*c*) or (*d*). (*f*) Single-core saturable transformer *ST*. (*g*) Alternative schematics of the *ST* of (*f*). (*h*) Multiple flux-path magnetic network: path ϕ_1 may have different core material and μ than path ϕ_2. (*i*) Multiaperture device, MAD, or transfluxor: additional holes may be added for other local flux paths.

In general, nonlinear magnetic circuits combine building blocks to accomplish a novel function[8], i.e., a function other than the transformation of voltage, current, and impedance levels. Novel functions include (1) *stabilization,* (2) *wave-shape conversion,* (3) *modulation,* (4) *frequency conversion* (multiplication or division), (5) *detection* of voltage or current thresholds, (6) *amplification,* (7) *switching,* (8) *logic,* and (9) *digital storage.*

The building blocks may be linear or nonlinear. Linear building blocks include the capacitor (C), the linear inductor (L), and the linear transformer (T). *Nonlinear noninductive* building blocks include the nonlinear varistor (R_N) and semiconductors such as the Zener, tunnel, and varactor diodes. The *nonlinear inductive* building blocks include the nonlinear inductor (L_N), the cross-field inductor,[9,10] the nonlinear transformer (T_N), the saturable reactor (SR), the saturable transformer (ST), and the multiaperture core device (MAD). The various nonlinear inductive building blocks are shown schematically in Fig. 1.4.

Some nonlinear building blocks, components, and devices were known to the engineering world before World War II. In this group are the SR, ST, magnetic amplifier (MA), the flux valve (or flux gate) compass, and the two-path flux network.[2] The last named (Fig. 1.4h) is the basis for the voltage-stabilizer transformer and the peaking transformer. The magnetic modulator and the static magnetic switch (static relay), among others, came into use later. An interesting application of nonlinear network theory[11,12] is the *rotator,* a circuit which rotates the *B-H* saturation curve counterclockwise to produce the more desirable steep slope of Fig. 1.3c.

By the creative combination of building blocks into more complicated circuits, engineers have evolved such devices as the frequency multiplier, which uses saturable transformers, and the magnetic pulse generator. The dc inverter transformer (a successor to the vibrator transformer) converts dc into square-wave ac in an oscillator circuit which uses transistors and a readily saturable core material. In the light of new memory and logic applications, the multiaperture core device, such as the transfluxor (Fig. 1.4i), is of particular interest. A special case of the multipath[1] magnetic network [which may have both saturable and nonsaturable flux paths (Fig. 1.4h)] is important because of its nondestructible memory. More complicated logic devices include the magnetic shift register and the magnetic bubble memory.* The field of nonlinear magnetics continues to grow. The

---

* Magnetic bubble memory can be fabricated on the scale of a chip. In such a device, microscopic bubbles, which are magnetic domains in a thin magnetic film, are sandwiched between two permanent-bias magnets.

magnetic amplifier[13,14] is now a subject in itself, and the application of digital magnetic devices is another growing area.[15,16]

In general, although the principal function of nonlinear inductive circuits is not transformation, *basic transformer behavior, magnetic induction, is evident during part of the periodic cycle to which the circuit is subjected.* Too, the physical structure and thermal characteristics of nonlinear magnetic components are the same as those encountered in conventional transformer design and construction. For these reasons the practical as well as analytic techniques of the mature transformer field are often enlisted for the design of nonlinear magnetic circuits.

1.6. THE ART OF TRANSFORMER DESIGN

In the United States, at least, the circuit engineer who draws up the specifications document is not expected to design the transformer. The component is more efficiently and economically produced when the problems of design and manufacture are referred to a reliable transformer vendor. This physical separation between the two members of the original engineering pair, the circuit and component engineers, occurred early in the evolution of the engineering team. However, with the growth of the field of power electronics, which relies heavily on magnetic circuits, the two specialties have tended to merge and the circuit engineer has become increasingly involved with the practical aspects of transformer design.

a. Concepts Used in Analysis and Synthesis

In the design of a transformer we draw on concepts from three broad subject areas: geometry, physics, and design procedures. Major topics in these subjects can be grouped as follows:

Geometry	Physics	Design Procedures
Field theory	Dielectrics	Optimization
Topology	Heat	Similitude
Linear network theory	Magnetics	Computer-aided design
Nonlinear network theory		Reliability

The complexity of the transformer engineer's problem is only partially indicated by the multiplicity of materials whose mechanical, electrical, magnetic, dielectric, thermal, and environmental characteristics must be known; this becomes clearer when the specifications include successful performance in a complex circuit. Whatever the difficulties presented by the design problem at hand, however, they are solved by an amalgamation of circuit analysis, intuition, and trial and error. The component engineer would like to synthesize the final design by the direct application of general principles to the specifications at hand, but in practice the design procedure is often influenced by previously proven designs. Thus, the design process may involve some "cut and try" in which repetitive calculations are carried out by means of a computerized program. Field mapping, using the *finite element model,* is sometimes helpful.[17,18] It is not inappropriate to view the total process as art as well as science; indeed, what transformer engineers confidently regard as within the range of their special skills, they describe as "the state of the art."

b. Optimization

The theme of optimization[19] runs through this book. Optimization is a process. It begins with a set of specifications, that is, specific objectives and one or more constraints on the transformer's performance. It ends with a design which, measured by explicit criteria, is deemed optimum. The objectives—principally *small size, low weight,* a *high degree of efficiency, low cost,* and *reliability*—are discussed at length at various points in this book, as are the tradeoffs that become necessary when these objectives conflict, a not infrequent situation.

The analysis of the transformer's geometry is an important branch of optimization. It is interesting to confirm our intuitive observation that, in the design of power transformers, not only does the optimum geometry change with the objective; but it also fluctuates with the relative cost of the iron, the copper, and the labor. The repercussions of this observation can be disconcerting: in principle, if the relative cost of copper and iron changes, then the geometry should also change. When low cost is the overriding objective, the optimum design is determined by the marketplace.

The synthesis of a design is considerably simplified by the use of the computer. One obvious important consequence of computer-aided design is that an "optimum" solution can be arrived at within a reasonable length of time; other advantages to be gained by using the computer are less obvious.

An interesting use of the computer has been developed within the framework of the design of the switching power supply (the power train), the governing objective being to achieve maximum efficiency.[20]

The variables are described in terms of a "multidimensional design-parameter space," and the behavior of one set of parameters is depicted in terms of the others by computerized graphics. The resulting three-dimensional display or contour plot is a visual gestalt that reveals the relationships more graphically than a set of equations.

As an illustration, consider the relation $z = F(x,y)$, which represents, in a converter transformer, the following variables: z = temperature rise ΔT; x = the volume v of the core (the product of A and l); and y = the switching frequency f_s. But z is a function of many variables, including the permeability of the core and the turns of the winding. A plot such as Fig. 1.5, which depicts the manner in which temperature rise varies with the volume of the core and with the frequency, is very helpful in the assessment of a series of designs. Computerized graphics thus enable us literally to see the degree to which one set of variables affects another. The changes in the contours (i.e., the partial derivatives or gradients) in different regions of the display show us the salient features of the design at a glance.

1.7. ORGANIZATION OF THE BOOK

The structure of this book reflects the author's view that, in the basic sense, the transformer circuit serves either to transmit power (Part 1) or to transmit information (Part 3). Because of the compelling importance of the transformer's reliability, dielectrics and heat are

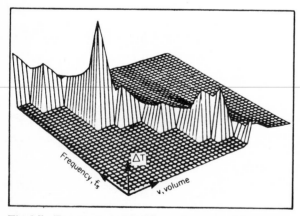

Fig. 1.5   Temperature rise ΔT vs. volume v and frequency f_s. *(From T. Wilson, Jr., et al., "DC-to-DC Converter Power Train Optimization for Maximum Efficiency," IEEE Power Electronics Specialists Conference Record 1981, Vol. 81-CH 1,552-7. By permission, Institute of Electrical and Electronics Engineers.)*

treated early in Part 1. The subject of the magnetic circuit and inductors (Part 2) has been placed midway in the book in order to stress its relevance to both power transformers and information-transmission transformers.

At the close of each chapter of the book there is a list of references. These include, of course, the customary citations of quoted information and equations, but they are intended to be more than that. Together they constitute a selected bibliography of sources of reasoning and concepts as well as reports of research and findings pertinent to the topics discussed in the book.

REFERENCES

1. G. Slemon, *Magnetoelectric Devices* (New York: John Wiley & Sons, 1966), Chaps. 1–3.
2. B. Jones, "Magnetoresistance and Its Applications," *Wireless World*, Vol. 76 (January 1970): 17–19.
3. C. Bajorek, D. Thompson, J. Cocke, "Active Transformers for Off-Hook Sensing in Telephone Applications," *IEEE Transactions on Magnetics*, Vol. MAG-14, No. 5 (September 1978): 1,062–64.
4. J. Watson, *Applications of Magnetism* (New York: John Wiley & Sons, 1980).
5. B. Jenkins, *Introduction to Instrument Transformers* (London: G. Newnes, 1967).
6. A. Wright, *Current Transformers* (New York: Halsted; London: Chapman & Hall, 1968).
7. T. Murata, R. Rikoski, "Mutator Simulated Floating Inductors," *International Journal of Electronics*, Vol. 39, No. 2 (1975): 229–32.
8. H. Storm, "Applications of Nonlinear Magnetics," *Transactions AIEE*, Vol. 77, Part I (July 1958): 380–88.
9. R. Heartz, H. Buelteman, "The Application of Perpendicular Superposed Magnetic Fields," *Transactions AIEE*, Vol. 74 (November 1955): 655–60.
10. T. Gross, "Revisiting the Cross-Field Inductor," *Electronic Design* (March 15, 1977): 82–85.
11. L. Chua, *Introduction to Nonlinear Network Theory* (New York: McGraw-Hill Book Company, 1969).
12. ———, "A Good Turn for Old Components," *Electronics* (May 29, 1967): 109–22.
13. H. Storm, *Magnetic Amplifiers* (New York: John Wiley & Sons, 1955).
14. A. Milnes, *Transductors and Magnetic Amplifiers* (London: Macmillan & Company, 1957).
15. H. Katz (ed.), *Solid State Magnetic and Dielectric Devices* (New York: John Wiley & Sons, 1959).
16. A. Meyerhoff (ed.), *Digital Applications of Magnetic Devices* (New York: John Wiley & Sons, 1960).

17. P. Moon, D. Spencer, *Field Theory for Engineers* (Princeton, N.J.: D. Van Nostrand & Company, 1961).
18. J. Dishman, D. Kressler, R. Rodriguez, "Characterization, Modeling and Design of Swinging Inductors for Power Conversion Applications," *Proceedings Powercon 8* (Dallas, 1981).
19. W. Cunningham, "Optimization Theory," Chap. 2 in Bell Telephone Laboratories, *Physical Design of Electronic Systems,* Vol. 4 (Englewood Cliffs, N.J.: Prentice-Hall, 1972).
20. T. Wilson, Jr., et al., "DC-to-DC Converter Power Train Optimization for Maximum Efficiency," *IEEE Power Electronics Specialists Conference Record 1981,* Vol. 81 CH1652–7.

PART 1

The Transmission of Power

CHAPTER **2**

The Power Transformer: Analysis

All transformers are power transformers when a load connected to the secondary winding consumes power, no matter how little. Traditionally, the term *power transformer* has been reserved for the transformer which transfers the components of power—voltage and current—at a single frequency and in amounts sufficient to cause the transformer to heat up.

The power transformer performs a basic task, and its properties are theoretically so simple that they are summarized in Eqs. (1.2) to (1.4). Nevertheless, as the user well knows, the power transformer often materializes as a black box which is too large, too heavy, and too hot.

It is perhaps just because this black box has so simple a function that the user may fail to realize the extent to which the circuit design dictates that it be large, heavy, and hot. In this context, an understanding of the factors affecting the size and heat of the power transformer can be of value to both the designer and the user of the transformer.

2.1. SCOPE

The number of variables which affect the performance, rating, size, and thermal characteristics of the power transformer is surprisingly large, as we shall see in the course of our four-part discussion. In this chapter we focus *qualitatively* on the geometry of the transformer and its function within the circuit. Considerations of leakage inductance, the magnetic characteristics of core materials, and heat losses have been postponed lest they obscure the outlines of our preliminary view. Many important decisions about size can be made without dwelling on the problems of heat and dielectrics, which are deferred to Chaps. 3 and 4. In Chap. 5 we approach our subject more totally, more closely, and more quantitatively, taking into account the complex of variables which determines the size of both the state-of-the-art and the limit-of-the-art power transformer. In Chap. 5 our first view is shown to be only approximate, but its validity and usefulness are confirmed.

2.2. GEOMETRY

Viewed physically, the transformer is a closed ferromagnetic *core* containing an aperture through which magnet wire is threaded, so that both magnetic flux and current travel a closed path and are linked with each other. So simple a requirement permits almost unlimited variation in the shape of core and coil or, to use another traditional expression, of iron and copper. Yet the size of any transformer may be conveniently characterized by means of its basic geometry, whatever its shape.

Each of the cores shown in Fig. 2.1 has a cross-sectional area, and so does the window or aperture associated with it, even though each core is constructed differently. Every core has an average length of magnetic path, l. The coil which passes through its window has an average length, the mean-length turn or *MLT*. The term *coil*, as used in this chapter, denotes the totality of the magnet wire windings; the primary and secondary windings—whether built separately or together—girdle the same aperture.

In practice the primary winding may be split, there may be more than one secondary winding, and any of these windings may consist of several layers of continuous magnet wire which has been wound either around a coil form or directly on the core. Each discrete, helically wound length of wire is called a *winding*. The magnet wire is usually copper, but it may be silver or aluminum; its cross section may be round or rectangular. It may be coated with a film or covered with a fiber, and these materials, as well as others which are used to insulate it, have appropriate dielectric, thermal, and mechanical properties.

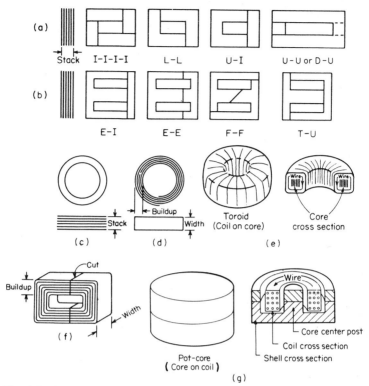

Fig. 2.1 Basic core geometries. (*a*) Laminations, single window. (*b*) Laminations, double window. (*c*) Stacked O rings (washers). (*d*) Spiral-wound strip. (*e*) Wound toroid and its cross section; core may be (*c*) or (*d*). (*f*) Cut C core. (*g*) Pot core and its cross section with cap of top shell removed.

a. Basic Core Construction

The four most important basic cores are known as laminated, toroid, C, and pot cores.

(1) Laminations. The most widely used core consists of thin sheets of iron alloy (e.g., silicon-iron, nickel-iron) which are readily stamped or punched into I, L, U, E, F, T, and O shapes. Such stampings or punchings are called *laminations,* and they are interleaved and piled up into a *stack.* Coils for such cores can be economically wound on a multiple winding machine.

A particularly important alphabet pair, the *scrapless* or wasteless E-I lamination set, can be stamped from a large sheet without any waste (Fig. 2.2). Standardized proportions and low cost have made

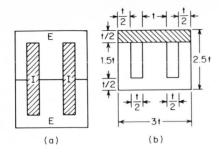

Fig. 2.2 Scrapless lamination. (*a*) Stamping to produce two scrapless E-I sets. (*b*) Dimensional proportions. The length of the magnetic path $l = 6t$ when the coil is wound over the center limb (tongue), *t*.

this set a part of almost every inventory, and transformers can be more quickly delivered when it is specified.

(2) **The Toroid.** When the O or washer-shaped rings are stacked, the result is the characteristic circular shape of the *toroid* core (Fig. 2.1*c*). A more popular form of the toroid, the *wound toroid*, is shown in Fig. 2.1*d*. This consists of a thin strip of iron alloy which has been concentrically wound into a tight circular spiral. It is geometrically equivalent to the stacked toroid if the width of the strip is equal to the height of the stack of washers and if the depth or "buildup" of the spiral equals the difference in radii of the washer. Both the stacked ring and the wound toroid have a rectangular cross section (Fig. 2.1*e*). Although the ideal shape, the classic *torus* (or doughnut), has a round cross section, it is not feasible to produce the torus from strip or rings and it is not considered economical to fabricate it from compressed powder particles or ferrite.

(3) **The C Core.** The disadvantages of the toroid are eliminated by the wound-and-cut core. This core is wound into a rectangular, rather than a circular, spiral and then cut to produce two C-shaped cores with a rectangular cross section. The C-C or C-core shape is shown in Fig. 2.1*f*. The coils can be wound on conventional machines, and the window area can be completely filled with copper.

Almost all the alphabet shapes can be economically fabricated from powdered iron, powdered permalloy, ferrites, and sendust. The *powder cores* are generally inferior when the operating frequency is below the kilohertz (kHz) range, but at higher frequencies they often cost less and perform better than the more popular laminations.

(4) **The Pot Core.** The toroid, which offers important advantages, has two practical disadvantages: the hole at the center of the coil represents an incomplete utilization of window area, and only one coil can be wound at a time. In principle, these disadvantages can be eliminated if the physical relationship between the iron and copper is interchanged. In the powder *pot-core* assembly (also called toroidal shell) this is in fact done. The coil is penetrated by, and completely enclosed

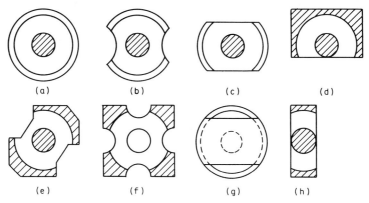

Fig. 2.3 The pot core and various modifications. (*a*) CC, *c*up *c*ore (as in Fig. 2.1*g*). (*b*) PM, *p*ot core *m*odule. (*c*) Q, cube core. (*d*) EP, high packing density. (*e*) RM, *r*ectangular *m*odule. (*f*) X, cross (square perimeter). (*g*) TT, *t*ouch *t*one; (*c*) is superimposed on (*g*). (*h*) E-C core.

within, two metallic cups (Figs. 2.1*g* and 6.6), and it may be wound on a conventional winding machine. However, the assembly does not permit the efficient transfer of heat from the internal coil to the outer core surface, and it is used chiefly for high-Q inductors and low-power transmission transformers.

Removal of a portion of the outer shell can overcome most of the disadvantages of the pot-core geometry (Fig. 2.3). In the PM, Q, EP, RM, X, TT, EP, and EC cores, the bobbin is exposed to the ambient air in varying degrees, which results in improved heat transfer and access to the terminations of a winding. In several instances (EP, RM, X), the external envelope has been contoured to facilitate the close packaging of cores on a circuit board.

b. Core-Type and Shell-Type Transformers

The two basic types of transformer assembly are depicted in Fig. 2.4. In the simple *core-type* structure, a single core with a single window provides a single magnetic path (Fig. 2.4*a*). When the coil is split, as in Fig. 2.4*b*, the assembly provides an economical alternative to the toroid.

In the *shell-type* structure, a single coil girdles two cores in such a way that the magnetic field becomes common to both cores, which divide the magnetic flux between them. The two cores may be physically fused, as in Fig. 2.4*c*. In Fig. 2.4*d* the coil has been assembled around the center limb, or tongue, which is twice as wide as any of the sides.   Since the two original cores were geometrically identical, the total cross-sectional area of the shell is equal to the product of the stack and tongue.   The structure may be achieved by pairing any of the combinations shown in Fig. 2.1*a*, as shown in the E-I, E-E, F-F, and T-U combinations of Fig. 2.1*b*.

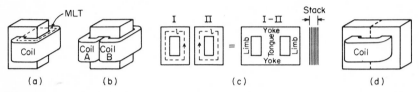

Fig. 2.4   Core- and shell-type transformers.   (*a*) Simple core type, with single coil, uses single window cores shown in Figs. 2.1(*a*), (*f*), or (*g*).   Dotted line indicates the *mean-length turn* (MLT) of the copper.   (*b*) Core type with two coils.   (*c*) Two single-window cores combined into a shell structure.   The shell may thus be any double-window structure shown in Fig. 2.1(*b*).   Dotted lines indicate length *l* of average magnetic flux path.   (*d*) Shell core and coil.

c. **Polarity**

At any given moment of an excitation cycle, the beginning or "start" of every concentric winding of a coil has identical polarity.   So does every ending or "finish."   The dot convention provides a convenient method for labeling the respective starts or finishes of the windings and thereby the polarity of each terminal.   Every start (S) or every finish (F) is assigned a dot on the schematic, as shown in Fig. 2.5*a*.   Current from the generator E_1 is seen as flowing into a dot marked on the primary terminal and as resulting in a current which flows out of a terminal marked as a dot on the secondary winding e_2 and then to the load R_L.   The respective starts or finishes of two concentric windings may thus be represented by separate dots.

When the polarity of two windings is not known, it is a simple matter to measure the output voltage of each, connect them in series, and then measure their combined output.   If the total equals the sum of the individual outputs (additive or boost polarity), the start of one winding has been connected to the finish of the other (Fig. 2.5*b*).   If the total is the difference between the two (subtractive or buck polarity), the two starts or finishes have been connected (Fig. 2.5*c*).

Paradoxical results are sometimes obtained when the test is performed on a transformer whose windings are not concentric.   In Fig. 2.5*d* two identical windings are shown on opposite limbs of the core.   One winding links the flux in an upward direction of flow, while the other links it in a downward direction.   When the two starts or finishes are connected, additive rather than subtractive polarity results.   This is a consequence of the fact that the flux flows from the start to

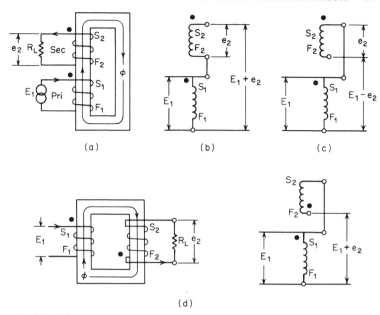

Fig. 2.5 Polarity conventions. (*a*) Concentric windings on core or shell structure. Current flows into the dot shown on the primary and out of the dot shown on the secondary. (*b*) Additive polarity (S_1-F_2 or S_2-F_1). (*c*) Subtractive polarity (S_1-S_2 or F_1-F_2). (*d*) Windings on opposite limbs of core structure. Additive polarity requires an S_1-S_2 or F_1-F_2 connection, in contrast to (*b*).

the finish of one winding and, following a continuous path, enters the other winding at the finish and leaves it at the start.

2.3. THE IRON-COPPER AREA PRODUCT

The most significant geometric characteristic of the core and coil assembly is the product of the area of the core cross section and the area of the window. Any core and coil combination can be compared with any other by means of the area product A_P:

$$A_P = A_{Fe}A_{Cu} \tag{2.1}$$

where A_{Fe} is the core cross-sectional area (stack × limb or tongue) and A_{Cu} is the window area (length × width).

a. VA and the Area Product

Assume, for the sake of simplicity, that the core of an isolation transformer (see Sec. 2.5*a*) has but two windings, a primary and a secondary, and that the insulation occupies a negligible part of the window area. Also assume that the area of both the primary and secondary magnet wire is proportional to the current in each winding. This assumption is consistent with the law of equal ampere-turns (A-turns):

$$N_1 I_1 = N_2 I_2$$

which is a simple transposition of Eq. (1.3).

The current per unit area of wire, or current density J, is usually made equal in each winding to provide uniform heat generation throughout the window. Half the window area is assigned to the primary and half to the secondary:

$$J = \frac{I_2}{K_{Cu} A_{Cu}/2N_2} = \frac{I_1}{K_{Cu} A_{Cu}/2N_1} \tag{2.2}$$

where K_{Cu}, the *copper space factor*, represents the fraction of the window area occupied by the copper.

The A-turn product is then proportional to the window area:

$$N_1 I_1 = N_2 I_2 = K_1 A_{Cu} \tag{2.3}$$

where $K_1 = K_{Cu} J/2$ combines into one constant both the copper space factor and the current density.

The fundamental transformer equation for a periodic sine wave, Eq. (1.8), may be expressed in terms of volts (V) per turn of the primary winding and transposed into:

$$\frac{E_1}{N_1} = K_2 A_{Fe} \tag{2.4}$$

where E_1 is the input voltage. We use the constant $K_2 (= 4.44 B_m f 10^{-8})$ because the flux density is constant when input voltage and frequency are fixed.*

The volt-ampere (VA) rating $E_1 I_1$ here equals $E_2 I_2$ (its usual definition) because our present discussion neglects power losses. To obtain the VA rating of the transformer in terms of its geometry only, we multiply the last two equations:

$$\text{VA}_T = E_1 I_1 = K_1 K_2 A_{Fe} A_{Cu} = K_3 A_P \tag{2.5}$$

---

*The actual core area A_{Fe} is less than the measured area because air spaces are inevitable. The fraction of available area actually occupied by core material, about 0.9, is called the *stacking factor*. Analogously, the *packing factor* describes the volume occupied by the magnetic particles of a powder core, as distinct from the volume occupied by the interspersed nonmagnetic materials.

where K_3 combines all the previous constants. The turns have canceled out, and their absence demonstrates that their number is not basic to the VA rating of the transformer.

We see from Eq. (2.5) that the transformer rating, the volt-ampere product, is directly proportional to the product of iron and copper areas. Even though the complications of temperature rise resulting from copper and iron losses have been disregarded, we are nevertheless able to confirm an important qualitative observation: *The higher the VA rating, the greater the area product.* The definition of K_3 also carries with it the implication that the area product can be reduced by increasing the *frequency, flux density, copper space factor, or current density.*

b. LI^2 and the Area Product

The VA rating can also be expressed in terms of the inductance L of a winding and its magnetic energy, LI^2. Thus, omitting subscripts:

$$EI = (\omega LI)I$$

$$VA = \omega LI^2$$

$$LI^2 = K_4 JBA_{Fe}A_{Cu} \qquad (2.6)$$

$$\text{where } K_4 = 10^{-8}/2\sqrt{2}$$

We see, therefore, that an inductor as well as a transformer can be specified in terms of core geometry.

The concept of the VA rating is basic to the art of transformer design. It is used in two senses. To the transformer designer it is the volt-amperes VA_T which is associated with the geometry of the transformer. To the user, it is the volt-amperes VA_L which is furnished to the actual load. A convenient figure of merit enables us to gauge the effectiveness of the transformer in any particular circuit:

$$\frac{VA_L}{VA_T} \leq 1 \qquad (2.7)$$

Unfortunately, the distinction between the two VA ratings is easily obscured. The user justifiably focuses on actual performance and on the load rating VA_L. Yet, as we shall see in Sec. 2.5, the VA_T necessary to furnish the desired VA_L is profoundly affected by specifications which may at first appear subsidiary or even irrelevant to the circuit engineer. Before we discuss this question, two preliminary points about weight and area should be discussed.

c. Area Product Equivalence

The area product of any particular core structure can be equated with that of the scrapless E-I lamination. The area product of a scrapless lamination with a square stack (i.e., stack height = tongue width) is $A_P = 1.5t(0.5t)t^2$, where t is the width of the tongue. Transposing:

$$t = \left(\frac{A_P}{0.75}\right)^{1/4} \tag{2.8}$$

Equation (2.8) (see Chap. 5 and Table 5.2), enables the user to estimate the VA_T of a nonscrapless core from the ratings assigned to standard scrapless laminations.

d. Rating and Weight

The weight of either the iron or copper is equal to the product of its volume and density and is proportional to the cube of its linear dimension. That is: Weight $\propto (A^{1/2})^3$, where A represents either the iron or copper area. Since the VA_T rating is proportional to A^2, $A \propto (VA_T)^{1/2}$ and:

$$\text{Weight} \propto (VA_T)^{3/4} \tag{2.9}$$

From this equation we can see that if VA_T is doubled, weight is increased by a factor of $2^{3/4} = 1.68$, an increase not of 100 percent, but of 68 percent. Again, as in Eq. (2.5) we have assumed that all other parameters are kept constant.

Equations (2.8) and (2.9) enable the engineer who knows the area product of any particular transformer to approximate both its VA_T rating and weight.

We are now ready to turn our attention to the problems created when the VA rating is specified along with concomitant specifications which increase the disparity between VA_L and VA_T. The reader is reminded that our hypothetical transformer is a two-winding isolation transformer whose window is completely occupied with copper. We now add the condition that it furnishes power to a standard resistive load.

2.4. MAJOR PARAMETERS

Certain parameters have a powerful bearing on both the VA_T rating and the size of the power transformer. These include: load power factor, regulation, efficiency, and dc polarization. A fifth, the copper space factor, is often affected by circuit requirements even though it is not explicitly specified. We elaborate on these topics in Chap. 5.

2.5. CIRCUIT FUNCTIONS

The size and geometry of the transformer are markedly influenced by the circuit functions. Important examples of transformers so influenced are the isolation transformer, the autotransformer, the regulating transformer, and the power-converter transformer. In each of these transformers, the VA rating is significantly affected by the winding connections, the load characteristics, and the waveforms of the current and voltage in the windings.

a. The Isolation Transformer

Isolation, which is defined as the absence of a metallic conductive connection between primary and secondary windings, is required in most transformers for electronic circuits and for any of various reasons. The need for conductive (sometimes called galvanic) isolation is clear when the hazards of shock or fire must be eliminated or reduced. However, there are important instances where, if extreme isolation is desired, it becomes necessary to obtain a marked reduction of capacitative and resistive (common-mode) current flow between windings and ground. One very effective technique is to place the primary and secondary windings on opposite legs of a U-I lamination, as shown in Fig. 2.4*b*, or on opposite sectors of a toroid.[1] Such transformers are often much larger than their VA ratings would indicate.

b. The Autotransformer

If the isolation (isol) property of a transformer is deliberately undone by connecting the secondary and the primary in series, we obtain the *autotransformer* (auto) connection shown in Fig. 2.6. As compared with the isolation connection, the autoconnection permits a higher

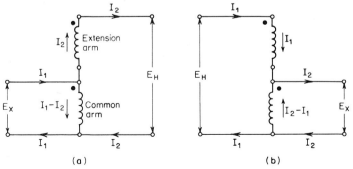

Fig. 2.6 Autotransformer. (*a*) Voltage step up: $n < 1$, $r > 1$. (*b*) Voltage step down: $n > 1$, $r < 1$.

input or output current. This is because the winding, which is common to both output and input, carries a current equal to their difference. The overall voltage is greater than that of either the common winding or the extension arm (i.e., the winding not common both to input and output).

The improvement in the performance of a given transformer is given by[2]:

$$\frac{VA_{L, auto}}{VA_{T, isol}} = \frac{\text{total voltage}}{\text{extension-arm voltage}} = \frac{E_H}{E_H - E_X} \qquad (2.10)$$

where E_H is the higher voltage and E_X the lower voltage. E_H is the secondary voltage in the step-up autotransformer and the primary voltage in the step-down autotransformer.

When the ratio definitions $n = E_H/E_X = 1/r$ are employed, Eq. (2.10) becomes:

$$\frac{VA_{L, auto}}{VA_{T, isol}} = \frac{n}{n-1} = \frac{1}{1-r} \qquad (2.11)$$

From Eq. (2.11) it can be seen that the improvement is greatest when the difference between input and output voltages is small. For example, the rating of a 100 VA_T isolation transformer is increased by the autoconnection to 200 VA_L if $r = 0.5$ or, more dramatically, to 1,000 VA_L if $r = 0.9$.

Regulation (reg) also improves as a result of the autoconnection:

$$\frac{reg_{auto}}{reg_{isol}} = 1 - r \qquad (2.12)$$

In practical terms, Eq. (2.11) means that we can use a core whose area product is smaller by a factor $1 - r$ than the isolation transformer core. With the autoconnection, the figure of merit defined by Eq. (2.7) can exceed unity, a weight-saving feature of importance to the circuit engineer.

c. Regulating Transformers

Transformers can be designated to regulate either output voltage or current when there are large variations in input voltage or load impedance.[3] We refer to three types of regulating transformers: *high-leakage reactance, ballast,* and *ferroresonant* transformers. Each of the three types is required to exhibit high series impedance and has a suitable lamination geometry by means of which leakage flux is established between the input and output windings.[4] In a typical configuration (Fig. 2.7), a magnetic shunt is used to obtain loose magnetic coupling between the windings.[5] The air gap between the stack of shunt

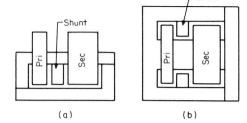

Fig. 2.7 High-leakage reactance geometry. (*a*) Core type. (*b*) Shell type, two shunt stacks.

laminations and the core limb establishes the magnitude of the leakage reactance.*

High-leakage reactance transformers are used in circuits which limit the current flowing to nonlinear loads. They are basic components in welding circuits[7] and battery chargers. Ballast transformers are widely used in lamp and magnetron circuits, in which current and power are stabilized within prescribed limits.[8] Typical loads are fluorescent and high intensity discharge (HID) lamps. The nonlinear volt-ampere characteristics of the mercury, metallic halide, and high pressure sodium lamp are not easy to model accurately, and circuit simulation may call for analog as well as digital design procedures.[9]

Ferroresonant transformers stabilize voltage applied to a resistive load, a capacitive load, or a rectifier load. The ferroresonant transformer is useful in many power conversion circuits and is discussed further in Sec. 5.6.

2.6. POWER CONVERSION CIRCUITS

The field of *power electronics* deals with the application of various devices and components to the conversion, control, and conditioning of electric power.[10] Here *conditioning* means any of several constructive alterations to the power circuit and its waveform. Examples are changing a voltage source to a current source, altering the form of the output wave from a square to a sine, and limiting harmonic distortion at the output of an inverter.

Inductive components are thus found in circuits utilizing a variety of electron devices such as the:

rectifier diode	trigger diode (diac)
avalanche diode	transistor
Zener diode	thyristor
Schockley diode	*tri*ode *ac* switch (triac)

---

* A high-leakage reactance geometry, in which two shunt paths are separated by a capacitor winding, is also used in a transformer to furnish sinusoidal constant current, such as the *transfilter*.[6]

Other solid-state components used in control circuits are the thermistor, varistor, magnetoresistance device, Hall effect device, and any of various light-sensitive devices such as the photovoltaic (solar) cell, photodiode, phototransistor, and light-emitting diode (LED).

The transformer and inductor are basic components in the ac-to-dc converter as exemplified by the rectifier circuit, the dc-to-ac converter circuit, and in dc-to-dc converter circuits. In the context of converter circuits, mention must be made of the power frequency changer, which provides a means for the direct conversion of ac to ac without an intervening dc link. However, the subject of this interesting technology lies outside the scope of our survey.[11]

Perusal of the literature on power conversion reveals a larger, perhaps bewildering variety of circuits. Because our initial task, prior to design, is to establish a VA rating for the transformer, we will focus on the following salient features of the conversion circuit:

1. The waveforms of current and voltage
2. The rms and peak values of current and voltage
3. The frequency, pulse duration, and duty cycle
4. The magnitude of unbalanced dc in the windings

In the sections which follow, we have selected representative examples of rectifier, inverter, and converter transformer circuits.

2.7. THE RECTIFIER CIRCUIT

Rectifier circuits require transformers with a larger VA_T rating than do circuits which do not rectify ac into dc. There are several reasons for this. (1) The current wave shape in the rectifier transformer becomes nonsinusoidal (Fig. 2.8) and its effective (rms) value changes. (2) The required voltages may necessitate larger primary and secondary windings, and thus a larger window area. (3) Certain filter circuits, and sometimes even the magnitude of filter components, have an adverse effect upon VA_T.

a. Utilization Factor

The comparison of VA_T ratings in different rectifier circuits is facilitated by the utilization (or utility) factor (UF):

$$UF = \frac{\text{average dc output power}}{\text{average } VA_T \text{ rating}} \qquad (2.13)$$

This definition, which assumes a filter without power losses, an ideal rectifier diode, and a resistive load, is a simple extension to the rectifier circuit of the figure-of-merit concept of Eq. (2.7).

The reciprocal of Eq. (2.13) is a more convenient yardstick because it permits us to assign the value of unity to the dc output power and thus obtain a normalized VA_T rating for each rectifier circuit.   Table 2.1 is based on the reciprocal.   To illustrate its use, assume that 100

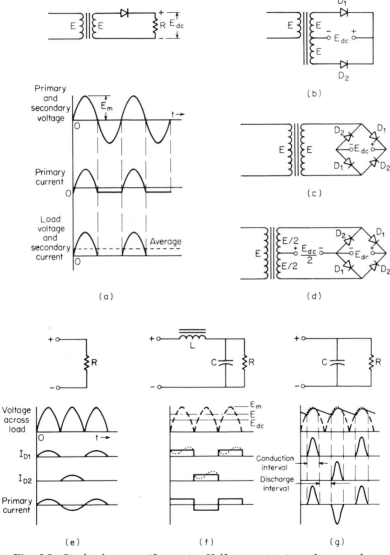

Fig. 2.8   Single-phase rectifiers.   (a) Half-wave circuit and wave shapes. (b) Full-wave center-tap (FWCT) with diodes D_1, D_2.   (c) Full-wave bridge. (d) Full-wave bridge with center tap to obtain $E_{dc}/2$.   Full-wave inputs. (e) Wave shapes for unfiltered R load.   (f) Wave shapes for L-input filter: currents are rectangular for infinite L; dotted line on diode current indicates ripple with finite L.   (g) Wave shapes for C-input filter.

dc watts are required from a rectifier whose average VA_T rating is found, from the appropriate table, to be 1.34. $UF = 1/1.34$, and the required transformer rating is 100×1.34, or 134 VA_T.

Although a high utilization factor is obviously desirable, it is only one of many factors among which the circuit engineer must effect a compromise. Some of the others, such as ripple, regulation, efficiency, and cost, also require analysis.

b. Transients

In the rectifier, the circuit leakage reactance, distributed capacitance, and load resistance form an *LCR* circuit. Thus, when the input or output is switched on or off, transient surges of voltage and current occur. The problems of transients is not new. It occurs in radiotelegraphy because keying is equivalent to switching a rectifier load (e.g., a modulated amplifier) off and on in accordance with a code.

The difficulties created by transient surges, especially when the current is switched off, have been aggravated by the introduction of solid-state switches. These popular semiconductors are impressively efficient and small compared with their predecessors, the selenium diode, the copper-oxide rectifier, and vacuum or mercury-vapor diode tubes. However, the selenium and copper-oxide diodes, which have large junction areas and high capacitance, can withstand surges lasting for milliseconds (msec), whereas the newer devices often burn out in microseconds (μsec).

When the primary of a power transformer is switched *on*, transient currents may blow out line fuses or open circuit breakers in a mysteriously sporadic fashion. The mystery can be traced to the fact that the transformer core was left in a state of residual flux density B_r when the input power was turned off. The particular B_r may be anywhere within the range $\pm B_m$ (see Fig. 1.2b). Let us assume that B_r is 0.75 B_m and therefore not far from B_s, the saturation flux density. If the power is again switched on at a moment of increasing input voltage, the flux change ($B_s - B_r$) is small and the core saturates. The transient inrush of magnetizing current, limited only by the low impedance of the primary, is very large and thus the line fuse blows out. The use of fuses with a thermal lag eliminates this particular misfortune.

There may also be problems on the secondary side. When there is a disconnected or inductive load, the sudden application of voltage at the instant that input voltage is at its maximum E_m may result in a transient oscillation whose peak value may be as great as $2E_m$.*

---

* See Sec. 10.3b.

The diode may break down if this surge exceeds its peak inverse voltage (PIV).

Transients are as much a circuit problem as a transformer problem, so the circuit engineer must exercise great care if an interruption of current in inductive, capacitive, and even resistive loads is anticipated. An additional "bleeder" resistor across a resistive load provides some protection because then the load cannot be entirely switched off, but some efficiency is thereby sacrificed. An initial reconsideration of the requisite diode PIV and the peak voltage rating of filter capacitors is often helpful, even to the extent of selecting those with ratings two or three times as large as might otherwise be deemed necessary. Where this is not feasible, transient suppressors such as the silicon carbide varistor or back-to-back Zener diodes may be indicated.

2.8. SINGLE-PHASE RECTIFIERS[12]

In the previous section we defined a utilization factor by means of which we can compare the VA_T ratings in different rectifier circuits. However, to compute the ratings which we wish to compare we must first obtain an estimate of the rms value of the nonsinusoidal transformer currents of the three basic single-phase rectifier circuits. Then, having derived VA_T ratings for these circuits when they do not contain filter inductors or capacitors, we proceed to investigate how the presence of these components affects the rating. We will thus be able to compare all the basic variations of single-phase rectifier circuits.

The principal characteristics of the three basic single-phase rectifiers are summarized in Table 2.1.

a. Inductor-Input Filter

The inductor filter, although of great utility in other rectifier circuits, is rarely employed in the half-wave circuit.

Values for the various currents and VA_T ratings in the infinite-inductor (L_∞) input circuits are listed in Table 2.1, which includes, for the convenience of the circuit engineer, corresponding values of diode current, diode PIV, and rms ripple voltages.

(1) **Critical Inductance.** If we now decrease the inductance which we had assumed to be infinite, the ripple current increases, manifesting itself in a deepening valley in the top of the rectangular wave. When the valley deepens sufficiently to meet the zero axis of current (Fig. 2.8f), inductance L has reached the critical value L_c. At the point where the ripple is zero, the peak value ($\sqrt{2}\ I_r$) of its lowest harmonic

TABLE 2.1 Single-Phase Rectifier Transformer Ratings*

	Half-wave	Full-wave, center-tapped		Bridge	
	R	R	L	R	L
Ripple frequency	f	$2f$	$2f$	$2f$	$2f$
Ripple V_{rms}	1.11	0.471	0.471	0.471	0.471
Ripple V_{rms} total†	1.21	0.482	0.482	0.482	0.482
Primary E_p	2.22	1.11	1.11	1.11	1.11
Primary I_p	1.21	1.11	1.00	1.11	1.00
Primary VA	2.69	1.23	1.11	1.23	1.11
Secondary E_s	2.22	1.11‡	1.11‡	1.11	1.11
Secondary I_s	1.57	0.785	0.707	1.11	1.00
Secondary VA	3.49	1.74	1.57	1.23	1.11
Average VA_T	3.09	1.49	1.34	1.23	1.11
PIV per diode	3.14	3.14	3.14	1.57	1.57
I_{pk} per diode	3.14	1.57	1.00	1.57	1.00
I_{av} per diode	1.00	0.50	0.50	0.50	0.50
I_{rms} per diode	1.57	0.785	0.707	0.785	0.707

* Voltages, currents, and VA are based on unity dc output voltage and current. Voltage drops in transformer, diode, and inductor are neglected.

† Includes second and third harmonics of ripple frequency.

‡ Secondary voltage at each side of center tap.

component, i.e., of the second harmonic current, is also equal to the average rectified current I_{dc}.

If the inductance is reduced further, below L_c, the crest of the ripple rises, and the valley would deepen and become negative if it were not blocked by the diode. A discontinuity in the ripple then results, as if a slice were removed from the valley portion of the ripple wave. The effect of this discontinuous current in series with the inductor is a momentary increase in output voltage during each half cycle of the power frequency.

If instead of decreasing the inductance we increase the load resistance, we encounter the same discontinuous behavior. If R varies over a range which includes the point where discontinuity occurs, as it well may, the result will be poor dc voltage regulation. The criterion for the continuous flow of current, and therefore for good regulation, is:

$$I_{dc} \geqq \sqrt{2}\, I_r \qquad (2.14)$$

and the value of critical inductance is given by:

$$\frac{\omega L_c}{R} = \frac{1}{3} \tag{2.15}$$

This equation may then be transposed into the familiar, practical criterion for the continuous flow of current when $f = 60$ Hertz (Hz):

$$L_c = \frac{R}{1,130} \tag{2.16}$$

The rms secondary current I_s increases as inductance L decreases from infinity to the critical value L_c. The increase in the popular FWCT (full-wave center-tap) circuit is stated as:

$$\frac{I_s}{I_{dc}} = 0.707 \sqrt{1 + \frac{a_L{}^2}{2}} \tag{2.17}$$

where $a_L = L_c / L = R/3\omega L$.

Values of I_s for typical multiples of L_c are given in Table 2.2. To illustrate, let us adopt for a moment the conservative practice of using an inductance twice the critical value. The secondary current (from Table 2.2) will be 0.75 I_{dc}, and the transformer rating will be about 6 percent higher than the value in Table 2.1, where L is infinite.

It is clear from Table 2.2 that transformer size is reduced when L is increased, but the advantage may be illusory. If a critical inductance is used instead of $2L_c$, the result is often a significant reduction in the weight of the choke, an advantage which may overcome the increase in transformer weight resulting from the VA_T increase of $0.87/0.75 = 1.16$, or 16 percent. The total situation should therefore be evaluated before a decision is made, for sometimes an inductance only 10 percent above L_c is the most practical choice.

(2) **Critical Inductance and Controlled Rectifiers.**[13] When the diodes of the FWCT circuit are replaced by thyristors, critical inductance becomes a function of the firing angle θ and the average-to-peak output voltage ratio E_{dc}/E_m. The addition of a *flyback diode* (FBD) across the input terminals to the LC filter makes it possible

TABLE 2.2   Secondary Current
with Finite L

L/L_c	a_L	I_s/I_{dc}
∞	0.00	0.707
4.00	0.25	0.72
2.00	0.5	0.75
1.33	0.75	0.80
1.00	1.00	0.87

TABLE 2.3 Firing Angle and
Critical Inductance

θ (degrees)	E_{dc}/E_m	$\omega L_c/R$
0	0.70	1/3
35	0.58	1/2
90	0.30	1
180	0	$\pi/2$

for L_c to remain finite in value all the way down to zero output voltage.* We indicate several limiting cases in Table 2.3. Critical inductance can be substantially larger than the value used in the standard rectifier circuit.

(3) **Ripple Reduction.** If capacitor reactance $X_c \ll R$, the common case, the ripple attenuation—that is, the ratio of input ripple e_r to output ripple voltage e'_r—is:

$$\alpha = \frac{e_r}{e'_r} = 1 - \omega^2 LC = 1 - \left(\frac{f}{f_0}\right)^2 \tag{2.18}$$

where f_0 is the resonant frequency of the LC combination and f the frequency of the second harmonic.

When ripple attenuation is substantial, the equation may be stated as:

$$\alpha \cong \left(\frac{f}{f_0}\right)^2 \tag{2.19}$$

Equations (2.15) and (2.19), although simple, are of great utility in the design of a ripple filter.

b. Capacitor-Input Filter

As we have already noted, diode current and secondary current are discontinuous when L is less than L_c. The filter behaves as a capacitor-input filter and in the limit, when $L = 0$, it has become a capacitor-input filter.

In the circuit which contains a capacitor-input filter (Fig. 2.8g) the capacitor charges during the brief conduction interval in which E_m exceeds E_{dc}. The current through secondary, diode, and capacitor has the wave shape of a narrow pulse, becoming higher and narrower as capacitance increases. If there were no series resistance in the circuit, the current peak would reach alarmingly high values, requiring

---

* The flyback diode also is used in the converter circuit. See Figs. 2.13 and 2.14.

drastic upward revision of the required diode and capacitor peak-current ratings. (Fortunately, the transformer's winding resistance is usually sufficiently high to limit the peak currents and the necessary revision or derating is typically 20 to 40 percent.) Between charging intervals, the capacitor discharges current through the load R at an exponential rate, and the resulting ripple voltage has a sawtooth wave shape.

Output voltage E_{dc} now depends on the time constant RC, and regulation will be poor when the load R varies, unless C is very large. Alternatively, the choice of C will determine both secondary current I_s and secondary voltage E for any given E_{dc} and load R. E and I_s also depend on the total series resistance R_s, which comprises the sum of all resistances in series with each diode:

$$R_s = a^2 R_1 + r_2 + R_d \tag{2.20}$$

where $a = 1/n = $ step-up turns ratio
$\qquad R_1 = $ primary resistance
$\qquad r_2 = $ secondary resistance*
$\qquad R_d = $ diode forward resistance

An analysis of the C input circuit is, not surprisingly, far more complicated than that of the L input circuit. O. Schade[14] summarizes his comprehensive analysis in a set of graphs in which normalized currents and voltages are plotted as a function of circuit regulation R_s/R and the dimensionless ratio $R/X_c = \omega CR$. The latter ratio is equal to the Q of the shunt (parallel) combination of R and X_c (capacitive reactance) and is a felicitous circuit parameter because it is directly related to the ripple factor γ_T, the quotient of total rms ripple and average dc voltage.

We use Gray's computation to illustrate this relationship:[15]

$$\gamma_T = \frac{1.58}{Q} = \frac{1.58}{2\pi fCR} \qquad Q > 20 \tag{2.21}$$

In this equation, f is the principal harmonic, i.e., the supply frequency if the circuit is half wave, or the second harmonic of the supply frequency if the circuit is full wave. Thus, a Q of 100 will produce a total rms ripple voltage of 1.58 percent of the dc output voltage.

Since we are interested in the effect of the capacitor on the transformer rating we have listed, in Table 2.4, some representative values of regulation and Q and their corresponding normalized secondary rms voltages and currents. The values are taken from Schade's graphs.

---

* In the FWCT circuit, r_2 is defined as half of the total secondary resistance.

TABLE 2.4 Capacitor-Input Filter Transformer Ratings
Secondary leg voltage E and current I_s as a function of the percentage
of resistance R_s/R and the load $Q = R/X_c$.*

| | | $R/X_c = 2\pi fCR$ | | | | | |
| | | 1 | | 10 | | 100 | |
	R_s/R (%)	E	I_s	E	I_s	E	I_s
Half-wave	2	1.91	1.75	0.94	2.4	0.83	2.60
	5	1.97	1.70	1.01	2.2	0.93	2.35
	10	2.02	1.68	1.14	2.1	1.09	2.15
	25	2.36	1.65	1.54	1.9	1.49	1.97
Full-wave, center-tapped†	2	1.07	0.86	0.81	1.3	0.78	1.5
	5	1.10	0.85	0.87	1.2	0.84	1.21
	10	1.15	0.85	0.95	1.1	0.93	1.15
	15	1.22	0.85	1.03	1.05	1.01	1.1
	25	1.35	0.83	1.17	0.95	1.16	0.98

* X_c = capacitative reactance at principal ripple frequency f.

† Primary current I_p is 1.414 times I_s in the full-wave center-tapped circuit. In the bridge and full-wave doubler circuits, I_s and I_p equal 1.414 times the value of I_s in the full-wave center-tapped circuit.

A comparison of this table with Table 2.1 reveals significant differences: (1) Circuit regulation and Q govern the variation of transformer voltage and current; (2) VA_T is higher for capacitor input than for inductor input; and (3) VA_T varies with capacitance.

The variation of VA_T with the capacitance raises a problem which we did not have to consider in discussing the inductor filter circuit. The exact value of inductance in the latter circuit, provided only that it exceed L_c, does not affect the secondary voltage and has only a moderate effect on the magnitude of the winding current. Omitting the value of L from the specifications results, at worst, in a slightly hotter transformer if the design is based on the value of current in Table 2.1 instead of Table 2.2.

On the other hand, the capacitor-input rectifier transformer cannot be accurately designed without a specification of the input capacitance. The circuit engineer is not always certain of the exact value of input C and may not determine it until the transformer has been received and the C input filter has been adjusted for the desired ripple, regulation, and voltage in the actual circuit. It is therefore common practice for the transformer designer to select for E_s and I_s a set of values which are cited by the diode manufacturers as representative of average practice. Typical values[16] of secondary VA_T of the FWCT would be $2EI_s = 2(0.85)(1.15) = 1.96$; of the bridge, $EI_s = 0.85 \times 1.63 =$

1.38; and of the half-wave circuit, $EI_s = 1 \times 2.3 = 2.3$. Primary VA_T of the FWCT would be

$$0.85(1.15)1.41 = 1.38$$

and of the bridge circuit, $0.85 \times 1.63 = 1.38$. If we compare these values with Table 2.1, we see that the average transformer rating is higher in the full-wave circuit than it is in either the R (unfiltered) or the L filter circuit.

Some practical comments are in order. The capacitor-input circuit results in poorer regulation and a larger transformer, but it costs less and weighs less. When low ripple, rather than good regulation, is of central interest, the CR circuit will do the job. It should also be noted that the FWCT circuit, which requires a larger transformer than does the bridge circuit, requires only two diodes (that dissipate less heat than four diodes). Such economic considerations, added to the comparatively low cost of electrolytic capacitors, account for the wide popularity of the capacitor-input FWCT circuit, especially in commercial applications.

c. The Voltage Multiplier[17,18]

There are occasions when a large step up of voltage $1 : n$ is desired, and the use of a multiplier circuit is indicated. The need for a multiplier may arise in high-frequency circuits where winding and circuit capacitance places a limit on the step-up ratio (see Sec. 9.2b). Such circuits (containing $2n$ capacitors and $2n$ diodes) multiply peak secondary volts by n, but they also multiply rms current in the secondary circuit of the transformer by a factor greater than $1/n$. Hence, the designer of the multiplier transformer must take care that the VA_T rating accurately reflects the higher-than-normal rms value of the secondary current.

2.9. THE INVERTER CIRCUIT[19]

A rectifier circuit converts ac at its input to dc at its output. The inverter circuit inverts dc at its input to ac at its output. Each of the rectifier circuits shown in Fig. 2.8 can be transformed into an inverter circuit by substituting an active switch, such as a transistor or thyristor, for the diode. Then, by applying dc voltage at the output (now called the input) and switching the transistors alternately on and off during the period $T = 1/f$, we obtain square-wave ac at the input (now called the output).

We see this in the transistor inverter circuit (Fig. 2.9a), where switch-

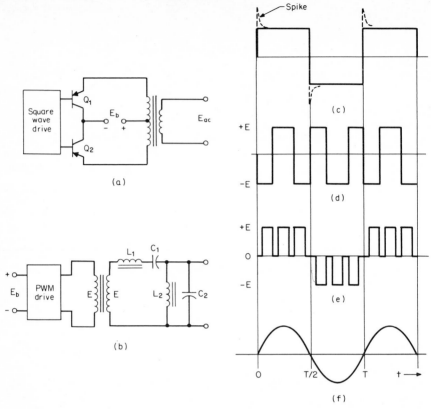

Fig. 2.9 Voltage-driven inverter circuits. (*a*) FWCT circuit. (*b*) PWM inverter circuit, thyristor drive. (*c*) Square-wave input and output waveform, circuit (*a*). (*d*) Two-level PWM waveform; circuit (*b*). (*e*) Three-level PWM waveform. (*f*) Sine-wave output waveform.

ing is initiated by square-wave drive to the base of each transistor, producing a square wave at the output (Fig. 2.9*c*). Since a flux change $\Delta\phi = 2\phi_m$ occurs at the end of each time interval $\Delta t = T/2$, the basic Faraday equation (Eq. 1.1) becomes, for a square-wave circuit:

$$E_{\text{rms}} = 4fNB_mA\,10^{-8} \qquad (2.22)$$

Comparison of this equation with that for the sine wave

$$E_{\text{rms}} = 4.44fNB_mA\,10^{-8} \qquad (1.8)$$

indicates a difference only in the coefficient. A significant conclusion, however, can be inferred from the relation between area product and VA rating Eq. (2.5). A transformer with square-wave excitation requires $4.44/4 = 1.11$ times the area product, and hence its size (in theory) is larger than a transformer with sine-wave excitation.

An important application of the inverter circuit occurs in the *unin-terruptable power supply* (UPS). Here there is a need for a sine wave-form in the output, but it is important that the filter components and transformer be kept small. There is a variety of ways to accomplish this. Some involve the use of three-phase circuits, in which stepped waveforms that approximate the contour of a sine wave are generated. Because the stepped waveform has a small amount of harmonic con-tent, little or no filtering is necessary.

a. Pulse Width Modulation

An alternative means of generating a sine waveform involves the use of *pulse width modulation* (PWM). Characteristic of PWM cir-cuitry is an active device (transistor or thyristor) which is switched at a rate higher than that of the power line. There are several types of switching circuits.

For the *programmed waveform* inverter, a periodic switching pat-tern is selected which will produce the best waveform for the number of switching operations per cycle. In the *carrier-modulated* PWM inverter, the power devices switch at the zero crossing of two signals. One signal is a reference triangular waveform at a frequency higher than the fundamental frequency. The other signal, which modulates the triangular signal, is that of the line power frequency. The degree of modulation determines the magnitude of harmonic distortion. A natural objective is to achieve optimum VA ratings of both the trans-former and filter components.[20] In one such system, the design makes use of a single-phase drive circuit (Fig. 2.9b), in which the pulse width is modulated to produce either a two-level waveform (+E, −E) or a three level waveform (+E, 0, −E). Either waveform, when filtered, results in a sine wave (Fig. 2.9f). But the three-level PWM circuit results in minimum size of the transformer. With the virtual elimina-tion of the third and fifth harmonics, LC filter components can be made very small.

Serious problems in the drive circuits can arise when the output load has a low power factor, e.g., an induction motor during transient or starting conditions. In such situations, the designer may elect to choose as the source a current generator instead of the customary voltage generator.

b. High-Voltage Electronic Ignition

The ignition circuit for the gas-combustion engine of an automobile converts dc from a battery to high-voltage ac applied to a spark-gap

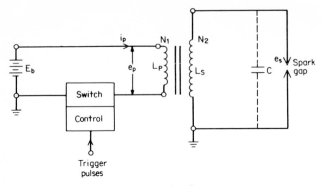

Fig. 2.10   Automotive ignition circuit.

load.   In the inverter circuit of Fig. 2.10,* breaker contacts in series with the primary are replaced by a semiconductor switch such as a transistor or thyristor.[21]   It then becomes feasible to step up low dc voltage E_b to kilovolts of ac (e_s) because the turns ratio N_2/N_1 is magnified by the stepped-up voltage $e_p > E_b$ which results from the opening of a switch in an inductive circuit.   Thus:

$$e_p = L_P \frac{di}{dt} \tag{2.23}$$

and

$$\frac{e_s}{e_p} = \frac{N_2}{N_1} \tag{2.24}$$

Alternatively, the primary voltage e_p during the off interval can be computed from an equation which equates flux linkages ϕ with the volt-second product:

$$\phi = E_b t_{on} = e_p t_{off} \tag{2.25}$$

With a six-cylinder engine and a shaft rotation of 1,200 to 3,100 r/min, the minimum circuit frequency is (1,200/60)6 = 120 Hz. The time between pulses would be 1/120 = 8.33 milliseconds (msec). Typical values in a *high-energy ignition* (HEI) circuit which satisfy Eqs. (2.24) to (2.26) would be as follows:

$E_b = 13.5$ V dc $L_P = 7.75$ mH (millihenrys)

$t_{on} = 8.0$ msec $dt = t_{off} = 0.31$ msec

$$\frac{N_2}{N_1} = 100$$

* Note the similarity to the flyback circuit, Fig. 2.13*b*.

$e_p = 350$ V peak $= (7.75 \times 14)/0.31$

$di = i_p = 14$ A max   (average $= 7$ A)

$e_s = 35,000$ V $= (13.5 \times 100 \times 8$ msec$)/0.31$ msec

2.10. CONVERTER CIRCUITS[22]

The addition of diodes to the inverter circuit of Fig. 2.9a results in the conversion of dc to dc (Fig. 2.11). The commutation from one switch to the other does not take place instantaneously. The finite interval (dwell time) between on and off periods produces a trapezoid dc waveform at the output (Fig. 2.11c). Filtering requires only a small shunt capacitor because of the brevity of the dwell time.

a. Switch-Mode Circuits

In Fig. 2.12 we depict three basic circuits used in a switch-mode regulated power supply (SMPS). Modulation of the pulse width occurs

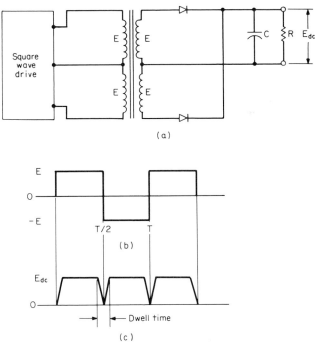

Fig. 2.11   Square-wave converter.   (a) Push-pull CT circuit. (b) Square-wave input voltage.   (c) Output dc volts.

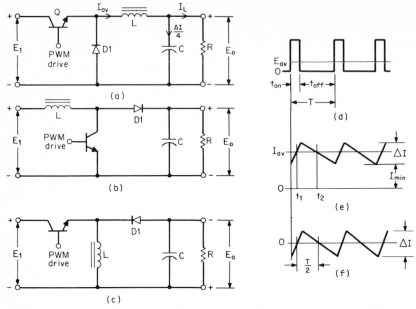

Fig. 2.12   Switch-mode circuits and waveforms without line isolation. (*a*) Series inductor step-down, forward converter circuit. (*b*) Shunt switch step-up, boost converter circuit. (*c*) Shunt inductor step-up, flyback converter circuit. (*d*) PWM-drive waveform. (*e*) Inductor ripple current of circuit (*a*). (*f*) Capacitor ripple current of circuit (*a*).

at the base of the transistor switch and results in the transformation of dc from one level to another.   Step down or step up of voltage is a function of the duty ratio *d*, defined as:

$$d = t_{on}/T \tag{2.26}$$

where t_{on} is pulse on time and $T = 1/f$ is the period corresponding to the frequency *f*.   Each of the three circuits behaves, in effect, like a *dc transformer* with a voltage ratio of $1:M$, as in:

$$M = E_o/E_I \tag{2.27}$$

In Table 2.5, we summarize the nomenclature and transfer ratio for each of the generic circuits shown in Fig. 2.12.   Note that the transformation from one circuit to another arises from a simple change in the location of the three elements—switch, inductor, and diode. Thus dc is stepped down in a forward (series inductor) circuit and stepped up by either a *boost* (shunt switch) circuit or a *flyback* (shunt inductor) circuit.

Attenuation of the triangular ripple current is achieved by designing inductor L and capacitor C to meet a limit specified for the peak-to-

TABLE 2.5 Converter Circuits

Topology	Figure	Nomenclature	M	Voltage ratio, E_o/E_1
Series L	2.12a	Buck (forward)	d	t_{on}/T
Shunt switch	2.12b	Boost	$1/(1-d)$	T/t_{off}
Shunt L	2.12c	Buck-boost (flyback)	$d/(1-d)$	t_{on}/t_{off}

peak ripple current ΔI (typically 0.4 I_o) and the ripple voltage ΔE. The basic equations for the *step-down* case are:

$$L\,\Delta I = \frac{E_o}{t_{off}} = \frac{E_o(1-d)}{f} \tag{2.28}$$

$$C\,\Delta E_c = \frac{\Delta I}{4}\frac{T}{2}\frac{\Delta I}{8f} \tag{2.29}$$

Equation (2.29) has been written to display the product of two terms: *average* ripple current ($\Delta I/4$) and the half period ($T/2$) during which it occurs (see Fig. 2.12f).

In the *step-up* circuit, the basic equations are:

$$L\,\Delta I = E_1 t_{on} \tag{2.30}$$

$$C\,\Delta E = I_o t_{on} \tag{2.31}$$

The equations for the two circuits have been cast in this simple parallel form so that we can more readily see how ripple current and ripple voltage relate to the magnitude of L, C, and pulse time. Two further comments are in order. In a regulated power supply, the values chosen for L and C are determined by constraints placed on the limiting values of input and output voltage and current. Also, if minimum speed of response to transient changes of load current is specified, L must satisfy an upper as well as a lower limit of inductance.

b. Transformer Coupling

The transformerless converters shown in Fig. 2.12 do not provide for isolation between output and input, a disadvantage which is overcome in the circuits shown in Fig. 2.13. The flyback circuit with a step-up transformer now yields a step-up ratio which is magnified by the turns ratio:

$$\frac{E_0}{E_1} = \frac{N_2 t_{on}}{N_1 t_{off}} \tag{2.32}$$

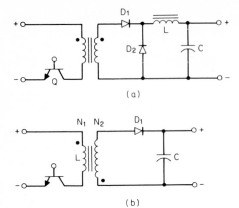

(a)

(b)

Fig. 2.13 Converter circuits with line isolation. (*a*) Forward converter; free-wheeling diode D₂ reduces secondary losses. (*b*) Flyback converter with step-up transformer.

Because these circuits are single-ended and are polarized with dc, they have the same disadvantages as the half-wave rectifier circuit. With flux density restricted to a limited excursion:

$$\Delta B = B_m - B_r < B_m \tag{2.33}$$

Thus, it becomes necessary to use a large iron-copper area product to produce a moderate VA rating.

Better utilization of the area product is achieved by using balanced circuits in which an equal flow of current produces symmetrical changes in flux over a complete switching cycle. This occurs in the bridge circuit, which requires two pairs of switches and diodes. Balance is also obtainable with the push-pull forward converter. This less costly circuit uses only one pair of switches and one pair of diodes. Consider the popular circuit shown in Fig. 2.14a, in which both windings are center-tapped and a PWM drive circuit is connected to the primary. This results in half-wave modulation (Fig. 2.14b). Upon rectification, we obtain a train of rectangular voltage pulses (Fig. 2.14c). The reactance of C is small and the waveform of the voltage across L is rectangular. The waveform of the current through L is triangular (Fig. 2.14d), resulting in a ripple voltage across load R which is also triangular in shape.

The single-ended converter circuit, in its conventional form, requires a large iron-copper area product. However, studies of the topology of converter circuits show that efficient utilization of the iron and copper of inductive components is possible even with a minimum of active elements. One example is a circuit in which unbalanced dc is eliminated by the addition of two capacitors, one in series with the primary

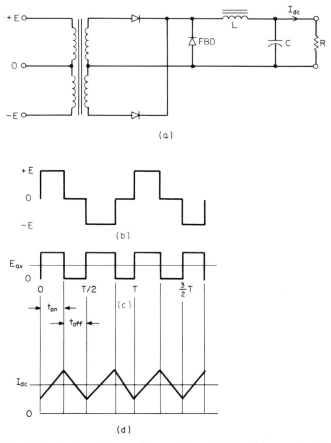

Fig. 2.14 Push-pull forward converter. (*a*) Transformer circuit, *CT* primary and secondary. (*b*) Half-cycle width modulation input. (*c*) Rectified output voltage to filter. (*d*) Triangular current waveform.

and another in series with the secondary.* The result is a magnetically efficient circuit with only one transistor and one diode.[24]

Operation at high frequency results in a reduction of the area product of the transformer, as we noted earlier in comments following Eq. (2.5). It follows that the operation of switching circuits in the supersonic band of frequencies should and does result in a drastic reduction in the size of both the transformer and the filter inductor, as we shall see later in Secs. 5.7 and 7.6.

---

* See Fig. 7.8. In the *resonant converter* circuit, a series input capacitor (lumped with the distributed capacitance) is chosen to resonate with a series inductor and the leakage inductance of the transformer.[23]

2.11. RELIABILITY

Two requirements dominate present-day specifications: miniaturization and reliability. *Miniaturization* is the design or redesign of a component for minimum size and weight. *Reliability* characterizes a component that we are confident will perform to specifications.[25]

Reliability is considered so important that it is often given priority over cost, delivery, and miniaturization. It is sought as a separate good, for it is recognized that reliability and quality, although closely related, are not synonymous. Neither does high cost necessarily ensure reliability. Moreover, reliability will be adversely affected by miniaturization unless the designer is very careful.

Dielectric considerations are relevant to reliability, and thermal considerations are relevant to miniaturization. We therefore interrupt our discourse on the power transformer to discuss dielectric design in Chap. 3 and thermal design in Chap. 4. In Chap. 5 we shall make use of all the concepts discussed in Chaps. 2, 3, and 4.

REFERENCES

1. W. Olschewski, "The Hybrid Compatible Transformer," *IEEE Transactions on Components,* Vol. CH MT-2, No. 4 (December 1979): 487–90.
2. M.I.T. Electrical Engineering Staff, *Magnetic Circuits and Transformers* (New York: John Wiley & Sons, 1943), pp. 396–98.
3. D. Paice, "Constant Voltage, Current, and Power: A Circuit for All Reasons," *IEEE Transactions on Industrial Electronics,* Vol. IEC I-25, No. 1 (February 1978): 55–58.
4. N. Grossner, "The Geometry of Regulating Transformers," *IEEE Transactions on Magnetics,* Vol. MAG-14, No. 2 (March 1978): 87–94.
5. H. Garbarino, L. Stratton, "Current-limiting Power Transformers," *Electrical Manufacturing* (October 1959): 116–18.
6. A. Kusko et al., "Transfilter," *IEEE Transactions on Magnetics,* Vol. MAG-12, No. 4 (July 1976): 389–92.
7. H. Zade, *Welding Transformers and Rectifiers* (London: Macmillan & Company, 1967).
8. A. Kusko, T. Wroblewski, *Computer Aided Design of Magnetic Circuits* (Cambridge, Mass.: MIT Press, 1969).
9. H. Lord, "Analog Equivalent Circuit Aided Design of Ferroresonant Transformers and Circuits," *IEEE Transactions on Magnetics,* Vol. MAG-13, No. 5 (September 1977): 1,293–98.
10. W. Newell, "Power Electronics," in D. G. Fink (ed.), *Electronics Engineers Handbook* (New York: McGraw-Hill Book Company, 1975), Sec. 15.
11. L. Gyugyi, B. Pelley, *Static Power Frequency Changers* (New York: John Wiley & Sons, 1976).

12. F. Connelly, *Transformers* (London: Sir Isaac Pitman & Sons, 1950), pp. 248–60.
13. R. Distler, S. Munshi, "Critical Inductance and Controlled Rectifiers," *IEEE Transactions on Industrial Electronics* (March 1965): 34–37.
14. O. Schade, "Analysis of Rectifier Operation," *Proceedings IRE*, Vol. 31 (July 1943): 341–61.
15. T. Gray, *Applied Electronics*, 2d ed. (New York: John Wiley & Sons, 1954), p. 328.
16. International Telephone and Telegraph Corporation, *Reference Data for Radio Engineers*, 6th ed. (Indianapolis: Howard W. Sams & Company, 1977).
17. J. Brugler, "Theoretical Performance of Voltage Multiplier Circuits," *IEEE Journal of Solid-State Circuits*, Vol. 6, No. 3 (1971): 132–35.
18. A. Ruitberg, "Design Techniques for Miniaturized Spacecraft High Voltage Power Supplies," *Solid-State Power Conversion*, Vol. 5, No. 4 (July–August 1979): 12–23.
19. E. Hnatek, *Design of Solid-State Power Supplies*, 2d ed. (New York: Van Nostrand Reinhold, 1981), Chap. 10.
20. S. Sriraghavan et al., "Voltampere Ratings of Components of PWM Inverter System," *IEEE Transactions on Industrial Electronics*, Vol. IEC I-25, No. 3 (August 1978): 278–84.
21. G. Huntzinger, G. Rigsby, *HEI: A New Ignition System*, Paper No. 750,346, Society of Automotive Engineers, Automotive Engineering Congress, February 1975.
22. A. Pressman, *Switching and Linear Power Supply: Power Converter Design* (Rochelle Park, N.J.: Hayden Book Company, 1977).
23. V. Vorpérian, S. Ćuk, "A Complete DC Analysis of the Series Resonant Converter," *IEEE Power Electronics Specialists Conference Record*, Vol. 82CH 1,762–4 (1982): 85–100.
24. S. Ćuk, "General Topological Properties of Switching Structures," *IEEE Power Electronics Specialists Conference Record*, Vol. 79CH 1,461–3 AES (1979): 109–30.
25. IEEE, *Recommended Practice for Achieving High Reliability in Electronic Transformers and Inductors*, Standard No. 392 (New York: Institute of Electrical and Electronics Engineers, 1976).

CHAPTER 3

Insulation and the Dielectric Circuit

Although the transformer has an enviable reputation for reliability and longevity, its major weaknesses have been known since its introduction to industry in the nineteenth century. The principal problems arise from the presence of *moisture, vibration, corona,* and *heat.*

The pattern of failure of transformers is similar to that of most passive components. There is an "infant mortality" period of brief duration, typically 10 to 50 hours (hr), when the rate of failure is irregular or erratic. During the major period of service, failures occur at a slow but constant rate, usually measured in thousands of hours. This interval is known as the *mean time between failures* (MTBF). During the final phase, the wear-out period, the rate of failure accelerates.

This chapter is devoted to a discussion and analysis of one of the most important causes of infant mortality of the small transformer: failure of the insulation in the dielectric circuit.

3.1. RELIABILITY OF THE INSULATION

In our survey we stress the importance of geometry and the problem of nonuniform electric fields. Our analysis will show that the presence of discontinuities in the dielectric circuit (voids and impurities, for example) can result in corona and, therefore, in premature breakdown of the transformer. We focus on the problem of corona, the bugbear of insulation, because all too often it is the cause of the infant mortality of a transformer. In consequence, we shall seek out those guidelines for design which will result in minimum levels of corona. If successful, we will enhance the probability that the insulation will endure for its normal expected 10,000 to 20,000 hr.

3.2. DIELECTRIC STRESSES

It is by no means an overstatement to say that the design of reliable insulation is as difficult, at times as complex and challenging, as other major transformer design problems. Our survey, which skirts a vast accumulation of data in an effort to focus on basic principles, of necessity cannot do justice to a subject which more naturally falls in the domain of the physical chemist and the dielectric engineer.

The major function of insulation is, of course, to insulate. The dielectric material should function as a nonconductor over the entire range of environmental stresses. We know that dielectrics such as impregnants and resin are used as a barrier against moisture. Solid insulation is used for mechanical functions such as the rigid support of splices and terminals with stable clearances. In the coil itself, the insulation of the layers must have sufficient compressive strength to withstand winding tension. The mechanical characteristics of the dielectric are just as important as its chemical, dielectric, and thermal properties.

When the transformer is energized with an applied voltage, it is subjected to thermal and dielectric stresses. The internal temperature rises, and immediately there are differences in potential almost everywhere in the transformer: (1) between turns, (2) between layers of wire, (3) between windings, (4) between terminals, (5) between a lamination or the core itself and the end turn of a layer (this clearance is called the *margin*), and (6) between the ground (the core or can) and each terminal.

The voltage of any one of these differences in potential is designated by V_{ab}. The designer's task is to satisfy, under all environmental conditions, the equation:

$$V_{ab} < V_t < V_B \qquad (3.1)$$

where V_B is the *breakdown* voltage and V_t the *test* voltage.

The *sparkover* voltage is V_S, and when visible sparkover occurs between terminals:

$$V_B = V_S$$

Breakdown is deemed to occur when there is an abrupt and large increase in the normally very small leakage current.*

Equation (3.1) is of deceptive simplicity. The transformer designer has a strong incentive to employ the least insulation necessary to satisfy Eq. (3.1) because, as we noted in Chaps. 2 and 5, the size of the transformer is a function of the copper space factor.† The less the insulation and the smaller the clearances, the smaller the transformer and (see Chap. 4) the better the heat transfer.

Let us rewrite Eq. (3.1) as:

$$C_a V_{ab} + A = V_t = \frac{V_B}{C_t} \tag{3.2}$$

where C_a, A, and C_t represent safety factors.

We may then define the designer's task as the determination of the minimum values of C_a, A, and C_t which provide: (1) a high degree of reliability, (2) the smallest transformer (least insulation and smallest clearances), and (3) efficient transfer of heat.

Now consider the effect of time and of wave shapes. In most types of solid and liquid insulation, the breakdown voltage V_B is a function of time:

$$V_B = F\left(\frac{dV}{dt}, \; t_i, \; \sum_1^N t_i\right) \tag{3.3}$$

where dV/dt is the rate at which the voltage V_{ab} is applied, t_i is the duration of the voltage, $\sum_1^N t_i$ represents the accumulated test time, and N is the number of tests of duration t_i.

Especially in solid insulation, V_B decreases progressively as the rate, duration, and number of tests increase. The peak value of V_B also depends on the wave shape of the voltage as well as on whether it is dc or ac. Thus, the value of V_B of a sine wave and of a train of pulses differs. When the voltage has a sine waveform, V_B varies with frequency; when the waveform is a pulse or a pulse train, V_B is a function of rise time, pulse duration, and pulse repetition rate.

Since breakdown is also a function of the geometry of the electrodes, the uniformity of the electric field and, of course, the nature of the dielectric, it is perhaps not surprising that different laboratories often differ about the value of V_B (by 30 percent is not unusual). Consequently, the physical chemist wisely employs the *intrinsic breakdown value* of a dielectric material as a way of isolating its intrinsic properties

---

* Breakdown is discussed further in Sec. 3.13.
† The copper space factor is analyzed further in Chap. 5.

from the effect of asymmetry of the electric field and electrode geometry.

Clearly, standardized test conditions are mandatory if valid correlations and reproducible results are to be achieved. In the *rapid-rise–short-time test* for insulation specimens, V is increased until V_B occurs. In the more stringent *step-by-step test,* V is raised in steps of brief intervals (usually 1 min) until the breakdown voltage V'_B is reached. V'_B is less than V_B.

The most common dielectric test used for electronic transformers is to apply an ac test voltage V_t of commercial line frequency between the windings and between the windings and the ground for 5 to 60 seconds (sec). Test voltages above 1,000 volts (V) root-mean-square (rms) voltage are applied gradually at a rate not exceeding 500 V/sec.

Interwinding tests do not reveal the adequacy of the insulation between layers or turns of the winding itself. The *induced voltage test* yields this information. A voltage two or three times higher than working voltage is applied to a winding at a frequency two or three times higher than normal to avoid magnetic saturation of the core. Such a test is valuable because it reveals defects which might eventually result in a shorted turn and excessive exciting current. The test is especially important when the layers of the transformer are not insulated, a condition common in the random-wound bobbin and the toroid.

3.3. ELECTRIC STRENGTH

The concept of the strength of the electric field, or intensity $E,$ defined as a voltage gradient, is helpful in evaluating the breakdown level of the insulation:

$$E = \frac{dV}{dx} \tag{3.4}$$

where x is the spacing between two points in a field.

The field consists of lines of electric flux terminating on the surfaces of each electrode, a and b. If the density of the flux is uniform, then E is constant between the electrodes. This condition obtains for flat parallel electrodes (Fig. 3.1*a*) which are separated by a small spacing $x = d$. If V_{ab} is the voltage between electrodes, the gradient is $E = V_{ab}/d$. The electric strength E_B sufficient to break down a dielectric may then be defined as:

$$E_B = \frac{V_B}{d} \bigg|_{x = d} \tag{3.5}$$

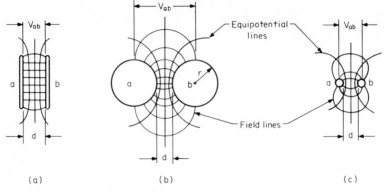

V_{ab}

V_{ab}

V_{ab}

V_{ab}

Equipotential lines

a   b

a   b

a   b

r

d

d

Field lines

d

(a)   (b)   (c)

Fig. 3.1 Symmetrical electrodes.

In the thin solid materials used for the insulation of the layers of a coil, the electrical conductivity is a linear function of E, and E_B varies inversely as an exponential function of d:

$$E_B \cong \frac{V_B}{d^{2/3}} \tag{3.6}$$

In a large number of solids, the value of E_B, in short-time tests, (typically 60 sec) is in the range of 800 to 3,000 volts per milli-inch (V/mil). In pure liquids, such as carefully filtered oils, E_B is almost independent of d, so that strength is usually expressed in kilovolts per one-tenth inch (kV/0.1 in). Typical values are in the range of 30 to 42 kV/0.1 in, i.e., 3,000 to 4,200 V/mil or 118 to 165 kV/cm.

3.4. ELECTRODE GEOMETRY AND NONUNIFORM FIELDS

One of the major difficulties of providing reliable insulation stems from the presence of nonuniform fields in various critical regions of the transformer.

a. Symmetrical Electrodes

It must not be assumed that identical or symmetrical electrodes are sufficient to ensure a uniform field. In the familiar parallel-plane configuration, for example, the field is uniform and symmetrical, but there is fringe flux at the edges and there are nonuniform fields in those areas. Even in a uniform region, it is possible for distortion of the field to occur when a discontinuity is introduced into an otherwise homogeneous dielectric. The nonuniform field is of fundamental importance and warrants an extended discussion.

First, let us recall that, between a pair of electrodes, there is an electric field which may be visualized in terms of lines of force. The total number of lines, or flux ϕ, is:

$$\phi = \int E \cos \theta \, da \qquad (3.7)$$

where θ is the angle between E and an element of area dA in the field.

Consider the general expression:

$$V_{ab} = \int_a^b E \cos \, dx$$

If the flux density $d \, \phi / dA$ is constant, and if $\cos \theta = 1$, then E is a constant:

$$V_{ab} = E \int_a^b dx \qquad V_{ab} = \frac{E}{b-a}$$

Thus, in the arbitrary case where two plane electrodes with equal areas are separated by a clearance d, as in Fig. 3.1a, we may say that $E = V_{ab}/d$, since the field is uniform.

The analysis of nonuniform fields is greatly facilitated by the use of graphical field mapping.[1] In this method, field lines and equipotential lines are drawn between electrodes in accordance with the following principles:

1. Field lines and equipotential lines always intersect at right angles.

2. A field line always meets an electrode normally (i.e., at a right angle) because an electrode is an equipotential surface.

3. Each flux tube is divided into three-dimensional field cells whose sides are curvilinear squares.

4. Electric flux ϕ over any cross section of a flux tube is the same.

5. Electric flux density D is inversely proportional to cell width ($D = \epsilon E$, where ϵ is the *permittivity* or dielectric constant).

6. Field intensity E is inversely proportional to cell length measured in the direction of the field line.

Now suppose that the electrodes are symmetrical spheres with large radii, as in Fig. 3.1b. When equipotential lines are drawn, it will be seen that, although the field is symmetrical, it is not uniform over the entire surface of the electrode. In the region covering a small surface area and an imaginary tube between a and b, however, the field is fairly uniform and reasonably constant. Even so, the flux density is greater than in a uniform field (Fig. 3.1a), and E is greater than V_{ab}/d. We characterize this behavior as *field enhancement*.

In Fig. 3.1c, which depicts much smaller spheres, there is an even smaller region in which the field is uniform. Here, $E \gg V_{ab}/d$. For practical purposes, the tiny spheres must be replaced by thin needles whose points have small radii of curvature. The behavior of small spheres is therefore simulated by the use of needle-shaped electrodes, described as *needle geometry*.

In each of these cases, thus, the field may be symmetrical but nonuniform.

b. Asymmetrical Electrodes

The field may be asymmetrical as well as nonuniform. If the surface areas of each electrode are unequal ($A_a < A_b$), and if their separation ($b - a = d$) is comparable to or greater than A, then the intensity of the local field E_N is the immediate vicinity of A_a (where $E_N = E_a$) will be greater than in the vicinity of the larger area A_b (where $E_N = E_b$).

Suppose that one electrode is a flat plane and that the other is shrunk in size to a sphere of very small radius r. We thus obtain the *point-plane electrode geometry* of Fig. 3.2, in which, because of the divergent field:

$$E_a \gg E_b > E_{ab} = V_{ab}/d$$

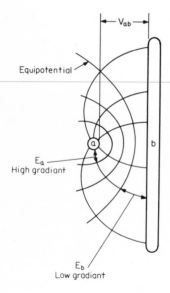

Fig. 3.2 Asymmetrical electrodes: point-plane geometry.

There is less field enhancement in the vicinity of electrode a, if d is substantially increased $(d \gg r)$. E_a may then be only slightly greater than E_b.

c. Corona: Partial Discharges

Let us suppose again that the electrodes are identical large planes (or spheres of very large radius) separated by 1 centimeter (cm) in dry air [at 1 atmosphere (atm) = 76 mm Hg, 20°C].

If V_{ab} is raised to $V_S = 30.3$ kV or 21.4 kV rms, sparking (breakdown of the air) occurs, and we say that $E_B = E_S = 3.03$ kV/cm. It should be noted that the E_S of air is not constant but varies inversely with gap spacing:

$$E_S = 24.22 + \frac{6.08}{\sqrt{d}} \qquad \text{kV/cm} \qquad (3.8)$$

It is also important to note that E_S rises sharply when the gap is less than 1 millimeter (mm).*

Now let us alter the electrodes so as to produce an asymmetrical geometry, for example, spheres of unequal radii or a point-plane configuration. The field is now asymmetrical as well as nonuniform. If the electrodes are separated by 1 cm, and if V_{ab} is raised to a value of V_i (let us say 20 kV, somewhat less than 30 kV, which is the approximate breakdown value of air), the field E_b in the vicinity of the plane may be 25 kV/cm. No breakdown occurs near this electrode. But the nonuniform field $E_N = E_a$, in the vicinity of the point or small sphere, may substantially exceed 30 kV/cm. A self-sustained, local, luminous, gaseous discharge, called *corona*, will then occur in the vicinity of that electrode. We call V_i (20 kV in our example) the *corona-inception voltage* or *corona-onset voltage*. If V_{ab} is dc, then the onset of corona is affected by polarity; that is, V_i is slightly less at a negative point than at a positive point.

Local gas discharges may also occur when there is inhomogeneity or distortion of the electric field as a result of discontinuities in the dielectric (see Sec. 3.9).

Corona is not easily detected. Even in air (where it produces nitrogen oxides and ozones), its luminosity may be marginal. And complete breakdown of the dielectric does not occur immediately, but may take place hours or even hundreds of hours later. Analysis of the statistics

---

* For a sine wave, $V_S = 21.4$ kV rms. It remains approximately constant up to 20 kHz, but then decreases steadily to a value 15 percent lower in the 5 to 15 MHz range. Above this, there is a further reduction of 30 to 40 percent.

of breakdown indicates that the life of insulation is inversely proportional to the ratio of corona inception to applied voltage:

$$\frac{\text{Corona life}}{\text{Normal life}} = \left(\frac{V_i}{V_{ab}}\right)^{\eta} \tag{3.9}$$

The exponent ($\eta = 3$ to 10) accounts for the sad fact that the life of the insulation shrinks from years to hours.[2]

3.5. FIELD ENHANCEMENT

It is difficult to calculate analytically the magnitude of E_N even for geometrically simple electrodes, and graphical field mapping, which may be used with success, is generally laborious.   In designing an insulation system, it is very helpful to estimate the *field-enhancement factor* (FE):*

$$\text{FE} = \frac{\text{max stress}}{\text{mean stress}} = \frac{E_N}{E_{ab}} = \frac{E_B}{E_i} \tag{3.10}$$

In the example we used for corona, the field was enhanced by a factor of $30/20$, or 1.5.

Field enhancement can occur in symmetrical fields in which there is no corona.   Thus, if both electrodes are small spheres or needle points, $E_a/E_{ab} = E_b/E_{ab}$ exceeds unity and breakdown occurs at lower voltages than in a geometry of large spheres or flat plane electrodes.

In the transformer, the physical manifestation of small spheres is those contours which have a small radius of curvature.   Examples are the sharp edges of a terminal or can, the ball of solder on a terminal, and even the tiny globules of solder in the vicinity of a splice.   The needle point is physically represented by loose strands from a lead wire or the sharp point on a poor solder joint.

In an oil-filled transformer, the surface of a flat or spherical electrode may develop (as a result of repeated sparkover tests) small but sharp points called *asperities*.   These asperities can produce a field enhancement as great as 5, which produces an effective change in the area of an electrode, described as the *area effect*.   It has been observed that the inception gradient E_i decreases with the logarithm of the area A.   For spherical electrodes with a radius r, an equivalent area $A \propto rd$ is defined, and the area effect is conspicuous when d is a small gap, i.e., 0.2 to 5 mil.

---

* Also described in the literature as *stress-intensification* or *stress-concentration factor*.

3.6. CURVATURE OF CONTOURS

An estimate of minimum radius of curvature in a gas can be arrived at through electrostatic considerations. The electrostatic potential at the surface of a sphere of radius r is:

$$V = \frac{1}{4\pi\epsilon_0}\frac{q}{r} \tag{3.11}$$

where q is the magnitude of charge on the surface and ϵ_0 is the permittivity of free space.

The maximum charge q_m that can be retained on the surface is limited by the breakdown gradient E_S:

$$q_m = 4\pi\epsilon_0 r^2 E_S \tag{3.12}$$

Therefore, the maximum potential V_i to which the sphere can be raised before corona occurs will be $V_i = rE_S$. Thus, the radius for corona inception is:

$$r = \frac{V_i}{E_S} \tag{3.13}$$

To illustrate, suppose the gas is air, for which $E_S = 30$ kV/cm, and that the maximum potential difference in a transformer is 3 kV. Then the minimum radius of any contour should exceed 0.1 cm *if corona is to be avoided.*

We will now compare this estimate of curvature with one obtained by considering the needle point-to-plane geometry, often regarded as inducing particularly severe field enhancement. This geometry may be mathematically represented by a semi-infinite hyperboloid with radius r, whose axis is normal to a plane electrode.[3] The maximum field enhancement at the tip is:

$$\frac{E_a}{E_{ab}} \cong \frac{2d/r}{ln(1 + 4d/r)} \tag{3.14}$$

where $d/r > 10$, E_a is the maximum gradient at the tip, and $E_{ab} = V_{ab}/d$ is the average gradient between electrodes. When d/r equals 10, 20, and 30, the average field is enhanced by factors of 5.4, 9.1, and 12.5, respectively.

If Eq. (3.14) is cast into a form similar to Eq. (3.13) by setting $E_a = E_S$ and $V_{ab} = V_i$, we obtain:

$$r \cong \frac{2}{ln(1 + 4d/r)}\frac{V_i}{E_S} \tag{3.15}$$

As we increase d/r from 10, to 20, and to 40, we obtain decreasing values of the coefficient: 0.538, 0.445, and 0.417, respectively. The radius of curvature predicted is about half the value predicted by the simpler Eq. (3.13), and we may therefore regard Eq. (3.13) as a conservative *criterion for minimum curvature* of sharp corners, edges, and contours.

3.7. PRESSURE AND TEMPERATURE: PASCHEN'S LAW

A generalized form of Paschen's law is that, in a uniform field, the breakdown voltage V_B of a gas is a function F of the product of pressure p and gap clearance d, but is inversely proportional to the absolute temperature T:[4]

$$V_B = F\left(\frac{pd}{T}\right) \tag{3.16}$$

A useful formula is:

$$V_B = 24.22\, \frac{293\, pd}{760\, T} + 6.08 \sqrt{\frac{293\, pd}{760\, T}} \tag{3.17}$$

where p is pressure in mm Hg (torr), T is absolute temperature, and d is spacing in cm;* that is, V_B decreases as altitude increases (or atmospheric pressure decreases) and as temperature rises.

In Fig. 3.3, the pd abscissa is based on 25°C, or 298°K.[5] The breakdown voltage above sea level can be estimated by using auxiliary altitude scales based on the mean temperature at high altitude.[6] A constant temperature of −68°C is assumed between 50,000 to 100,000 ft. If a different ambient temperature is assumed, the abscissa must be shifted by the ratio of absolute temperatures. An increase from 25°C to 85°C, for example, would result in a displacement to the left by a factor of 358/298, or 1.2, with a consequent reduction in breakdown voltage.

When a field is nonuniform because of the asymmetrical geometry of the electrodes, Paschen's law is not strictly valid. However, an application of the *law of similitude* shows that if all the linear dimensions of such a configuration are increased by the same factor by which the gas pressure is decreased, then the VA characteristic of a given gas discharge system remains constant.

---

* Some useful definitions: 1 torr = 1 millimeter of mercury (mm Hg); 1 atmosphere (atm) = 760 mm Hg = 1.0133 $(10)^5$ newton (N)/square meter (sq m); 1 bar = 10^5 N/sq m.

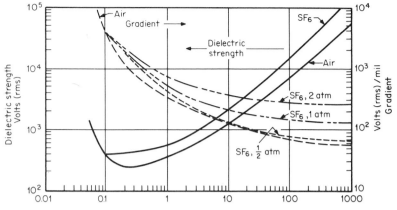

Fig. 3.3 Paschen's law curve: air and SF₆ (25°C). Solid curves show dielectric strength vs. $p \times d$ (atm × mils); dotted curves show gradient vs. d (mil).

The major implications of this important principle often pose difficult problems to the designer of transformers, as well as of all other components which must operate reliably at high altitudes. If we wish to maintain the same safety factors at high altitude as we do at sea level, then we should keep pd constant. For example, if p drops to about 33.6 mm Hg at 70,000 ft, then d should theoretically be increased by a factor of 760/33.6, or 22.6. Unfortunately, this value is much too large to be practical in small transformers: a ⅛-in (3-mm) gap between terminals would have to be increased to about 2¾ in (70 mm).

It is also important, if a component is to operate at high altitudes, to evaluate the sharp contours. The sparking gradient of air drops to about 2.5 kV/cm (when $pd = 10 \times 33.6$, or 336) at 70,000 ft. If we employ Eq. (3.15) to estimate the maximum exposed radius of curvature, we obtain:

$$r \cong 0.5 \frac{V_i}{E_s} = 0.5 \frac{500(1.4)}{2,500} = 0.14 \text{ cm}$$

If the maximum potential of the transformer is increased to 2.5 kV peak, then r should be increased to 0.5 cm.

The simple addition of a "service loop" on a lead wire to a terminal may be sufficient to distort the electric field and lower the corona-onset voltage. For example, a reduction of 20 percent has been noted at 4.5 mm Hg (about 113,000 ft).

A number of interesting anomalies occur at very low pressures, since a discrete minimum breakdown voltage—about 300 V—occurs when pd equals about 5. Thus it becomes possible for breakdown to occur in a long gap at a lower voltage than in a short gap! If pressure is reduced sufficiently, as in an almost perfect vacuum, gradients in the order of 100 kV/mm can be obtained.

3.8. THE GAS-FILLED TRANSFORMER

From Eq. (3.16) we may infer one of the advantages of the pressurized gas-filled transfomer; i.e., even when clearances are small, breakdown voltages can be very high. There are other advantages.

Consider the equation of state of an ideal gas:

$$pv = nR_0T \qquad (3.18)$$

where v is the volume, n the number of moles of the gas, and R_0 is the gas constant per mole.* If pv/nR_0 is substituted for T in Eq. (3.16), it may be restated as:

$$V_B = F\left(\frac{dn}{v}R_0\right) \qquad (3.19)$$

In this restatement, the internal breakdown voltage is independent of temperature—an advantage when high temperatures are anticipated. And since n/v is proportional to the density of gas, which even under high pressure is less than that of most solids or liquids, the gas-filled transformer will weigh less than one with a solid or liquid filling.

In principle, therefore, the only limitation to achieving an extremely high degree of resistance to breakdown when pressurizing a metal-enclosed transformer is the strength of the seams and their ability to remain leak-proof.

It is customary to rate the breakdown strength of gases in relation to nitrogen N_2, the breakdown strength of which is set at unity. Most suitable gases, other than air and hydrogen,† have strengths ranging between 2.2 and 2.8.

N_2 has a very low boiling point, $-195.8°C$, but poor thermal conductivity. Consequently, a gas such as sulfur hexafluoride SF_6, which boils at $-63.8°C$ and remains stable up to 150°C, is often employed. When an even higher operating temperature is necessary, perfluorethane, which boils at $-81.6°C$ and remains stable up to 350°C, is used. Its breakdown strength (relative to air) is 1.6.

---

* R_0 is related to Boltzmann's universal constant K_0 and Avogadro's number N_0, by $K_0 = R_0/N_0$.

† Hydrogen, which has excellent thermal conductivity and a low boiling point, $-252.8°C$, cannot be used because of its flammability.

Since corona may decompose some gases or decrease their electric strength, particular care is taken in the design of internal electrode geometry to minimize field enhancement.

3.9. DIELECTRIC DISCONTINUITIES

Solids and liquids are about a thousand times as dense as gas. Because electric strength is roughly proportional to density, the voltage gradients required to break down solids and liquids at atmospheric pressure are 10 to 100 times those required to break down gases. The E_S value of air is about 80 V/mil at 1 cm; the E_B of solid and liquid dielectrics ranges from 800 to 8,000 V/mil.

Such attractively high values are not easily exploited, however, because it is almost impossible to eliminate completely the discontinuities of dielectrics in a practical insulation system. Such discontinuities occur whenever two or more dielectrics (with different dielectric constants) are in an electric field. They also occur when there are cavities, cracks, and voids either within a solid insulation material or at the interface of an electrode and the insulation. In oil, discontinuities arise from bubbles of air, hydrogen, or water, and from suspended impurities.

a. Corona Inception in Solids[7,8,9]

If there are voids (e.g., trapped air bubbles) or cracks in a solid or at the interface of a solid and an electrode, an air gap exists within the solid, at its surface, or at the interface. We represent this condition by an equivalent circuit, Fig. 3.4a, in which C_a, the capacitance of the air gap, is in series with C_e, the capacitance of the solid. Bridging the two is C_d, the capacitance of the remaining solid. Across the air gap is a voltage $V_a = V_{ab}C_e/(C_a + C_e)$, which may also be written:

$$V_a = \frac{V_a}{1 + (d/\epsilon d_a)} \cong \frac{d_a}{d} V_{ab} \qquad (3.20)$$

where ϵ is the dielectric constant of the solid, d the electrode spacing, and d_a the depth of the void.

Since $V_a = V_{ab}$, a discharge in the gap can occur when there is an inception voltage $V_i = V_{ab}$ which is less than V_B, the normal breakdown value of the solid. When V_{ab} rises to that value at which the gap voltage V_a equals the air saturation potential V_S, (Fig. 3.4b) a gas discharge, i.e., corona, begins. The accumulated charge on C_a flows through the external circuit in a brief interval (in the order of 10^{-7} sec). Since the voltage across C_e rises, the V_a drops to an extinc-

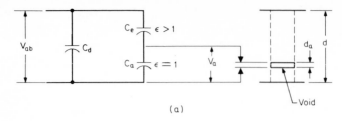

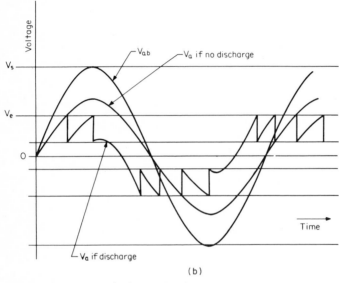

Fig. 3.4 Corona: gas discharge circuit.

tion voltage V_e. The V_e now rises at a sinusoidal rate to V_S. The sawtooth train of pulses which results may repeat several times each half-cycle of the sine wave. The Fourier spectrum of such a waveform is in the radio frequency (RF) range and is commonly detected by employing a circuit comprising a high-pass RF filter and a wide-band oscilloscope.

At $V_a = V_S$, corona starts in the gap:

$$V_i = \left(d_a + \frac{d}{\epsilon}\right)\frac{V_S}{d_a} < V_B \qquad (3.21)$$

where V_S, the sparking potential of air, varies as a nonlinear function of the air gap [see Eqs. (3.8) and (3.17)].

Empirical values of V_S [obtained from Eq. (3.17)] or plots of E_S vs. air gaps may be substituted into Eq. (3.21). If V_i is then plotted against

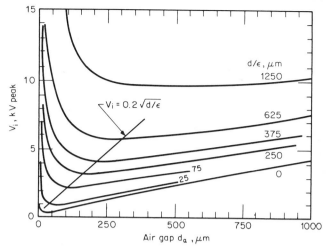

Fig. 3.5   Discharge inception vs. air gap.

d_a with d/ϵ as a parameter, a family of curves (Fig. 3.5) is obtained.[10] Note that each curve has a minimum value of corona-onset voltage V_{im} and that each minimum increases as d/ϵ increases.

b. Field Distortion and Inclusions[4]

The series air gap circuit just analyzed is only a special case of dielectric discontinuity.   Further insight may be obtained by the analysis of the more general case, a dielectric material immersed in a region which has a different dielectric constant.

Assume that an originally uniform field E_o, in a medium with a dielectric constant of ϵ_1, has placed in it a dielectric sphere of ϵ_2 with radius r.   It can be shown that the field E is enhanced by inclusions such as gas bubbles or solid particles.   If there is a gas bubble inside the sphere (Fig. 3.6), the field is constant and the enhancement ratio:

$$\text{FE} = \frac{E}{E_o} = \frac{3\epsilon_1}{2\epsilon_1 + \epsilon_2} \tag{3.22}$$

can reach a maximum of 1.5 when $\epsilon_1 \gg \epsilon_2$.

When the inclusion is a solid, enhancement occurs at its surface and depends on its shape, which can be defined in terms of the axial ratio γ (long diameter/short diameter) of the particle.   Analysis shows that for a sphere FE = 3, but for a prolate spheroid, it can be larger:

γ	1	2	3	5
FE	3	5.8	7.5	18

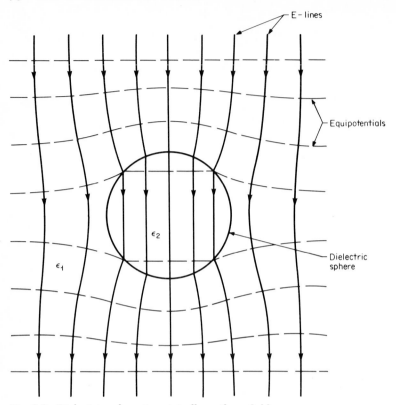

Fig. 3.6   Dielectric sphere in originally uniform field.

Since the inside and outside dielectrics each have different electrical conductivity, there will be a temperature difference between the inside and outside of the sphere.   If E_o is close to E_B, local hot spots can occur.

If the sphere is a gas, a gaseous discharge—corona—may occur. In a typical vacuum of 7.6 mm (i.e., one-hundredth atmospheric pressure), a trapped bubble shrinks to one-hundredth its normal volume. But it retains a fifth of its normal radius.   Thus, *even after vacuum impregnation, air bubbles may be sizeable enough to cause corona.*

An important problem occurs in randomly wound coils with many thousands of turns or in tight coils of considerable thickness (called *deep coils*).   It is difficult to impregnate such coils thoroughly, even in a nearly perfect vacuum.   If the entrapped sphere is a solid impurity or a liquid one (e.g., undissolved moisture), either E_B is reduced or local hot spots may initiate chemical deterioration of the ambient dielectric.

3.10. DESIGNING TO AVOID CORONA

If the equivalent circuit shown in Fig. 3.4a is a valid model of corona inception, then it is possible, in principle, to ensure the absence of corona by selecting in Fig. 3.5 that value of d/ϵ whose corona inception voltage V_{im} exceeds the peak working voltage V_{ab}; that is, a criterion for avoiding corona in solids in uniform fields is:

$$V_{ab} < V_{im} = F(d/\epsilon) \tag{3.23}$$

In accordance with this approach, we plot minimum corona inception voltage V_{im} vs. d/ϵ using a range of air gaps d_a as a parameter. Note that the smallest value of V_{im} is consistent with the Paschen curve in Fig. 3.3 and with Eq. (3.21), since V_{im} is approximately 300 V when $d/\epsilon = 0$ (see Fig. 3.5).

An empirical formula for the line intersecting the minima of the curves in Fig. 3.5 is of the form:

$$V_{im} = K'(d/\epsilon)^{1/2} \tag{3.24}$$

where K' depends on the sparking gradient E_s of the gap and on its permittivity. Most important, in solids, is the problem of the air gap or void. The coefficient in this example depends on which system of units d (electrode spacing) is expressed in:

K'	V	d
1,000	volts	milli-inches
200	volts	microns
6.32	kilovolts	millimeters

We can conclude, from our discussion, that the choice of electrode spacing will depend on the insulation permittivity and will be markedly affected by the degree of local field enhancement. For solid insulation we can formulate an anticorona design criterion in terms of a minimum rms gradient E_i:

$$E_i = \frac{V_i \text{ rms}}{d} = \frac{K'/\sqrt{2}}{\sqrt{d\epsilon}} \tag{3.25}$$

$$d = 2\epsilon \left(\frac{V_i \text{ rms}}{1,000}\right)^2 \text{mil} \tag{3.26}$$

Let us apply Eq. (3.26) to the specification of interwinding insulation. If the maximum working voltage between a pair of windings is 1,000 V rms and an organic insulation with a value of $\epsilon = 3.2$ is selected, then d should be not less than 6.4 mil (163 m) thick. E_{min}, in this case, would be 156 V/mil. In actual practice, d must also be of sufficient thickness to withstand a number of applications of the dielectric test voltage, 3,000 V rms.

We know from experience that, in the evaluation of porous materials (e.g., glass fiber cloth) or materials known to contain voids or pinholes, we need to derate the short-term electric strength substantially. Design values of 75 to 100 V/mil are therefore in use, about a third of the values used for films which are relatively free of pinholes. When film or other sheet materials are used in a transformer, it is not unusual for voids to result because of the evaporation of solvents in the impregnant* or when impregnation during the vacuum cycle is incomplete. Air gaps such as cracks can also occur as a result of thermal shock during the baking cycle of an impregnating or potting procedure.

Since the diameter and nature of a void are difficult to determine by tests for corona, we may very well be justified in adopting a "worst case" design approach, using Eq. (3.26) and curves such as Fig. 3.5. We would then characterize a solid insulation *system* that contains material whose value of ϵ is low as having a higher probability of withstanding corona than one that contains material whose value of ϵ is high.

Natural mica appears to be an exception to this line of reasoning. This classic insulation, which has an ϵ of approximately 6 and which breaks down at 1,000 to 3,000 V/mil, has a proven history of outstanding resistance to corona. Mica also has the exceptional property of maintaining a high intrinsic strength at temperatures as high as 600°C, which pulverize most synthetic dielectrics. Thus, even if corona occurs at its surface, this material suffers very little chemical degradation. Unfortunately, natural mica has a slippery surface; moreover, it is difficult to use as layer insulation (the insulation between windings) without fracturing it. However, many of its desirable properties have been retained in a variety of synthetic mica products and composite laminates. For example, integrated phlogopite mica, silicone-bonded ($\epsilon = 4.5$), has an electric strength of 400 V/mil at room temperature which drops only 30 percent after 1,000 hr at 600°C.

It is possible to use several layers of different types of insulation in place of one layer and obtain a surface which facilitates the mechanical winding of a coil, but this procedure increases voids. The need has

---

* The use of solventless resins (e.g., epoxy) as impregnants and Parylene as a conformal coating is noteworthy. In these materials, polymerization does not produce voids. Potting resins (e.g., urethane) will usually require vacuum degassing (removal of gas in a vacuum chamber) to minimize voids.

been met by bonded, composite laminates, in whose manufacture particular care is taken to avoid trapped voids.

In Table 3.1 we have listed some insulation materials typically used in the construction of small transformers.[10] Note that there is a marked difference between short-term electric strength and the recommended maximum stress.

There are several ways of *controlling or minimizing dielectric stresses.* For example, the equivalent circuit for corona (Fig. 3.4a) may be extended to an asymmetrical electrode geometry exposed to

TABLE 3.1 Recommended Voltage Stress and Temperature Limits for Various Insulations

Type	Insulation material	Dielectric constant ϵ 100 Hz	Dielectric strength, V(rms)/ 0.001	Max stress, V(rms)/ 0.001	Hot-spot max, °C
Gases	Air	1.0	—	—	
	Sulfur hexafluoride (SF_6)	1.0	—	—	180
Liquids	FC75 or FC 77*	1.9	350	150	105
	Coolanol 20†	2.5	350	150	105
	Askarel	5.5	350	150	105
Solids	Epoxy resin	4.2	>500	200	130
	Polyester resin	4.0	550	150	130
	RTV	3.0–3.5	>300	150	200
	Laminate, epoxy glass	5.2	500	150	130
	Diallyl phthalate (DAP)	4.2	>300	150	130
Films	Mylar‡ (0.001-in thick)	3.2	>3,000	200	130
	Kapton‡ (0.001-in thick)	3.5	>5,000	250	200
	Teflon (0.002-in thick)	2.1	>900	200	180
Sheet	Kraft paper, unimpregnated	2.0	>250	75	105
	Kraft paper, resin-impregnated	3.5	>500	200	130
	Nomex,‡ unimpregnated	2.7	150	75	180
	Nomex,‡ liquid-impregnated	2.8	>850	200	180
	Nomex,‡ resin-impregnated	2.9	>350	150	155
	Pressboard	2.0	>200	100	130
Other	Mica	5.8	3,000	250	200
	Wire enamel	3.2	>800	200	105

* 3M Company trademarks for fluorocarbon liquids.

† Monsanto Company trademark for silicate-ester-base fluid.

‡ Trademark of E. I. du Pont de Nemours & Company, Inc.

SOURCE: E. Henry, L. Wilson, in C. Harper (ed.), *Handbook of Components for Electronics* (New York: McGraw-Hill Book Company, 1977), Chap. 9, pp. 9–25.

air or a gas. Assume that corona is present around the sharp edge of a terminal. The field gradient in the area of the corona will be reduced if the terminal is covered with a material having $\epsilon > 1$ (e.g., epoxy or silicone rubber). In terms of the equivalent circuit, the capacitance C_a in the corona area has been increased by this factor ϵ and the local gradient has been reduced by the same factor. If the field-enhancement factor is 2, then corona can be quenched by using a material with a dielectric constant greater than 2.

Other techniques, using corona rings, equipotential shields, and semi-conductor coatings, are described elsewhere in the literature.[11,12,13]

3.11. COMPOSITE LIQUID AND SOLID DIELECTRIC SYSTEMS

The electric strength of an oil dielectric is markedly affected by contaminants such as water, gas bubbles, and solid particles. The presence of water or a conducting globule denotes a condition $\epsilon_2 \gg \epsilon_1$, and one analysis shows that the electric field is given by:[4]

$$E = 600 \sqrt{\frac{\pi \sigma}{\epsilon_1 r}} \, GH \tag{3.27}$$

In this expression, GH is a function of the axial ratio γ (see Sec. 3.9b), and σ is the pressure due to surface tension on the globule of radius r. Maximum GH corresponds to $\gamma = 1.85$, and the critical field for an unstable globule is:

$$E_B = 487.7 \sqrt{\frac{\sigma}{\epsilon_1 r}} \qquad \text{V/cm} \tag{3.28}$$

For example, when $\sigma = 43$ dyne per centimeter (dyn/cm), the peak breakdown voltage is a function of permittivity of the oil and radius of the globule:

ϵ_1	$r(\mu m)$	kV/cm	kV/mm	kV/0.1 in
2	1	226	22.6	57
2	25	45.2	4.52	11.5

Theoretical analysis is fairly consistent with experience.

In insulating oils, suspended solid particles as small as 0.1-mil (2.5-μm) diameter are sufficient to cause field enhancement. It has also been observed that an increase in water from a mere 5 to 50 parts per million (ppm) by volume is sufficient to reduce E_B by a factor of 3. If only a small number of particles is present, derating the oil's voltage gradient by a factor of 3 would be sufficient precaution against hot spots. If, however, the contamination of the oil has reached moderate proportions, then each enhanced field at the surface of each particle results in a vectorial addition of enhanced fields. This behavior, together with the area effect, perhaps accounts for the observed fact that the breakdown gradient of oil is usually an inverse function of the clearance d between electrodes when d is small. Even if dielectric breakdown does not immediately occur, the life expectancy of the insulation system is reduced. Because very small inclusions have a marked effect on the electric strength of the liquid, quality standards for insulating oil will specify vacuum degassing (e.g., 0.1 torr) and a high degree of purification (e.g., 50-μm pore size of the filters).

In a composite liquid-and-solid dielectric system, the discharge gradient is a function of the thickness of the insulation as well as the properties of the liquid. An empirical formula may take the following form:[14]

$$E_i = 3.6/d^{0.58} \quad \text{kV/mm} \tag{3.29}$$

for a system in which $\epsilon_1 = 2$, $\epsilon_2 = 5$, and d is in the range of 0.3 to 1 mm.

Many designers of high-voltage transformers have consistently preferred a composite liquid-and-solid dielectric system because of its proven field record of long life expectancy, even though tests for corona indicate the presence of measurable local discharges.

3.12. SURFACE DISCHARGES AND TRACKING[15]

Dielectric discontinuity also occurs at the surface of a dielectric exposed to air. Examples are the surface area between two exposed terminals on a coil; the surface area between the lug and ground of a terminal insulated by porcelain, glass, or Teflon; and the unimpregnated area of the margin between the end turn of a winding and the lamination. Again, it is difficult to ascertain the exact distribution of the electric field lines without resorting to graphical mapping of the field.

Here, the significant spacing between electrodes is not the shortest distance but, rather, the shortest length which continuously follows the contour of the surface. This distance is usually called the *creepage clearance,* since breakdown (now called flashover) is possible along the surface of the dielectric.

Flashover, a surface discharge on a solid or liquid, is thus distinguished from *sparkover,* which occurs in air, especially since flashover often occurs at a voltage substantially lower than V_S.[4] When humidity is great, the sparkover voltage may drop by 60 percent and the flashover voltage may drop 50 percent (ice and snow are not troublesome).

It is common practice to make the creepage clearance between terminals at least twice that of the sparking clearance, but not less than
$\frac{1}{16}$ in (1.6 mm).

In an adverse climate, the flashover voltage may depend entirely
on the surface resistance rather than on the electric strength of the
insulation or the air. Fungus lowers the surface resistance, a problem
which can be substantially avoided by using inhibiting sprays, varnishes,
and paints. However, certain paints, when used in the immediate
region of the terminal area of a metal-encased transformer, tend to
lower the surface resistance. As a simple precaution, paint (especially
spray) is not used on the terminal surface of the metal-encased transformer.

The role played by water, when it forms a film on the surfaces between electrodes, is that of a catalyst in many types of chemical reac-
tions. Thus, an adsorped layer of moisture will dissolve water-soluble
ionizable materials such as salts, acids, and alkalis. If corona is present
in the immediate area, the nitrous acids and salts may be dissolved
in the film of moisture. And, when the surface temperature rises
sufficiently to boil away the moisture, the residue of ionizable salts is
conductive. Actually, there are local discrete regions of contaminants.
Thus, tiny surface arcs, called *scintillation,* produce local heating and
erosion of the dielectric, as well as conductive residues. The presence
of carbonized channels on the surface is called *tracking.* Flashover
is then only a matter of time.

It is important to note that tracking is accelerated if there is a difference of dc potential between terminals. The electrodes and the water
solution (which behaves like an electrolyte if it contains ions of a salt)
provide the necessary conditions for electrolysis and, therefore, for
the flow of dc leakage currents. Since electrolysis is accelerated by
the magnitude of the dc voltage as well as by heat, tracking occurs
more often in dc than in ac circuits. It is also interesting to note
that *rectification of ac is possible* with asymmetrical electrodes.
In a point-plane electrode system, unequal polarization of the dielectric
occurs at the two electrodes, resulting in unequal diffusion rates and,
therefore, rectification.

We are thus faced with a type of breakdown which is difficult to
assess quantitatively and difficult to eliminate without providing impractically long creepage paths. Since some moisture and some surface
contamination are inevitable over a sufficiently long period of time,
one solution to the problem is to inhibit the formation of a continuous
film by applying a chemical such as silicone oil to the surface in order
to create discontinuous droplets of water. Another solution is to use
the so-called *track-resistant dielectrics.*

It is believed that resistance to tracking is increased when the dielectric material has a small carbon content, and that halogenated organic materials are undesirable. Phenolics, when the voltage is above 600 V, are more susceptible to tracking than certain track-resistant compositions containing an appropriate filler material. A resin such as polymethylmethacrylate, which depolymerizes at high temperature but without carbonization, is highly track resistant (6 kV). Other equally track-resistant materials include TFE-fluorocarbon (Teflon) as well as time-honored porcelain, although the latter cracks or chips when flashover is severe. Many of the track-resistant dielectrics are proprietary.

To summarize, the problem of tracking is most serious under two conditions: (1) the presence of dc voltage in the transformer; and (2) adverse environments, such as coastal and marine areas where the air contains much salt vapor and industrial areas where the air is chemically contaminated. Under these conditions, it is simple prudence to avoid using the "open" style transformer.

3.13. THEORY OF BREAKDOWN[16]

There are a number of theories concerning the mechanism of breakdown in dielectrics. Studies of electron avalanche in solids, of streamer theory in fluids, and of breakdown in solids and liquids caused by heat and erosion are of great theoretical, as well as practical, importance to the physical chemist. There are two important reasons why the theory of thermal breakdown is of particular interest to the transformer designer: (1) the theory provides certain guidelines for reliability and (2) the art of miniaturization inevitably creates thermal problems which, in turn, create dielectric problems.

According to the theory of thermal breakdown, the breakdown of solid and liquid dielectrics is initiated by a temperature rise in the insulation when the electric field exceeds a critical value. When this happens, heat is generated by either the flow of ionic currents, the interaction of electronic currents with the lattice or, if the field is ac, the displacement of bound ions or dipoles. If the heat cannot be dissipated at the rate at which it is generated, the material melts, decomposes, or becomes a dissociated electrolyte.

The analysis is based on the principle that heating a unit volume of insulation results either in a rise in temperature or in a flow of heat out of the unit volume. This concept is summarized by the equation for the continuity of heat:

$$E^2\sigma = C_v \frac{dT}{dt} + \frac{d}{dx}\left(k\frac{dT}{dx}\right) \tag{3.30}$$

where E is the intensity of the local field, σ is the electrical conductivity, C is the constant-volume heat capacity of the dielectric (mass times specific heat or calories/degree temperature rise), dT/dt is the rate of change of absolute temperature, k is the thermal conductivity, and dT/dx is the temperature gradient in a slab of thickness x. The coefficients σ, C_v, and k are functions of the temperature, and σ may also vary with E.

The analysis is considerably simplified if one of the two differential terms of Eq. (3.30) predominates over the other. *Impulse breakdown,* which occurs in the order to microseconds, is attributed to the first term. *Long-time breakdown,* in the order of seconds but sometimes hours, is attributed to the second term. Long-time breakdown should not be confused with breakdown associated with thermal aging. In discussing small power transformers, we are usually concerned with the second term, long-time breakdown.

The principle of continuity of current flow requires that the current density $J = \sigma E$ be constant throughout the dielectric. To keep track of the components of the conductivity which generate heat and those which do not, it is convenient to define the complex dielectric constant:

$$\epsilon = \epsilon' - j\epsilon'' \tag{3.31}$$

where ϵ' is the dielectric constant and ϵ'' is the loss factor. J may then be written for dc and ac as follows:

For dc:
$$J = \sigma_{dc} E$$

For ac:
$$J = j\omega C_o \epsilon' E + \omega C_o \epsilon'' E \tag{3.32}$$

Ac conductivity σ_{ac} equals $\omega C_o \, \epsilon''$, where C_o is the capacitance. The rate of heat generated, or power W, is the scalar product JE, so that:

For dc:
$$W = \sigma_{dc} E^2$$

For ac:
$$W = \omega C_o \epsilon'' E^2 \tag{3.33}$$

We note that σ_{dc} and ϵ'' both increase exponentially with temperature T and that ϵ'' is a function of frequency. In practice, the reciprocal of σ_{dc}, or resistivity, is often used as a measure of the quality of a dielectric.

a. Thermal Instability in Solids

Now consider an inhomogeneous dielectric, of thickness d, whose conductivity is not uniform because of the presence of impurities or chemical contamination. For simplicity, assume a series circuit with unequal conductivities $\sigma_1 > \sigma_2$. This is done in order to analyse the consequences of any incipient weaknesses resulting from field asymmetry and corona.

In the analysis of instability, three conditions are recognized:

(a) V_{ab} is dc
(b) V_{ab} is ac, and the dielectric is good ($\epsilon' \gg \epsilon''$)
(c) V_{ab} is ac, and the dielectric is poor ($\epsilon'' \geq \epsilon'$)

In (a), $J = \sigma_{dc}E$ is constant and independent of any local distance d_1. If, when dc voltage V_{ab} is applied, σ_{dc} increases with T, then E drops in order to maintain a constant J. The power dissipated, $W_1 = \sigma_1 E^2$, results in a local hot spot, remains constant, and there is thermal stability.

In (b), $J = j\omega C_o \epsilon' E$, and it is essentially constant since ϵ' decreases only slightly as T rises. To illustrate, ϵ' of Teflon drops from 2.1 to 2.04 as T rises from 25°C to 150°C. ϵ'' is small, but it exists, and it accounts for a heat-producing component, $J'' = \omega C_o \epsilon'' E$. When ϵ'' *increases as temperature rises*, the principal component of current density, $J = j\omega C_o \epsilon' E$, is unaffected by J''. Therefore, E does not drop in order to maintain constant J, as it does in the first case (a). And, since heat increases in accordance with Eq. (3.33), thermal instability becomes possible and thermal runaway may occur. *Thermal runaway* describes the dynamic process in which a rise in temperature in a local region results in an increase of power loss in the dielectric and hence a further increase in heat at a rate faster than it can be dissipated.* When this happens, a critical temperature T_c, at which chemical decomposition begins, is reached. In due time—seconds or hours— dielectric breakdown occurs in a local region where $\sigma_1 > \sigma_2$.

If the field is uniform and there is no contamination or impurity, $\sigma_1 = \sigma_2$, and breakdown of the total insulation usually results in puncture of the insulation. When this occurs, the breakdown voltage may simply drop to V_S, the sparking potential of air.

In (c), $J'' = \omega C_o \epsilon'' E$, which is constant. When ϵ'' rises with T, E decreases to maintain constant J and constant $W = \omega C_o \epsilon'' E^2$. As in (a), the dc case, there is no thermal runaway.

Thus, thermal runaway is much more likely to occur with ac and good dielectrics than with dc or poor dielectrics.

It is normal practice to choose a "good" dielectric with a low angle of power loss, defined by:

$$\tan \delta = \epsilon'' / \epsilon' \qquad (3.34)$$

---

* Thermal runaway can be viewed as positive feedback of power in a closed system.

However, to reduce the likelihood of thermal runaway (in the event local hot spots develop) it is desirable to choose an insulation with a *low temperature coefficient of the loss angle.*

There are some types of transformer geometry where local field enhancement is unavoidably high and local breakdown is virtually inevitable. Then the designer may find it expedient to coat the local surface with a semiconductor material (e.g., silicon carbide-filled paint). This technique will increase the rate of local heat dissipation and reduce the temperature of a hot spot, thus avoiding thermal runaway and premature breakdown of the insulation.[11]

Other important generalizations have been made concerning thermal breakdown in solids. It can be shown that the temperature rise in a thick slab is independent of the thickness, depending only on the applied voltage, the electrical conductivity, and the thermal conductivity. That is:

$$V^2{}_{ab} = \frac{4K}{\sigma}(T - T_a) \tag{3.35}$$

where T is the hot-spot temperature at the center of the slab and T_a is the ambient temperature at the electrode.

It may then be shown that, in all thick materials:

$$V^2{}_m \sigma_a = \text{constant} \tag{3.36}$$

where V_m is the maximum stable voltage. An important consequence of this relation is that we cannot indefinitely increase the thickness of a material and expect it to withstand an arbitrarily high applied voltage. There is an upper limit of voltage which is independent of thickness, known as the *maximum thermal voltage V_T* because it cannot be exceeded without causing thermal instability. In electronic transformers, V_T can be expected to occur only when temperature of frequency is high.[4]

T is uniform in thin insulation and:

$$V_{ab} \propto \sqrt{\frac{d(T - T_a)}{\sigma}}$$

where $T - T_a$ now represents the temperature drop at the electrode interface. If σ is proportional to field strength* V_{ab}/d, we obtain:

$$V_{ab} \propto d^{2/3} (T - T_a)^{1/3} \tag{3.37}$$

This equation provides theoretical confirmation of the frequently observed dependence of E_B on the two-thirds power of the thickness

---

* This dependence of σ on E has also been observed in oils when d is small.

[see Eq. (3.5)]. If k is high (ordinarily it is not), then $V \propto d$. Thus, it is possible to say of thin solids that, in general:

$$V \propto d^m \tag{3.38}$$

where $m = \frac{1}{2}$, $\frac{2}{3}$, or 1.

It may also be shown that the critical decomposition temperature T of a hot spot may be lower, as well as higher, than the maximum average stable temperature T_m. In consequence, the time required for breakdown may be shorter (if $T_c < T_m$) than the time required for thermal runaway (when $T_m < T_c$).

The time interval preceding thermal breakdown depends on the frequency ω. This relationship, which may be inferred from Eq. (3.33), justifies the practice of reducing the dielectric test time by increasing the frequency. If 60 sec is the normal test time when the frequency is 60 Hz, then the time may be reduced to 9 sec when the frequency is 400 Hz. In practice, test time is not reduced below 5 sec. The fact that thermal runaway requires seconds (or even hours) to occur explains one of the advantages to be gained by using a nonsinusoidal waveform such as a rectangular pulse train with a low repetition rate. A microsecond pulse radar power transformer can safely be built with thinner insulation than if the waveform were sinusoidal.

b. Breakdown in Liquids

Oil is the traditional dielectric in high-voltage circuits (usually defined as 10 kV and higher) because its intrinsically high electric strength permits the clearances in the coil margins and between terminals to be small.

According to the thermal theory of breakdown, suspended particles (impurities) will enhance (increase) the electric field. This may, in turn, result in a hot spot which initiates the formation of a bubble and a discharge. Gassing may then occur, leading to a "bubble avalanche." However, oil has a major advantage in that a momentary breakdown need not result in permanent damage, as it usually does in solid insulation. A momentary or local breakdown in oil creates convection currents which carry the vapor bubble or the discharge away from the local hot spot. This self-healing behavior is advantageous in many circuits in which continuity of service is critical.

The liquids most frequently used include mineral oil, chlorinated hydrocarbons (askarels), fluorocarbons, and silicones. Great pains are taken to keep oils free from contamination. And since oils, to a varying extent, may act as a solvent, additional precautions are taken in the construction of a coil to eliminate materials which might contaminate the oil.

The vapor bubble is the analog of a void in a solid. In one analysis of breakdown, this is taken into account by deriving an equation using such appropriate variables as the mass M, the average specific heat C_p, the boiling point of the liquid T_b, and the latent heat of vaporization l_b (the calories per gram to vaporize a liquid without a change in the temperature):

$$AE^n\tau_i = M[C_p(T_b - T_a) + l_b] \tag{3.39}$$

where A is a constant, τ_i is the pulse width, and the exponent n (3/2) indicates that pressure on the liquid and bubbles is a function of temperature.

The advantageousness of small pulse widths may be readily inferred from this equation. In liquids, impulse breakdown progressively increases below 1 μsec, as τ_i decreases. However, in the 1 to 1,000 μsec range of pulses, the breakdown voltage is fairly constant, about 25 percent higher than the breakdown voltage of dc. (Breakdown is also a function of the rate of rise of a pulse, but this is not relevant to the thermal theory.)

We may also infer from Eq. (3.39) that a high boiling point is advantageous and that the breakdown voltage falls sharply as the ambient temperature approaches the boiling point. Although the heat of vaporization decreases as temperature rises, the specific heat rises. Consequently, E_B is most likely to decrease at either very low or very high temperatures.

At high altitudes, where atmospheric pressure is very low, the can of a liquid-filled transformer may expand, thus reducing the internal pressure and lowering E_B. A reduction to 86 mm Hg, about 50,000 ft, has been observed to reduce the breakdown voltage to 75 percent of its value at sea level.

3.14. DESIGN OF THE DIELECTRIC CIRCUIT

We conclude our survey with a brief discussion of design principles. It is possible to distinguish two approaches to the design of the dielectric circuit: *quality-control design* and *worst-case design*.

a. Quality-Control Design

A widely used design approach begins with the economic objective of using the minimum amount of insulation. The insulation (preferably of low cost) is carefully chosen to meet specified limits of quality, typically for parameters such as:

1. Electric strength of insulation, oil, and impregnant
2. Dissipation factor (loss angle)
3. Temperature coefficient
4. Purity (freedom from moisture, trapped gas, and inclusions)

Much care is taken to achieve a specified degree of vacuum during the process of impregnation. Tests for partial discharge specify a limited corona discharge [e.g., 20 picocoulombs (pC), 100 pC, 1000 pC] based on the dielectric's projected life expectancy.[17]

This approach relies heavily on vigilant enforcement of standards of quality control, inspection, and testing. To the extent that these conditions are met, the design results in a transformer of minimum cost and few returns from the field.

b. Worst-Case Design

An alternative approach is to design for maximum reliability. The designer assumes that the materials used will not be of consistent quality; that is, it is assumed at the outset that the dielectrics will have, or will in time develop, discontinuities: voids, cracks, bubbles, moisture, and inclusions. The designer focuses on *geometry* and on *derating the nominal electric strength* of the insulation. The procedure can be summarized as follows:

1. Establish a geometry with minimum asymmetry and, where necessary, alter the geometry so as to reduce local gradients by appropriate techniques of stress control.[11,12] If potentials are at the kilovolt level, split coils into narrow pie sections ("tires") to obtain low volts per layer and low volts per coil.[18] Sharp contours are to have a minimum radius (see Sec. 3.6). If necessary, coat critical surfaces exposed to air with a material of high permittivity.

2. Derate the electric strength of all materials on the supposition that dielectric discontinuities are unavoidable (see Table 3.1). Select materials of low permittivity. Strive for a compatible dielectric system. When solid insulation is used, the designer employs "anticorona" formulas such as Eqs. (3.13) and (3.24). When insulating oil is used, assume a field-enhancement factor of at least 3 (see Sec. 3.10).

3. If the transformer is deemed too large, consider using a composite liquid-and-solid dielectric system.

4. If the transformer is deemed too heavy, consider using a gas dielectric.

The worst-case design approach does not, in general, result in a transformer of minimum size or insulation of minimum cost. It does, however, increase the probable reliability of the component.

REFERENCES

1. J. Krauss, K. Carver, *Electromagnetics,* 2d ed. (New York: McGraw-Hill Book Company, 1973), Chap. 3, pp. 86–93.
2. E. Kuffel, M. Abdullah, *High Voltage Engineering,* (Oxford: Pergamon Press, 1970), pp. 89–90.
3. J. Burks, J. Schulman, *Progress in Dielectrics,* Vol. 1 (New York: John Wiley & Sons, 1959), p. 17.
4. L. Alston, *High Voltage Technology* (New York: Oxford University Press, 1968), Chaps. 3, 7, 10.
5. E. Henry, L. Wilson, in C. Harper, (ed.), *Handbook of Components for Electronics* (New York: McGraw-Hill Book Company, 1977), Chap. 9.
6. W. Starr, "High Altitude Flashover and Corona Problems," *Electro-Technology,* Vol. 69, No. 5 (May 1962): 124–27; No. 6 (June 1962): 128–34.
7. J. Mason, "Discharges," *IEEE Transactions on Electrical Insulation,* Vol. EI-13, No. 4 (August 1978): 211–48.
8. J. Devins, A. Sharbaugh, "Electrical Breakdown in Solids and Liquids," *Electro-Technology,* Vol. 68, No. 4 (October 1961): 98–116.
9. R. Bartnikas, E. McMahon, *Corona Measurement and Interpretation,* Engineering Dielectrics, Vol. 1 (Philadelphia: American Society for Testing and Materials, 1979).
10. M. Halleck, "Calculations of Corona Starting Voltage in Air-Solid Dielectric Systems," *Transactions AIEE,* Vol. 75, Part 3 (April 1956): 211–16.
11. L. Medina, "Prevention of Ionization in Small Power Transformers," *Proceedings IRE Australia,* Vol. 15, No. 5 (May 1954): 114–15.
12. H. Weber, "Transformer Design Concept to Increase the Dielectric Breakdown Strength," *IEEE Electrical Insulation Conference,* Vol. 75CH 1,014 (November 1975): 183–85.
13. S. Hirabayashi et al., "A New Corona Suppression Method for High Voltage Generator Insulation," ibid.: 139–42.
14. N. Parkman, "Some Properties of Solid-Liquid Composite Dielectric Systems," *IEEE Transactions on Electric Insulation,* Vol. EI-13, No. 4 (August 1978): 289–307.
15. K. Mathes, E. McGowan, "Electrical Insulation Tracking," *Electro-Technology,* Vol. 69, No. 4 (April 1962): 146–51.
16. J. Devins, A. Sharbaugh, "The Fundamental Nature of Electrical Breakdown," *Electro-Technology,* Vol. 67, No. 2 (February 1961): 104–22.
17. J. Burnham, E. Wong, H. Ota, "Corona Characteristics of Flaws in a Solid Encapsulated High Voltage Inverter Transformer," *IEEE Power Electronics Conference,* Vol. 74CH 0,863-1-AES (1974): 40–50.
18. E. Brown, "High Voltage RF Transformers," *Solid State Power Conversion* (May 1979): 54–58.

CHAPTER 4

The Thermal Circuit

In this chapter we are concerned with the factors responsible for a rise in the temperature of the transformer and the control of the temperature rise by appropriate design techniques.

4.1. SIZE, RATING, AND TEMPERATURE RISE

We were able, in Chap. 2, to stress certain basic relationships among the volt-ampere rating VA_T (or P_o), volume, and area product A_p of a transformer by deliberately ignoring power losses.

In practice, because of power losses W in the transformer, the input power E_1I_1 exceeds the output power E_2I_2. Power losses in the core (i.e., iron losses W_{Fe}) and power losses in the coil (i.e., copper losses W_{Cu}) are directly related to magnetic flux density B and current density J, and since we are now prepared to deal with them, it is desirable to reformulate Eq. (2.5) in terms of these variables and the frequency f:

$$VA_T = P_o = C_1 K_{Cu} K_{Fe} f J B A_{Fe} A_{Cu} \qquad (4.1)$$

where $C_1 = k_1 k_q \, 10^{-8}$; $k_1 = \frac{1}{2}$ (or the fraction of window occupied by the primary); and $k_q = 4.44$ (sine wave) or 4.0 (square wave). The constants K_{Cu} and K_{Fe} are the copper and iron space factors (i.e., the fraction of gross area occupied by copper and iron, respectively). J is expressed in amperes (A) per square centimeter (sq cm) (or amperes per sq in), B is in gauss (G) (or lines per sq in), and both core area A_{Fe} and coil area A_{Cu} are expressed in sq cm (or sq in).

Total power losses W in the transformer may be stated functionally as:

$$W = W_{Fe} + W_{Cu} = F(B, J, \text{vol}, f, \rho_{Cu}) \qquad (4.2)$$

where ρ_{Cu} is the resistivity of the wire in the coil.

The hottest point in the transformer is, we assume, at its center. But the temperature of the iron core is of less interest to the designer than that of the more vulnerable coil of copper wire, so the highest internal operating temperature T_o is taken to be that of the hottest part of the coil (typically, this is the innermost part of the coil, in direct proximity to the core). The differential between this temperature T_o and that of the ambient air T_a is θ, and both θ and T_o vary with the power losses and the size of the transformer:

$$\theta = T_o - T_a = F(W, \text{vol}) \qquad (4.3)$$

Viewed in this form, θ is an increase over the ambient temperature and is called the *temperature rise*. Between T_o and T_a there are other, lower, temperatures—in the film of air immediately surrounding the transformer, at the surface of the transformer, in the impregnant or compound, in the secondary, and so forth. Each of these temperatures represents a rise, a contribution to the total differential θ, when we view θ as progressively increasing as we proceed toward the innermost part of the coil. However, each of these temperatures becomes a drop if we proceed outward, from the center to the ambient air. In our discussion, we treat these progressive changes in temperature as rises.

In Chap. 3, we saw that life expectancy is intimately connected with operating temperature and insulation class and that reliability is inversely related to temperature rise. From the user's point of view, it is not the densities of the magnetic flux and current which are fundamental, but rather the relationships among size, rating, and temperature rise. To establish these relationships, we will consider the connection between θ and W stated in Eq. (4.3), first asking: How does size relate functionally to temperature rise and VA?

The procedure necessary to answer this question entails regarding the transformer rating as the dependent variable:

$$P_o = F(\theta, \text{vol}) \qquad (4.4)$$

We begin with a qualitative analysis of the relationships among the major parameters. In Chap. 5, we deal with the practical solutions to Eq. (4.4), that is, how the rating may be computed from the temperature rise and the size. Here we consider the solution to Eq. (4.4) in two different situations. The first solution involves the low-frequency transformer, in which the magnitude of B is limited by the need to avoid saturating the iron. The second involves the high-frequency transformer, where both B and J can be freely selected to yield maximum VA rating.

a. Saturation

We make the following assumptions:

1. B and J are substantially independent of each other.[1]   While B is bounded by magnetic saturation, J is bounded only by the maximum permissible temperature rise resulting from current flow in the coil.

2. The VA furnished to the load is to be varied by adjusting the load current.

3. Power loss is related to temperature rise by an equation of the form:

$$\psi = h\theta \tag{4.5}$$

where ψ is the power density or average power loss per unit area dissipated from the surface of the transformer and h is the coefficient of heat transfer.

Power dissipated in the coil is proportional to the square of the current. Power loss in the copper is:

$$W_{Cu} = K_{Cu} A_{Cu} l_t \rho J^2 \tag{4.6}$$

where ρ is resistivity of copper and l_t is the mean length of turn in the coil.

Power loss in the copper is a fraction α of the total loss W and, in the iron, a fraction β of W:

$$W_{Cu} = \alpha W_{Fe} = \beta W$$

But the latter's contribution to the temperature rise in the coil is somewhat less in the laminated transformer. If we define γ_1 as the fraction of power loss in the iron to be added to the loss in the copper, we have:

$$W'_{Fe} = \gamma_1 W_{Fe}$$

Values of γ_1 depend on the thermal coupling between core and coil and are a function of the geometry:

Geometry	γ_1
Laminations	0.5–0.75
Ferrite E and pot cores[1]	0.64–0.67
Toroid	0.7–1.0

We now equate effective power loss in the copper W'_{Cu} with the temperature rise, as follows:

$$W'_{Cu} = \alpha_1 W_{Cu} = h\theta A_s$$

where $\alpha_1 = 1 + \gamma_1\beta/\alpha$.

If we solve for J in terms of θ,

$$J^2 = \frac{h\theta A_s}{\alpha_1 \rho l_t K_{Cu} A_{Cu}} \tag{4.7}$$

and then substitute J into Eq. (4.1), we obtain:

$$P_o = C_1 fBK_{Fe}A_{Fe}\left(\frac{K_{Cu}A_{Cu}A_s h\theta}{\alpha_1 \rho l_t}\right)^{1/2} \tag{4.8}$$

This cumbersome equation can be simplified if it is rephrased in terms of the geometry, that is, the volume of the iron (vol_{Fe}) or the geometric mean volume of core and coil vol_m. To do this, we make use of the following definitions:

$$vol_{Fe} = A_{Fe}l \qquad vol_m = (vol_{Fe}vol_{Cu})^{1/2} \tag{4.9}$$

We also introduce the coefficients g_0, g_1, and g_6, which combine various dimensions of the transformer:*

$$\left.\begin{array}{l}
g_0 = \left(\dfrac{A_{Fe}A_s}{ll_t{}^2}\right)^{1/2} \\[2ex]
g_1 = \left(\dfrac{A_s A_{Cu}}{ll_t{}^2}\right)^{1/2} \\[2ex]
g_6 = vol_{Fe}g_1
\end{array}\right\} \tag{4.10}$$

We then obtain:

$$P_o = C_2 fBg_6(h\theta)^{1/2} \tag{4.11}$$

where $C_2 = C_1 K_{Fe}(K_{Cu}/\alpha_1\rho)^{1/2}$.

---

* The geometric coefficients defined here and in the next section are discussed at greater length in Secs. 5.4 and 7.7.

The rating P_o can also be expressed either in terms of the volume of iron or the mean volume of core and coil:

$$P_o = C_2 fB(\text{vol})_{\text{Fe}} g_1(h\theta)^{1/2} \tag{4.12}$$

$$P_o = C_2 fB(\text{vol})_m g_0(h\theta)^{1/2} \tag{4.13}$$

From the preceding equations, we can infer that the VA rating of a low-frequency power transformer is more responsive to an increase in frequency and magnetic flux density than it is to an increase in the temperature rise. In practice, however, f is prescribed by circuit conditions and B is limited by our choice of a practical core material. With f and B thus fixed, small size and high VA rating are facilitated by operating the transformer with a large temperature rise.

b. Optimum BJ

We follow the analysis of Judd and Kressler, choosing that BJ product which makes the rating P_o a maximum.[2] Power loss in the iron P_S can be expressed as:

$$P_S = \frac{W_{\text{Fe}}}{M} = bf^n B^m \tag{4.14}$$

where $M = K_{\text{Fe}} A_{\text{Fe}} l\delta$ is the weight of the core in kilograms (kg); δ is the density of the iron; and b, n, and m are empirical coefficients (see Secs. 5.3c and 6.1c).

In the laminated power transformer of open construction, thermal coupling between core and coil is poor, and we will want to assess the extent to which temperature rise is affected by the ratio of the power losses in the copper and iron. It is convenient to express temperature rise θ in the form:

$$\theta = r_{\text{Fe}} W_{\text{Fe}} + r_{\text{Cu}} W_{\text{Cu}} \tag{4.15}$$

so that the r-coefficients can be determined by empirical tests.

Alternative expressions, in terms of the ratio λ of coil and core thermal resistance (or conductance) are given by:

$$\left. \begin{array}{l} \theta = r_{\text{Fe}}(W_{\text{Fe}} + \lambda W_{\text{Cu}}) \\[2mm] \theta = r_{\text{Cu}}(\gamma W_{\text{Fe}} + W_{\text{Cu}}) \end{array} \right\} \tag{4.16}$$

where $\lambda = 1/\lambda = r_{\text{Cu}}/r_{\text{Fe}}$.

When core and coil are energized separately, it is possible to confirm, by means of graphical plots, that the coefficients of Eq. (4.15) are constant and independent of B and J.*

Analysis shows that the optimum value of current density (opt J) varies inversely with the size of the coil ($A_{Cu}l_t$):

$$\text{opt } J^2 = C_3\left(\frac{\theta}{r_{Cu}}\right)\left(\frac{1}{K_{Cu}A_{Cu}l_t}\right) \tag{4.17}$$

where $C_3 = \dfrac{m}{(2+m)K_{Cu}\rho}$

The optimum value of flux density (opt B) can be expressed as:

$$\text{opt } B = \left\{\frac{(\theta/r_{Fe})[2/(2+m)]^{1/m}}{Mbf^n}\right\} \tag{4.18}$$

and varies inversely with frequency.

If optimum values of J and B are substituted into Eq. (4.1), it is possible to obtain an expression for the *maximum power rating*. For the sake of clarity in this discussion, we simplify the result somewhat by assuming a single heat coefficient h and a single composite surface area A_s.† Further, if we choose the integer $m \cong 2$ as the exponent of B, we obtain:

$$\text{opt } P_o \cong C_5 f^a\left(\frac{A_{Cu}A_{Fe}}{l_t l}\right)^{1/2}(hA_s)\theta \tag{4.19}$$

where $C_5 = \dfrac{C_1 K_{Fe}}{2}\left(\dfrac{K_{Cu}}{b\delta\rho}\right)^{1/2}$

and the exponent $a = 1 + n/2 - n/m$.

Here, l is the length of the magnetic path (in centimeters), and a, b, and δ (the density of the iron in grams per cubic centimeter) are constants of the core material. We can simplify Eq. (4.19) by rephrasing it with the geometric coefficients g_2, g_3, and g_4. To do this we define them in a format similar to Eq. (4.10):

$$g_2 = \left(\frac{A_{Fe}A_{Cu}}{l l_t}\right)^{1/2} \tag{4.20}$$

$$g_3 = g_2 A_s \qquad g_4 = A_s/l l_t \tag{4.21}$$

We then obtain:

$$\text{opt } P_o \cong C_5 f^a g_3(h\theta) \tag{4.22}$$

---

* The effect of loss of power in the core on the temperature rise in the coil is discussed further in Sec. 4.2.

† This has been done by using the substitution: $hA_s = 1/(r_{Fe}r_{Cu})^{1/2}$.

Its equivalent, in terms of the mean volume of core and coil, is:

$$\text{opt } P_o \cong C_5 f^a (\text{vol})_m g_4 (h\theta) \tag{4.23}$$

The value of the exponent a depends on the choice of the core material (Sec. 6.1c).

Comparison of the saturation-limited (low-frequency) case, Eq. (4.11) and the optimum BJ (high-frequency) case, Eq. (4.22), is instructive. We infer the superiority of the high-frequency transformer, whose power rating per unit volume is substantially higher than that of the transformer operating at commercial power frequencies.

Provided also that reliability is not jeopardized, a large temperature rise is actually desirable. We will eventually see that a small size, high rating, moderate rise, and reliability are *all* simultaneously feasible, provided the coefficient of heat transfer is maximum and the transformer is operated at a low duty cycle.

Accordingly, our objectives are to ascertain the heat transfer coefficients in order to predict temperature rise and to consider those basic principles of thermal design which permit us to control and increase the coefficient of heat transfer h. We have discussed elsewhere the conditions under which transient heating may help us to miniaturize the transformer and increase its VA rating without adversely affecting its life expectancy and reliability.[3]

4.2. THERMAL PARAMETERS
AND CIRCUITS

Heat transfer in the transformer is basically a field phenomenon. Suppose the transformer's geometry is simple—a sphere or a cylinder, for example. When we have a homogeneous material of thermal conductivity k and of uniform power density W_v (power generated per unit volume), the task is to solve Poisson's equation:

$$\nabla^2 T_o = -\frac{W_v}{k} \tag{4.24}$$

for the internal temperature T_o, under suitable boundary conditions.[4]

The transformer is a complex component which contains two heat generators (core and coil) with materials of different volumes and thermal conductivity. In laminated cores, the flow of heat is not isotropic, and field mapping is used to plot the equipotential surfaces and to ascertain the temperature gradients.[5] In this chapter, we shall carry forward the attempt to translate the results of field mapping into equivalent circuits so that electric circuit theory can be employed. It is hoped we can thus clarify and simplify the relationships and that problems will become more susceptible to calculation.

When we use the circuit approach, we treat the flow of heat as if it were analogous to the flow of current I in a bar of metal whose area is A and whose length is x. According to Ohm's law:

$$I = GV \qquad G = \frac{kA}{x}$$

where G is the conductance, V is the difference in potential in volts (V), and k is the conductivity.

If the field form of Ohm's law is used, then we may write:

$$\frac{I}{A} = k\frac{V}{x} \qquad \text{or} \qquad J = kE \tag{4.25}$$

where J is the current density and E is the voltage gradient or electric field.

The thermal counterparts of Eq. (4.25) are:

$$W = \frac{kA}{x}\theta \qquad \psi = k\frac{\theta}{x} \tag{4.26}$$

Analogous to current flow is the heat flow rate, energy per unit time or flux W in watts (W). The difference in temperature θ, in degrees, corresponds to the difference in potential V, and θ/x is called the thermal gradient. Thermal flux density ψ,* the analog of J, is also analogous to magnetic flux density B in the equation:

$$\frac{\phi}{A} = \mu\frac{NI}{x} \qquad B = \mu H \tag{4.27}$$

where NI/x is the ampere-turns (A-turns) gradient, or magnetizing force, and permeability μ is the analog of conductivity.

Because of certain peculiarities in the historical development of the statements of relationships among thermal, mechanical, and electrical energy, there is an unduly large variety of heat units in the English as well as the metric system. Table 4.1 is designed to facilitate the conversion of the most frequently encountered units. The data in the table are divided into five major groups of thermal concepts:

1. Energy or heat: in calories (cal), joules (J), or British thermal units (Btu).

2. Power or heat rate: in watts (W), calories per second (cal/sec) and Btu per hour (Btu/hr).

---

* The term *flux density* is used in preference to the more simple *flux* in order to avoid ambiguity.

TABLE 4.1 Conversion of Thermal Units

	Multiply	By	To obtain
Energy or heat	Calories	4.186	Joules (W/sec)
	Joules	10^7	Ergs
	Btu	252	Calories
	Btu	1,055	Joules
	Btu	788	Ft-lb
Power or heat rate	Watts	1	Joules/sec
	Watts	0.239	Cal/sec
	Watts	3.413	Btu/hr
Power density	Watts/sq in	491.4	Btu/(hr) (sq ft)
	Watts/sq cm	6.452	Watts/sq in
Thermal conductivity or conducibility	Watt in/(sq in)(°C)	22.7	Btu ft/(hr)(sq ft)(°F)
	Watt in/(sq in)(°C)	2.54	Watt cm/(sq cm)(°C)
	Cal cm/(sec)(sq cm)(°C)	4.186	Watt cm/(sq cm)(°C)
	Cal cm/(sec)(sq cm)(°C)	10.6	Watt in/(sq in)(°C)
Coefficient of heat transfer or external conducibility	Cal/(sec)(sq cm)(°C)	27.0	Watts/(sq in)(°C)
	Btu/(hr)(sq ft)(°F)	0.00367	Watts/(sq in)(°C)
	Watts/(sq in)(°C)	273	Btu/(hr)(sq ft)(°F)

3. Power density (thermal flux density): in watts per square inch (W/sq in), watts per square centimeter (W/sq cm), and Btu per hour per square foot [Btu/(hr)(sq ft)].

4. Thermal conductivity k or internal conducibility.*

5. Coefficient of heat transfer h, or external conducibility, i.e., thermal conductivity times unit distance.[6]

The analogy between Eqs. (4.25) and (4.26) enables us to construct a thermal equivalent circuit which resembles an electric equivalent circuit. In the series circuit depicted in Fig. 4.1, a heat generator which delivers a power of W watts is analogous to a constant-current electric generator delivering I amp. A constant rate of heat flowing through three thermal resistors produces the thermal drops $\theta_1 = R_1 W$, $\theta_2 = R_2 W$, and $\theta_3 = R_3 W$. The total temperature rise θ equals $\theta_1 + \theta_2 + \theta_3$.

Now consider the far more complex thermal circuit of an enclosed transformer. In Fig. 4.1b, heat flow W_{Cu} originates in both the primary and secondary windings. Heat generated in the core W_{Fe} takes several paths, the principal one being that which joins the flow W_{Cu} from the surface of the coil and flows through the solid compound or liquid impregnant to the outer surface of the case. Thence, the heat W

---

* Fourier proposed the terms *external conducibility* and *penetrability* to distinguish them from *interior conducibility*, now called *thermal conductivity*.

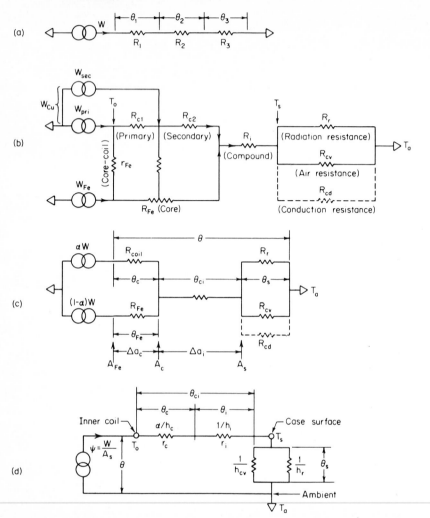

Fig. 4.1 Thermal equivalent circuits. (a) Simple series circuit. (b) Complete equivalent circuit of the transformer. (c) Simplified transformer circuit; $W = W_{Fe} + W_{Cu}$, α = copper fraction of loss. (d) Simplified series transformer circuit; $h_{cv} + h_r = h_s = 1/r_s$.

flows outward from the case, by radiation and convection, to the surrounding medium of air and, by conduction, to the chassis on which the transformer is mounted.

It is highly advantageous to simplify this circuit, but care must be taken so that accuracy is not unduly affected. The use of various simplifying procedures, in this and other treatments of heat transfer, is partly justified by the fact that the most basic empirical constants are almost never known to better than two significant figures.

The situation is indeed made immediately simpler when the copper and iron power losses are not too dissimilar or when, because of poor coupling (r_{Fe} in the figure has a large value), there is a negligible interchange of heat between coil and core. If copper power loss is denoted by αW (where α is the fraction of total loss W), such a transformer may be represented by Fig. 4.1c.

We wish to determine maximum temperature T_o in the coil, so that we can ascertain the total temperature rise from:

$$\theta = T_o - T_a = \theta_c + \theta_i + \theta_s \qquad (4.28)$$

where the individual thermal drops are represented by:

θ_c, the rise in the coil through R_c over the radial distance Δa_c between the surface of the core and the surface of the coil.

θ_i, the rise in the impregnant or potting compound through R_i over the radial distance Δa_i between the coil surface A_c and the case surface A_s.

θ_s, the rise at the surface (*surface rise*) through the parallel combination of thermal resistances R_r, R_{cv}, and R_{cd} between the case area A_s and the surrounding air and chassis at an ambient temperature T_a.

We can gain further clarity and simplicity by viewing the transformer thermal circuit as basically a series circuit. We stress the mathematical similarity of the three modes of heat transfer—radiation, convection, and conduction—by expressing each mode in the same form, rewriting Eq. (4.5) as:

$$\psi = \frac{W}{A_s} = h\theta$$

A_s, the external area of the case, is used as the reference surface for all calculations. A compelling reason for this is that, of the several concentric surfaces in the transformer, only the external surface is readily available to the user for measurements or estimates.

Heat originating inside the transformer and penetrating the case must equal the heat emerging from the case. This conforms to the principle of continuity of heat, the analog of Kirchhoff's law of conservation of current at a junction or node. If the transformer is viewed as a series thermal circuit with the surface as a node, as in Fig. 4.1d, then we may write the heat flow equations in the form:

$$\psi = \frac{h_c}{\alpha}\theta_c = h_i\theta_i = (h_r + h_{cv})\theta_s \qquad (4.29)$$

i.e., generated internal heat equals dissipated external heat. The subscripts c and i refer to coil and impregnant, respectively. The subscripts r, cv, and cd refer, respectively, to radiation, convection, and conduction.

The total temperature rise of Eq. (4.28) may now be divided into two components: one inside the case, $\theta_{ci} = \theta_c + \theta_i$ and one outside the case, θ_s. Thus:

$$\theta = \theta_{ci} + \theta_s = r_{ci}\psi + r_s\psi = \frac{\psi}{h} \qquad (4.30)$$

where r_{ci} is the combined coil and impregnant thermal resistivity and r_s is the surface thermal resistivity.

TABLE 4.2 Insulation Class and Temperature Rise

IEEE*	MIL-T-27	Operating temperature, T_o	Temperature rise, $T_o - T_a$	Typical ambient temperature, T_a
....	Q	85	45	40
O	...	90	50	40
A	R	105	65	40
		105	50	55
		105	40	65
		105	30	75
B	S	130	55	75
		130	45	85
F	V	155	80	75
		155	70	85
....	T	170	85	85
....	U	>170	...	85
				125
H	...	180	95	85
		180	55	125
220	...	220	95	125
C > 220	...	>220	...	125
		300	100	200
		500	150	350
		600	100	500
		650	150	500

* IEEE column from *General Principles for Temperature Limits in the Rating of Electric Equipment,* IEEE, Standard No. 1, March 1969.

Temperature rise is a dominating specification, and Table 4.2 contains, for reference, the various values of θ which are likely to result with standard ambient temperatures and standard classes of insulation.[7]
Total thermal resistivity $r = 1/h$ is:

$$r = r_{ci} + r_s = \left(\frac{\alpha}{h_c} + \frac{1}{h_i}\right) + \frac{1}{h_r + h_{cv}} \tag{4.31}$$

and the total thermal resistance R_T, in degrees Celsius (°C) per watt,* is the sum of the internal and external resistivities divided by the area:

$$R_T = \frac{r_{ci}}{A_s} + \frac{r_s}{A_s} \tag{4.32}$$

4.3. EXTERIOR AND INTERIOR HEAT TRANSFER

The principal modes of heat transfer are radiation, convection, and conduction.

Our initial task in this section is to determine the magnitude of the various individual coefficients of heat transfer. In Sec. 4.4, total h is ascertained so we may predict the temperature rise when W/A_s is known. Temperature rise may then be calculated from $\theta = \psi/h$. As will be seen, however, the problem is complicated by the fact that the exterior coefficients for radiation and convection are not constants but nonlinear variables which vary with the ambient temperature and with the temperature rise at the surface.

Stated briefly, the problem is first to solve nonlinear equations such as:

$$h = F(\theta_s, T_a) \qquad \theta_s = F(\psi, T_a) \tag{4.33}$$

and then, for each mode of transfer, to recast the equations in the standard form:

$$\psi = h\theta \tag{4.5}$$

a. Radiation

Stefan's law of radiated heat, in terms of radiation flux density, is:

$$\psi_r = \sigma\epsilon(T_s{}^4 - T_a{}^4) \tag{4.34}$$

---

* Since volts per ampere are electrical ohms, degrees Celsius per watt may be referred to, analogously, as *thermal ohms*.

where ϵ is the emissivity of the surface material and the constant σ equals $3.68(10)^{-11}$ if W/A_s is expressed in watts per square inch.

If the surrounding surface area is not much larger than the transformer, a small correction in the value of ϵ may be necessary to include the effect of the emissivity of the larger surface. It is important to note that high emissivity depends more on the reflectance of the surface than on its color. Thus, a glossy or polished black surface has less ϵ than a nonglossy light gray surface. It has been noticed that, at very high ambient temperatures, ϵ decreases.*

To restate Eq. (4.34) as a function of surface temperature rise, we use the convenient approximation:

$$T_s{}^4 - T_a{}^4 \cong 4T_a{}^3\theta_s\left(\frac{1+1.5\theta_s}{T_a}\right)$$

Stefan's equation may now be restated in the standard form:

$$\psi_r = h_r\theta_s \qquad h_r = F(T_a, \theta_s) \qquad (4.35)$$

where $h_r = b(1 + 1.5\theta_s/T_a)$ and $b = 4\sigma\epsilon T_a{}^3$, which is constant at a specified ambient temperature.

It is useful, particularly so we can compare it with other coefficients, to calculate h_r under standard conditions. In this section we employ the following arbitrary, though not atypical, conditions: $\theta_s = 40°C$, $T_a = 313°K$ (or $40°C$), and $\epsilon = 0.9$.

This yields, for the radiation parameters:

$$\left.\begin{array}{l} h_r = 0.00487 \text{ W}/(\text{sq in})(°C) \\ \psi_r = 0.195 \text{ W}/\text{sq in} \end{array}\right\} \qquad (4.36)$$

Table 4.3 is a compilation of calculated values of h_r over a wide range of values of ambient temperature and surface temperature rise. Of particular significance, and discernible in the table, is the extent to which thermal flux density (watts per unit area) changes with T_a and θ_s. Note that h_r increases gradually with θ_s but rapidly with ambient temperature T_a. This may be seen by comparing the lower right quadrant of the table with the upper left. Thus, when θ_s and T_a are substantially higher than $40°C$, h_r and ψ_r may grow by as much as an order of magnitude.

b. Natural Convection[8,9,10]

In the convection mode of heat transfer, the fluid in contact with the hot surface is in motion. The most familiar and typical example

---

* This decrease (which can be appreciable at $350°C$) has been compensated for by coating the surface with sodium dichromate.

TABLE 4.3 Radiation Constants: Conductivity h_r and Thermal Flux Density ψ_r.*

Ambient temperature, T_a

θ_s	313°K 40°C		328°K 55°C		338°K 65°C		348°K 75°C		358°K 85°C		398°K 125°C		473°K 200°C		623°K 350°C		773°K 500°C	
	h_r	ψ_r	h_r	ψ_r	h_r	ψ_r	h_r	ψ_r	h_r	ψ_r	h_r	ψ_r	h_r	ψ_r	h_r	ψ_r	h_r	ψ_r
25	.....	.....	.....	.....	5.7	0.143												
30	4.67	0.140	5.33	0.16	5.82	0.175												
35	.....	.....	5.45	0.191	5.92	0.207	6.45	0.226										
40	4.87	0.195	5.55	0.222	.....	.....	6.55	0.262										
45	4.96	0.223																
50	5.06	0.253					.....	.....	7.4	0.37								
55	.....	.....					6.94	0.382	.....	.....	10.1	0.555						
60	.....	.....					7.05	0.423	7.63	0.458	.....	.....	16.7	1.0				
65	.....	.....					7.15	0.465	7.75	0.503	10.4	0.667						
70	.....	.....					.....	.....	.....	.....	10.6	0.742	17.1	1.2	37.6	2.63	70.	4.9
100	6.05	0.605	6.9	0.69			.....	.....	8.67	0.867	11.6	1.16	18.5	1.85	39.9	3.99	73.1	7.31

* Multiply h_r by 10^{-3}. Values are based on $\sigma = 3.68(10)^{-11}$; $\epsilon = 0.9$.

of convection is the natural flow of air about the external surface of the transformer. Forced convection occurs when air, water, oil, or other fluid is constrained, by means of a fan or pump, to flow at greater than natural velocities. Here we consider air as the fluid medium, and convection is natural.

In natural or free convection, the thickness of the boundary layer of air varies inversely with the height of the component. This layer, or film, of air also becomes thinner as the differential θ_s between surface temperature and ambient temperature increases. Consequently, the transfer of heat by convection h_{cv} varies as some power of θ_s:

$$h_{cv} \propto \theta^n \qquad n > 1$$

The shape and the proportions of the transformer, as well as its proximity to neighboring components, affects the magnitude of the conductance coefficient h_{cv}. Indeed, this coefficient incorporates the effects of a rather large number of factors. Consider the equation for natural convection ψ_{cv}:

$$\psi_{cv} = \left(C_a S \frac{p^{1/2}}{L^{1/4}} \right) \theta_s{}^{5/4} \qquad (4.37)$$

where the coefficients (in parentheses) of the surface rise θ_s include a complex parameter C_a, a shape factor S, the relative barometric pressure p, and the significant dimension L. C_a depends on the basic properties of the particular fluid. It varies directly with the thermal conductivity k, the volume coefficient of expansion γ, density d, specific heat C_p, and the acceleration of gravity g. It varies inversely with viscosity μ_1.

Three of these factors—k, d, and μ_1—are temperature dependent.[*] These relationships are embodied in the expression:

$$C_a = F(T_a) = K a_{cv}{}^{1/4} = \frac{9.2(10)^{-3}}{T_a{}^{1/4}}$$

where a_{cv} combines the various factors just mentioned and is constant only at a specific ambient temperature T_a.

However, since a_{cv} varies only slowly with temperature, the coefficient is usually computed at an ambient temperature of 40°C, resulting in the important value $C_a = 0.0022$.

The effect of shape and proportions on the shape factor:

$$S = \frac{A_{eff}}{A_s}$$

---

[*] If the fluid is other than air, C_p is also temperature dependent. The h_{cv} of helium is larger than that of air.

T A B L E   4 . 4   Convection Shape Factor

Shape	Significant dimension, L	Shape factor, S
Sphere	Radius	0.63
Cylinder, horizontal	Diameter	0.45
Cylinder, vertical	Height < 2 ft	0.45–0.55
Cube	Height	0.636
Small parts	. . . .	1.45 (approx.)

is summarized in Table 4.4.   The large variation of S may be construed to be equivalent to a large change in the effective surface area A_{eff}.* Note that even a change in the orientation of the axis with respect to the azimuth may alter the significant dimension L if the shape is oblong or thin.   (Geometric design considerations are discussed in Secs. 4.5 and 5.4.)

If the convection coefficient is defined as:

$$h_{cv} = C_a S \frac{p^{1/2}}{L^{1/4}} \theta_s^{1/4} \qquad (4.38)$$

then the natural convection equation may finally be cast in a form consistent with Eq. (4.5):

$$\psi_{cv} = h_{cv} \theta_s$$

Table 4.5 contains convection values for h and ψ over a wide range of ambient temperatures and surface temperature rise.   Our interest in miniaturization accounts for the additional column, $L = 1$ in. Note the only moderate extent to which a fourfold reduction in the linear dimension improves convection.† Our data, which are based on $S = 0.63$ and A = total area, produce results roughly equivalent to assuming that $S = 1$ and A = the vertical area only.

It will be informative to compute a reference value for h_{cv} and to compare it, under standard conditions, with the reference value of the radiation coefficient h_r of Eq. (4.36).   Again, let us assume that $\theta_s = 40°C$ and $T_a = 313°K$, or $40°C$.   Also assume that $p = 1$ at sea

---

* Rippin, Harms, and Walters, in their analysis of a rectangular transformer (Ref. 4), assign different shape factors and coefficients to the top area and to the total vertical area.   The simpler treatment in this section leads to substantially the same results for a cube.

† An expression for the convection coefficient which is widely used for the large transformer ($L > 1$ ft) is: $h_{cv} = 0.00134 p^{1/2}\theta^{1/4}$.

TABLE 4.5 Convection Constants: Conductivity h_{cv} and Thermal Flux Density ψ_{cv}*

θ_s		313°K, 40°C		348°K, 75°C		398°K, 85°C		473°K, 200°C		623°K, 350°C		773°K, 500°C		Mean ψ_{cv}
		$L=1$	$L=4$	$L=1$	$L=4$	$L=1$	$L=4$	$L=1$	$L=4$	$L=1$	$L=4$	$L=1$	$L=4$	
25	h	3.08	2.18	3.0	2.12	2.99	2.12	2.79	1.97	2.60	1.84	2.46	1.75	
	ψ	....	0.054	....	0.053	....	0.053	....	0.049	....	0.046	....	0.044	0.051
30	h	3.23	2.28	3.14	2.22	3.12	2.2	2.91	2.06	2.72	1.93	2.58	1.83	
	ψ	....	0.068	....	0.067	....	0.066	....	0.062	....	0.058	....	0.055	0.064
35	h	....	....	....	2.3									
	ψ	....	......	....	0.081									
40	h	3.47	2.46	3.37	2.38									
	ψ	....	0.098	....	0.095									
45	h													
	ψ													
50	h	3.66	2.59											
	ψ	....	0.13											

55	h	….	….	….	….	….	….	3.62	2.56	….	….	….	….
	ψ	….	….	….	….	….	….	….	0.141	….	….	….	….
60	h	….	….	….	….	3.74	2.64	….	….	….	….	….	….
	ψ	….	….	….	….	….	0.158	….	….	….	….	….	….
65	h	….	….	….	….	….	….	….	….	….	….	….	….
	ψ	….	….	….	….	….	….	….	….	….	….	….	….
70	h	3.98	2.82	3.89	2.75	3.86	2.73	3.60	2.55	3.36	2.38	3.19	2.26
	ψ	….	0.198	….	0.193	….	0.191	….	0.179	….	0.167	0.19	0.158
100	h	….	….	….	….	….	….	3.93	2.78	3.67	2.6	3.48	2.46
	ψ	….	….	….	….	….	….	….	0.278	….	0.26	0.29	0.25
Mean h_{cv} ±0.3		…	2.5	…	2.49	…	2.47	…	2.31	…	2.31	…	2.15

* Values are based on $S = 0.63$, $\rho = 1$ (sea level) and $\psi = W/A$ where W is watts and A is total area. L is the significant dimension (in inches). Mean ψ_{cv} values are based on ambient temperatures between 40°C and 200°C. Multiply h_{cv} by 10^{-3}. Its mean deviation is ±0.3.

level, $L = 4$ in, a representative dimension for electronic transformers, and $S = 0.63$. We then obtain:

$$\left.\begin{array}{l} h_{cv} = 0.00246 \text{ W}/(\text{sq in})(°\text{C}) \\[2mm] \psi_{cv} = 0.098 \text{ W}/\text{sq in} \end{array}\right\} \qquad (4.39)$$

These values are somewhat lower than the radiation parameters of Eq. (4.36).

If we compare the convection table with the radiation table, we note that the convection coefficient (in contrast to the radiation coefficient) varies only slightly with the surface temperature rise and only moderately with the ambient temperature. On the other hand, a drastic reduction in convection results from an increase in altitude. An ascent to 50,000 ft (15.2 km) is sufficient to reduce p by a factor of 10 and h_{cv} by $\sqrt{10}$, a threefold decrease. In space vehicles, when g (the acceleration due to gravity) falls to zero, h_{cv} vanishes even when the normal density of air is maintained.

From the various foregoing considerations, it should be clear that the convection coefficient of heat transfer cannot be computed with great, or even moderate, accuracy.

c. Surface Temperature Rise

The sum of radiation and convection coefficients is the surface conductivity h_s.* Surface temperature rise should now be ascertainable from a knowledge of power dissipation from the surface, in accordance with:

$$\psi_s = (h_r + h_{cv})\theta_s = h_s\theta_s$$

Expansion of this expression yields:

$$\psi_s = a\theta_s{}^{1.25} + b\theta_s + bc\theta_s{}^2 \qquad (4.40)$$

The coefficients a, b, and c may be equated with parameters represented in Eqs. (4.35) and (4.37).

Unfortunately, the presence of the fractional and second-order exponents prevents a direct analytical solution for surface rise θ_s as a function of the power density ψ_s. Consequently, either a graphical or numerical solution is employed.

We have obtained h_s (Table 4.6) by adding those radiation coefficients (from Table 4.3) and those convection coefficients (from Table 4.5)

---

* Strictly speaking, the surface coefficient does not have the dimensions of conductivity. The terms *conductivity* and *exterior resistivity* are used because of their succinctness and familiarity.

TABLE 4.6 Surface Constants: Conductivity h_s, Resistivity r_s, and Thermal Flux (Power) Density ψ_s *

Ambient temperature

θ_r	313°K 40°C			328°K 55°C			338°K 65°C			348°K 75°C			358°K 85°C			398°K 125°C			473°K 200°C			623°K 350°C			773°K 500°C		
	h	r	ψ	h	r	ψ	h	r	ψ	h	r	ψ	h	r	ψ	h	r	ψ	h	r	ψ	h	r	ψ	h	r	ψ
25	.....	.....	.....	.....	.....	.....	7.8	128	0.195																		
30	6.95	144	0.209	7.83	128	0.235	8.02	125	0.24																		
35	.....	.....	.....	7.92	126	0.278	8.22	122	0.288	8.95	112	0.314															
40	7.33	136	0.294	7.97	125	0.319	.....	...	.....	9.04	111	0.362															
45	7.46	134	0.336																								
50	7.65	131	0.382										9.89	101.0	0.495												
55	.....	...	.....							9.43	106	0.517															
60										9.69	103	0.581	10.12	98.7	0.607	.....	...	.....	19.2	52	1.15						
65										9.75	102.5	0.633	10.24	97.6	0.666	13.0	77	0.845	51								
70										.....	...	.....	.....	...	.....	.....	...	.....	19.6	51	1.37	40	25	2.8	72.3	13.8	5.06

* Values are based on $L = 4$ in. ψ_s is in watts per square inch. Multiply h_s by 10^{-3}.

which cover a wide range of ambient temperatures and which make up a representative set of surface temperature rises. Also computed in the table are the surface heat transfer coefficient h_s, surface thermal resistivity r_s (the reciprical of h_s), and surface heat dissipation ψ_s, the last in watts per square inch.

Examination of the table reveals that there is a progressive nonlinear *decrease* in surface resistivity (and a corresponding increase in the capacity for heat dissipation) when either the ambient temperature or the rise at the surface is increased. A comparison of the tables for radiation and convection shows, also, that convection becomes progressively less important as the ambient temperature and the rise at the surface increase. In the so-called high-temperature transformer (high T_a, high θ), radiation and conduction are the principal modes of heat transfer.

Combining the standard radiation and convection coefficients of Eqs. (4.36) and (4.39) yields the standard surface constants:

$$\left.\begin{array}{l} h_s = 7.33 \text{ mW/(sq in)}(°\text{C}) \\ r_s = 136 \text{ sq in } (°\text{C})/\text{W} \\ \psi_s = 0.293 \text{ W/sq in} \end{array}\right\} \qquad (4.41)$$

Raising the ambient temperature from 40°C to 75°C increases the exterior conductivity to 9 mW/(sq in)(°C), and the thermal density to 0.36 W/sq in. Note that over the most commonly specified range of ambient temperatures (40°C to 85°C), $r_s = 120 \pm 20$.

The desirability of operating the transformer at very high temperatures can be readily inferred from Table 4.6. As much as an order of magnitude increase in h_s and ψ_s is obtainable by increasing the ambient temperature to 500°C and by increasing the permissible rise at the surface to 70°C.

d. Conduction through Coil and Compound

Calculations of the interior rise in temperature of the transformer, from its external surface to its innermost coil—i.e., proceeding from the surface, through the impregnant, oil, or potting compound, and through all the windings—can be involved, tedious, and of dubious precision. Nevertheless, it is possible to obtain useful estimates with only modest effort by employing two reasonable assumptions:

1. In most instances, the temperature rise in the coil θ_c and in the impregnant θ_i are substantially less than the surface rise, which is the differential between the temperature of the ambient air and the

temperature at the external surface of the transformer. That is: $\theta_c + \theta_i < \theta_s$.

Also, $r_c + r_i < r_s$.

2. The surface areas of the coil, of the coil and core, and of the enclosure can be translated into equivalent areas of cylinders, spheres, or cubes.

Since the rise in the temperature of the coil and compound are, in a good design, only a minor component of the total rise in the temperature of the transformer, small errors in their computation should not seriously affect the accuracy of the overall estimate.

In the schematic cross section of a transformer (Fig. 4.2) our problem is fourfold: to estimate the relative size of the several equivalent surface areas; to estimate the equivalent depth of coil insulation Δa_c; to estimate the thickness or depth of the compound or oil Δa_i; and to estimate the equivalent concentric radii a_1, a_2, and a_3.

In this treatment we can simplify the computation of heat by assuming that the effective exterior surface A_s is the area of either a circumscribing prism (e.g., a rectangular can) or a cylinder A_{cyl}. We then assume that the core-and-coil assembly has an area A_{sph} equivalent to that of a sphere. We also take the area of the mean surface in the coil to be A_{cm}.

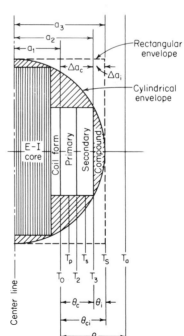

Fig. 4.2 Schematic cross section of coil and compound in cylindrical envelope.

Since the thermal flux density is not uniform over the entire surface of the transformer, it is difficult to estimate the effective surface area, especially for the open (unpotted or unmolded) transformer.

In the E-I transformer, thermal flow in the laminations is anistropic. Hence, heat flows more readily in the direction of the lamination edges than it does at right angles across the surface oxide or coating of each lamination.[4] The effective surface area is therefore lower than the value computed by summing the area of all surfaces. Thermal flow in ferrite cores is usually assumed to be isotropic (the same in all directions), and the computation of surface area is fairly straightforward. A variety of solutions is therefore used by designers. For the scrapless E-I transformer, we may use the formula:[11]

$$A_e = (7.71 + 11s)t^2 \qquad (4.42)$$

which equals $18.71t^2$ when the stack-tongue ratio s equals 1.

Other equivalent surfaces can be computed. We may attempt to construct an object, such as a sphere, whose contour approximately follows a thermal equipotential (i.e., a surface of constant temperature). If the transformer has the shape of a rectangular prism, we might approximate its equivalent diameter by taking the average of its height, width, and breadth. Then, for the scrapless lamination, we obtain a diameter of $2.5t$ and an area of $6.25\pi t^2$, or $19.7t^2$, if $s = 1$. Such a sphere is approximately inscribed in the prism. In Sec. 4.4a we compute an area A_{sph} of $23.1t^2$ for a sphere whose volume equals that of core and coil [Eq. (4.46)].

The area of a cylinder which circumscribes a scrapless transformer with square stack of t in is $10.6\pi t^2$, or $33.3t^2$. Hence:

$$A_{cyl}/A_{sph} = 33.3/19.7 = 1.68$$

The principal path of heat flow is assumed to be normal to the concentric surfaces of layer insulation, and the flow parallel to the axis of the coil is assumed, for the present, to be negligible.

The thermal resistivity of coil and compound assembly, defined earlier in Eq. (4.31), may now be stated as a more explicit function of geometry and the conductivities of coil and compound, k_c and K_i:

$$r_r = \frac{a(a_2/a_1)\,\Delta a_c}{k_c}\frac{A_s}{A_{cm}} + \frac{(a_3/a_2)\,\Delta a_i}{k_i}\frac{A_s}{A_{sph}} \qquad (4.43)$$

The area quotients appear in the equation because the exterior surface (mold or can) is the reference area for all computations. The form $(a_2/a_1)\,\Delta a_c$ is a convenient approximation of the more exact $a_2\ln(a_2/a_1)$, since $\ln(a_2/a_1) \cong \ln(1 + \Delta a_c/r_1)$.

We have already proposed standard values for the radiation and convection parameters. A similar procedure is now employed for coil and compound. In the next section, we arrive at standard thermal data for the transformer as a whole.

To arrive at standard data for coil and compound, we posit a specific and typical model. Our coil is mounted in the popular, scrapless E-I 100 core, molded with a ⅛-in (3-mm) thick, filled resin, whose thermal conductivity is 0.01 W in/(°C)(sq in). Thermal resistivity r_i of the compound is estimated to be:

$$r_i = \frac{1 \times 0.125}{0.01} 1.68 = 21 \text{ sq in } (°C)/W$$

The factor $1.68 = A_{cyl}/A_{sph}$ results from the assumptions that the un-molded transformer surface approximates the area of a sphere and that the reference area is that of a cylinder. The value of unity for the radii quotient is valid except for the very small transformer, for which it may increase to about 1.2. The mean thickness of the mold is assumed to equal the median thickness.

Employing the standard surface thermal density given in Eq. (4.41) to compute the temperature drop through the compound, we obtain:

$$\theta_i = 21 \times 0.293 = 6.1°C$$

To arrive at standard thermal data for the coil, we consider a two-winding transformer whose primary and secondary windings occupy equal depths in the window. Heat flows outward from the innermost primary winding, through the secondary, and to the outer cylindrical surface.

If we substitute the typical value of $k_c = 0.005$ into Eq. (4.43), the first fraction of which pertains to the coil, we arrive at a characteristic value for thermal resistivity of the coil:

$$r_c = 16.7 \text{ sq in } (°C)/W$$

Given the same standard value of ψ_s used in computing the temperature rise in the compound, the rise in the coil alone is estimated to be:

$$\theta_c = 16.7 \times 0.293 = 4.9°C$$

Using the values of r_c and r_i which we obtained in the foregoing discussion, we can now estimate the sum of coil and compound thermal resistivity, in the small (3-in high), scrapless, electronic transformer, to be:

$$r_{ci} = 35 \pm 15 \text{ sq in } (°C)/W \tag{4.44}$$

Note that the temperature rises in both the compound and the coil are of the same order of magnitude and that each is substantially less than the surface drop θ_s. These observations, which confirm one of our initial assumptions, provide some justification for the various approximations used in this section.

4.4. TOTAL TEMPERATURE RISE

The total resistivity of the transformer was stated earlier as:

$$r = r_{ci} + r_s = \left(\frac{\alpha}{h_c} + \frac{1}{h_i}\right) + \frac{1}{h_r + h_{cv}} \qquad (4.31)$$

The total temperature rise is the product of this equation and the thermal density, stated as:

$$\theta = \theta_{ci} + \theta_s = r\psi \qquad (4.45)$$

Since r decreases if the copper-power loss factor α decreases, there would seem to be an advantage to violating the traditional design practice of equating copper and iron power losses. In fact, this is not infrequently done. The magnetic flux density B is increased so that the loss of copper in the core is more than half the total loss. This practice improves copper regulation but at the expense of overall efficiency. The consequent deviation from $\alpha = 0.5$ compels us to reconsider our simplifying assumption that the interchange of heat between core and coil is negligible, but the problem of how best to distribute the power losses is not easily resolved. It is reconsidered at the beginning of Sec. 4.5 and in Sec. 5.3.

It is desirable to have at our disposal an estimate of the total resistivity r of a sealed or an open transformer. On the basis of the standard construction and conditions referred to in the previous section, an encapsulated transformer has a thermal resistivity of about 150 ± 30 sq in (°C)/W, and an open transformer has a resistivity of about 130 ± 27 sq in (°C)/W.

It is also useful to construct a table which predicts the total temperature rise for a wide range of θ_s and T_a. If we combine Table 4.6 with the most important data of the previous sections, we obtain Table 4.7. Here we have posited a transformer with a significant dimension of $L = 4$ in, and assume it to be compound covered, i.e., either canned and filled with a compound or encapsulated with a resin. The value of $r_{ci} = 35$ is assumed to be approximately constant over a wide range of ambient temperatures. Each of the maximum internal temperatures T_o corresponds closely to the standard IEEE temperatures[12] which serve to define the class of the insulation system (e.g., 105, 130, 155, 180, 220°C).

An examination of Table 4.7 reveals certain important trends:

1. Total resistivity r_T, the sum of surface and coil-compound resistivities in col. (7), is nonlinear and decreases when ambient temperature T_a and final temperature T_o increase.

2. As T_a and T_o increase, the coil-compound temperature drop θ_{ci} becomes a larger fraction of the total rise θ.

3. Total temperature rise is the product of the nonlinear variables r and ψ. Consequently, it is possible to keep θ constant by increasing the dissipation ψ, provided the ambient temperature is also increased.

T A B L E 4 . 7 Total Temperature Rise*

Ambient, T_a (1)	Surface rise, θ_s (2)	Surface resistivity, $r_s = 1/h_s$ (3,4)	$\psi_s = h_s\theta_s$ (5)	r_{ci} (6)	$r_T = r_s + r_{ci}$ (7)	$\theta_{ci} = r_{ci}\psi_s$ (8)	Total rise, $\theta = r_T\psi_s = \theta_s + \theta_{ci}$ (9)	Operating temperature, $T_o = T_a + \theta$ (10)
40	40	136	0.294	35	171	10.3	50.3	90.3
40	45	134	0.336	35	169	11.7	56.7	96.7
40	50	131	0.382	35	166	13.4	63.4	103.4
55	35	126	0.278	35	161	9.7	44.7	99.7
55	40	125	0.320	35	160	11.2	51.2	106.2
65	25	128	0.195	35	163	6.8	31.8	96.8
65	30	125	0.241	35	160	8.4	38.4	103.4
75	40	111	0.362	35	146	12.7	52.7	127.7
75	55	106	0.517	35	141	18.2	73.	148
75	60	103	0.581	35	138	20.3	80.3	155.3
85	60	98.7	0.607	35	134	21.2	81.4	166.4
85	65	97.6	0.666	35	132.6	23.3	88.3	173.3
125	65	77	0.845	35	112	29.6	94.6	219.6
200	60	52	1.15	35	87	40.3	100.3	300.3

* Column numbers correspond to steps in sequence to arrive at final operating temperature T_o. Column 1 is from Table 4.2; cols. 2 through 5 are from Table 4.6.

ψ_s = surface power (thermal flux) density (in watts per sq in); r_{ci} = coil-compound resistivity arbitrarily set to 35 (see text); r_T = total resistivity; and θ_{ci} = coil-compound temperature rise.

We also note, by way of recapitulation, that the total temperature rise increases with the significant dimension L and with the altitude.

a. Temperature Rise in an Equivalent Sphere

Not infrequently, it is desirable to estimate the weight of a transformer when both the permissible temperature rise and watts to be dissipated are given. A convenient relationship among θ, dissipation W, and weight M is obtained if we employ the concept of the equivalent sphere. We assume that the performance of any arbitrarily shaped transformer may be equated with a hypothetical sphere of radius ρ if it dissipates the same power loss W over its surface area $A_s = 4\pi\rho^2$.

First we estimate the volume v_t of a scrapless shell-type transformer with a square stack. We then find the radius of a sphere having the same volume and average density. The calculations proceed as follows:

$$v_t = 6t^3 + 4.5t^3 = 10.5t^{3*}$$
$$M = 0.276(0.9)6t^3 + 0.321(0.3)4.5t^3$$
$$= 1.49t^3 \text{ (core)} + 0.433t^3 \text{ (coil)}$$

$$\rho = \left(\frac{3(3 + 7.5s)}{4\pi}\right)^{1/3} t$$

Since $s = 1$ in a square stack, $\rho = 1.357t$.

The area of such a sphere is:

$$A_{sph} = 4\pi\rho^2 = 23.1t^2 \tag{4.46}$$

i.e., intermediate between $19.7t^2$, the area of the sphere inscribed in a prism, and $33.3t^2$, that of the circumscribed cylinder.

The volume of a sphere, $4\pi\rho^3/3$, equals the quotient of its weight M (in pounds) and its density d (in pounds per cubic inch). When we rewrite the basic relationship of Eq. (4.5) in terms of weight and density, we obtain:

$$\theta = \frac{W}{hA_s} = r\frac{W}{4\pi(3M/4\pi d)^{2/3}} = \frac{rW}{4.83}\left(\frac{d}{M}\right)^{2/3}$$

We estimate the density of the scrapless core and coil to be 0.19. The substantial amount of insulation in the coil dilutes the density of the copper (0.321) and iron (0.276) in the transformer. This equation then becomes:

$$\theta = r\frac{W}{14.6M^{2/3}} \tag{4.47}$$

---

* For the nonsquare stack: $v_t = 6st^3 + 1.5(2 + s)t^3 = (3 + 7.5s)t^3$.

In a standard open transformer, $r \cong 134$. (Ambient temperature is assumed to be 65°C and the surface temperature rise θ_s is moderate.) Equation (4.47) may then be further simplified:

$$\theta \cong 9.18 \frac{W}{M^{2/3}} \qquad (4.48)$$

According to this approximate formula, the temperature of a 2-lb scrap-less E-I transformer will rise about 40°C if it dissipates 6.9 W of heat.

In many cases, the initial computation of temperature rise need not be precise. It is then convenient to use an approximate expression relating temperature rise to the estimated metric surface area in square centimeters (sq cm):

$$\theta = r \frac{W}{A_s} \qquad (4.49)$$

Here r, the reciprocal of thermal coefficient h, will equal about $134 \times 6.45 = 864$, in typical convection cooling.*

b. Measurement of Temperature Rise

Calculating the temperature rise of windings by measuring resistance is usually the favored method because the data are more repeatable than when other, more direct methods, such as the embedded thermocouple, are used.

The resistance method is based on the assumption that over a moderate range of temperature, say 100°C, the change in the resistance of a winding is proportional to the change in its temperature:

$$\frac{R_h - R_c}{R_c} = \frac{T_h - T_c}{T_c - T} = \frac{\theta}{T_c + 234.5}$$

The subscripts h and c distinguish the hot and cold resistances and temperatures (in degrees Celsius). Because of the assumed linearity, the extrapolated temperature for zero resistance is −234.5°C rather than −273.2°C (absolute zero on the Kelvin scale). This inferred zero resistance temperature is computed from the temperature coefficient of standard annealed copper at 20°C, which is $0.00393 = 1/(20 + 234.5)$.

The ambient temperature may increase during the test, rising from

---

* The value $r = 850$ is used in estimates of temperature rise in high-frequency inductors (see Table 7.3). Values between 833 and 950 are encountered in the design of ferrite transformers.[13]

an initial value T_c to a final value T_a. Consequently, the average rise is computed with a more accurate expression:

$$\theta_M = \frac{\Delta R}{R_c}(234.5 + T_c) - (T_a - T_c) \qquad (4.50)$$

where $\Delta R = R_h - R_c$. A standard test condition is that $T_a - T_c$ not exceed 5°C.

A reasonably accurate measurement of θ is assured only if the hot resistance measurement is made quickly, that is, as soon as possible following interruption of power to the transformer. Speed of measurement is most important with a small transformer because of its brief thermal time constant.

c. The Hot Spot

When we measure the temperature rise of a winding, we obtain its average rather than its maximum temperature. Thus, a measurement of θ in the primary coil of the schematic of Fig. 4.2 would yield the average temperature T_p and not the maximum, or hot-spot, temperature T_o. We define the difference $T_o - T_p$ as the *hot-spot rise* θ_H, and θ_m as the average rise $T_2 - T_3$ between the coil surface and the center of the coil.

Since it is difficult to measure θ_H directly, it is rarely attempted in the small electronic transformer. It is important, however, to have a definite idea of its magnitude, especially since T_o has a direct bearing on life expectancy and reliability.

To gain a useful estimate of θ_H, we assume that the primary and secondary windings have the same accumulated thickness of insulation and that only half the heat from the copper flows through the primary, i.e., inner winding. We then write

$$\theta = \frac{T_o - T_2}{2} = \frac{1}{2}\frac{T_2 - T_3}{2}$$

and we infer, since $\theta_m = 4\theta_H$ and $\theta_H = \theta_c/6$, that:

$$\theta_m = \tfrac{2}{3}\theta_c \qquad (4.51)$$

Thus, the arbitrary assumptions of a symmetrical coil geometry and of a linear heat flow imply that the temperature rise in the hot spot is about one-sixth of the coil rise θ_c. But the rise in the coil rarely exceeds 12°C, except in a very deep coil or when the heat dissipation far exceeds the typical range of 0.2 to 0.4 W/sq in. Consequently, a normal hot-spot rise of 2°C, or even 5°C, will not ordinarily justify the special instrumentation (e.g., imbedded thermocouple) necessary to ascertain the actual maximum operating temperature T_o directly. [The rigorous analyses by M. Jakob and T. J. Higgins[14] apply more strictly to a coil which is not heated by a core. In that case, the maximum temperature is at the center of the coil. The average coil rise θ_m is two-thirds of the maximum rise θ_c only when the aspect ratio (coil length/coil build) is large. When the aspect ratio is 2.5, which signifies a more nearly square cross section, $\theta_m/\theta_c = 0.52$.] In large distribution transformers, an accurate assessment of the hot-spot temperature has an important bearing on the maximum permissible power rating. Studies using the "multiflow principle" indicate that hot-spot temperatures are substantially lower than most estimates.[15]

4.5. THERMAL DESIGN[16,17,18]

Ambient temperature is one of the prescribed parameters. The astute designer then endeavors to design power transformers for operation at as great a total temperature rise as is practical, since this will help achieve either a small transformer or one with a high VA rating—see Eqs. (4.11) and (4.22). Considerations of life expectancy and reliability, of course, set an upper bound on the permissible temperature rise θ; moreover, not infrequently, the proximity of components whose performance is adversely affected by a high surface temperature serves to limit θ to an even lower value.

If θ, T_a, and the VA rating of a transformer are prescribed, then its normal size may be ascertained. *Normal size* is defined as that size which results when natural, practical, and economical techniques, of minimum complexity, are utilized in the manufacture of the transformer. The normal transformer is built with standard, readily available materials, has one coil, wound with copper magnet wire, and contains solid insulation throughout. (The size and rating of the electronic transformer are the subjects of Chap. 5.)

Thermal design techniques comprise all possible procedures for (1) increasing the total thermal conductance hA [i.e., decreasing the thermal resistance (r/A)] and (2) when θ is prescribed, for maximizing the thermal power density, $\psi = h\theta$. Success in increasing h or ψ can increase the VA rating or reduce the size of the transformer.

We may be guided in design by the approach which results in *uniform power dissipation per unit volume*. When core and coil volumes are unequal, this approach is consistent with the field theory principle that *uniform heat generated per unit volume* produces undistorted thermal flux lines. This should then result in a uniform distribution of temperature gradients and the absence of hot spots. An effort may also be made to minimize total volume by equating the volumes and average perimeters of core and coil.

There is no universal agreement about what is natural, normal, or economical in techniques and materials. Consequently, in this section, we shall survey a wide range of thermal techniques, even though many will be deemed impractical at the present writing. Only a survey is attempted in this section, which deals with the following major topics:

Conduction circuit techniques (e.g., the heat sink, the conductive shield, conductive plates, and the heat pipe)
Improved interior conduction (e.g., coil geometry, foil winding)
Geometry and convection (coil proportions and fin cooling)
Forced convection
Cooling by vaporization (nucleate and film boiling)
The liquid-filled transformer (oil and fluorochemical cooling)
Total thermal design (e.g., combined fluorochemical and conduction techniques)

a. Techniques of Conduction

The common practice of ignoring exterior conduction as a mode of heat transfer results either from the assumption that it is a minor mode compared with radiation and convection or from the difficulty of devising a standard procedure for its measurement. Yet it is a well-observed fact that simply bolting a power transformer to a chassis plate or heat sink causes some reduction in operating temperature.

The magnitude of this decrease depends on the size of the plate and the intimacy of its contact with the transformer. To estimate the maximum decrease in operating temperature, therefore, assume that the plate area is very large and that contact is perfect, i.e., that there is no air gap or insulating film between the base and the plate. For simplicity, we will also assume that the transformer is a cube. If the plate is maintained at a constant T_a by means of a heat exchanger such as a cold plate or thermoelectric cooler,[19] then we expect the thermal flux density $W/A_s = \psi$ to be reduced by one-sixth at most. In terms of an equivalent circuit, Fig. 4.3a, the effect is equivalent to adding an external shunt resistivity r_{cd}, thereby reducing resistivity by one-sixth, or about 17 percent.

Significant decreases in r and θ have been demonstrated in many tests.[20] The use of internal conductive shields and plates can be effective.[21,22] An even more dramatic improvement results when a cold plate is substituted for the extended surface and maintained at a constant temperature by a heat exchanger. The θ of a high-temperature three-phase airborne transformer, for example, drops 72 percent,

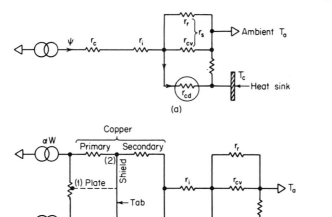

Fig. 4.3 Conduction techniques: schematics. (*a*) Heat sink may be extended surface (e.g., mounting base) or cold plate. (*b*) Conduction by tab to mounting base or cold plate: (1) conductive plate *f* (dashed) between coil form and lamination; (2) conductive shield (solid line) between primary and secondary windings brought out by means of tab.

from 103°C to 27°C when the transformer is mounted on a cold plate maintained at 125°C. The behavior is equivalent to diverting a large fraction of the coil heat to a heat sink rather than to the ambient air.

It is important to recognize the limitations of the conductive shield or plate techniques, especially in the absence of an effective heat sink. Such techniques tend to reduce the effective copper area of the coil and may, especially in a small transformer, increase copper power loss and regulation unduly.

Another drawback is the difficulty of eliminating the air gap in series with the heat path. Yet it is crucially important to keep this interface resistance to a minimum. Low gap resistance, in the order of 0.5°C/W, is obtained if the contact surfaces are clean, flat, free from burrs, and have a minimum of surface oxides. It is sometimes desirable to displace the air in a gap with a highly conductive, viscous lubricant.*

---

* Silicone grease (such as Dow Corning DC200, DC340) has been used to reduce contact resistance. Beryllium oxide ceramic wafers are reported to provide good thermal conductivity in situations where electrical isolation is necessary.

If an aluminum heat exchanger is used, treatment with a light chromate (Irridite) should be used in preference to anodizing, which creates appreciable thermal and electrical resistance.

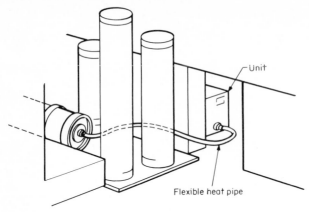

Fig. 4.4   Flexible heat pipe around obstructions.   (*By permission, Hughes Electron Dynamics Division, Torrance, Calif.*)

There are circumstances in which the mechanical layout of components in a restricted space results in cramped configurations of transformer and semiconductors (see Fig. 4.4).   An efficient thermal conductor, the heat pipe, is sometimes used to conduct heat to a heat sink located in a less congested region.[23,24]   The *heat pipe* is a sealed tube containing a liquid (typically water or methanol) surrounded by a wick structure and enclosed by a copper or stainless steel shell.   Heat at one end causes the liquid to vaporize and flow to the cooler end, where it condenses.   After condensation, the liquid returns, by capillary action in the wick, to the hot end, completing the cycle of heat transfer.   In one airborne application, the use of 0.250-in (6.4-mm) diameter heat pipe results in a flow of heat with 20 times greater conductivity than that of a solid copper rod of the same size.   The heat pipe is one-fourth the weight of the rod.[25]

Conductive techniques, like all other thermal techniques, must be evaluated in terms of their cost.   It is hard to justify the additional expense of production unless the reduction in the operating temperature or the increase in rating is substantial.

b. Improved Interior Conduction

Creative manipulation of the three variables in the formula for thermal resistance:

$$\frac{\Delta a}{kA}$$

has resulted in a number of successful techniques of thermal conductivity.

One technique, based on coil geometry, involves altering the aspect ratio (window length/window width) of the core so as to produce a thin or shallow coil.   Fewer layers of insulation are then required and the thermal path is shortened.   The use of either one or two coils, as in Fig. 2.4, entails a larger surface area.   A long, narrow window is readily achieved with certain L-type laminations.   Figure 4.5 shows the thin type of transformer which results from this approach to coil geometry.

A more novel approach is to improve thermal conduction by means of techniques such as foil winding and wafer-coil construction.[26] Maximum thermal conductance in the coil and maximum copper space factor are obtained by substituting copper or aluminum foil for the traditional circular magnet wire.   The use, on the conductor, of extremely thin insulation such as aluminum oxide further improves the conductor space factor (but also makes the coil vulnerable to scratches and shorted turns).

High thermal conductivity (as well as electric strength) in the coil requires an absence of trapped air in the interstices of the windings and insulation.   This is achieved by a combination of high-vacuum impregnation, vacuum molding, and the use of conductive fillers (e.g., silica).   It is desirable, from a thermal as well as dielectric point of view, to use compatible or homogeneous insulation throughout the entire transformer.

c. Geometry and Convection

In the previous section, we showed, by a heuristic argument, that a thin coil of large surface area is geometrically qualitatively superior.

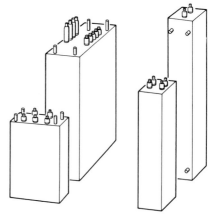

Fig.   4.5 Thin   transformers, constructed with L-type laminations and solenoid-type coils. (*By permission, Electronic Research Associates, Inc., Cedar Grove, N.J.*)

When surface convection is taken into account, however, we find that it is not at all an obvious matter to determine the best proportions for the thin coil or the long window.

(1) **Convection Geometry.** One solution to this problem involves first defining a "wiped surface" area for the popular configuration of a single coil and either two C cores or the E-I core (see Fig. 2.4c).[27] A method of finite increments is used to find those geometrical relationships which yield the maximum input VA rating per unit volume. Two conditions are that the temperature rise be restricted to a maximum of 40°C and that there be no regulation limitation. The optimum geometry is an overall shape that is rectangular, not square, and a core with a long narrow window. (See Table 5.4 and the discussion in Sec. 5.4c). Experience indicates, however, that an attempt to obtain exact optima is not justified since substantial variation in proportions produces only small changes in the VA rating per unit volume.

Problems may arise in the mounting of the thin transformer, depending on whether convection is natural or forced. When convection is natural, the convection coefficient h_{cv} (and therefore surface conductivity h_s) tends to decrease with height. The designer should be assured that a sought-for reduction in r_{ci} is not offset by an increase in surface resistance r_s. This requires that a stationary transformer be positioned with its thin dimension in the plane of elevation and that the large surface be placed in intimate thermal contact with the chassis.*

(2) **Fin Cooling.** Another geometrical procedure involves reducing the thermal flux density and the operating temperature by modifying the exterior geometry so as to substantially increase the effective surface area of the transformer. The most familiar application of this idea is fin cooling. Although a variety of fin shapes is feasible, the triangular (washboard), radial (concentric washer), and star types are those most commonly used for transformer enclosures. Figure 4.6 shows a star fin–shaped, high-power, pulse transformer.

Some idea of the efficacy of fins is gained if we visualize a cylindrical surface on which thin fins are equally spaced at a distance equal to their radial depth. Since the area in contact with the film of air is tripled, h_{cv} should also triple. Because the film thickness is about 0.2 in (for 3-in-high components), the fins should not be closer together than 0.4 in. (Similar reasoning would require that *hot components be spaced no closer than 0.4 in.*)

---

* Forced convection, however, may favor a different orientation of the transformer. It may be preferable to place the thin surface in contact with the chassis, since this permits a more effective channeling of air currents past the larger surface of the transformer.

Fig. 4.6 Fin transformer. (*By permission, General Electric Co., Schenectady, N.Y.*)

Certain desirable features of fin design may be inferred from the analysis of a straight fin with a rectangular profile. The effectiveness η_f of the fin (effective area/actual area) may be shown to be:

$$\eta_f = \frac{\tanh ma_f}{ma_f} \tag{4.52}$$

where $m = \sqrt{2h_{cv}/k\delta}$ and a_f is the radial depth, δ is the fin thickness, and k is the thermal conductivity of the fin.

Since the function defined in Eq. (4.52) increases as ma_f decreases, it is desirable to use a thick fin with a shallow radial depth. A substantial increase in area is more readily obtained by increasing the number of fins than by increasing a_f. Experience indicates that fin cooling is most effective and most economical when forced convection is available.

(3) Discussion. There is an important difficulty in the prediction of the temperature rise of compound-filled metal-encased transformers. In accordance with our simplified equivalent series thermal circuit, we expect that the measured rise of an open transformer will increase by $\theta_i\,°C$ when the transformer is placed in a can and filled with compound. Yet, under a variety of circumstances, experimenters have observed θ to: (1) increase, (2) show no change, and (3) decrease. Such anomalous behavior is not always easy to explain.

However, it can be inferred from the discussion of extended surfaces and the shape factor of transformers that an increase in r_i (the thermal resistivity of the compound) may or may not be offset, in varying amounts, either by a decrease in W/A_s due to an effective increase in the surface area or by an alteration in the shape of the enclosure. The variability of θ_i may also depend on the type of internal mounting brackets, on the shape and proportions of the can,* and on how it is mounted on the chassis. Thus, enclosing an open transformer may improve or degrade thermal performance.

d. Forced Convection[28]

Typical values of the natural convection coefficient h_{cv} are in the range 0.0025 to 0.0037 if the fluid is in air. Since the concomitant radiation coefficient is of the same order of magnitude at moderate ambient temperatures, it takes a large change in h_{cv} to produce a moderate increase in the surface coefficient h_s and a reduction in the surface temperature rise θ_s.

When equipment contains many heat-generating components producing a density exceeding 300 W/cu in (18 W/cu cm), it is possible to justify the expense of forced convection using a fan or pump. It is desirable to have a cooling duct, which permits a predictable rate and amount of fluid (e.g., air, water, oil) to transfer heat from the surface of the transformer or other component. Convection is further improved if the laminar flow is changed to a turbulent flow by using closely spaced ducts of short length in the direction of flow. Turbulence may also be promoted by adding irregularities such as louvers, ruffles, or slits to the surface of the ducts.

Factors which determine the forced convection coefficient h_{cf} include the rate of flow (volume, or weight per unit time); the heat transfer properties of the fluid, or Prandtl number (which combines the specific heat, thermal conductivity, and viscosity); and the cross-sectional flow area A_x of the duct.

The forced-convection coefficient h_{cf} may be expressed in the following form:

$$h_{cf} = C_f \omega^m \tag{4.53}$$

The coefficient C_f is a complicated function. It takes into account thermal properties of the fluid as well as certain geometric constants, including a roughness factor which depends on the total configuration of the components in the path of flow. ω, the weight-flow rate, equals dvA_x and is determined by the density d (0.07 lb/cu ft for air) and the product of fluid linear velocity v in lineal feet per minute

---

* It is sometimes necessary (especially if k_i is low) to paint the *inside* surface of the can in order to reduce the internal thermal rise θ_i.

(lin ft/min) and the duct area A_x.* A representative value of the exponent m for air is 0.8.

Surface temperature rise θ_{sf}, redefined in terms of the input and the output temperatures of the duct, is:

$$\theta_{sf} = T_s - T_a = T_s - \left(T_\alpha + \frac{\Delta T}{2} \right) \tag{4.54}$$

where the ambient temperature T_a is the average temperature of the fluid in the immediate vicinity of the hot component and ΔT is the difference between the output and input duct temperatures T_β and T_α, respectively.

Except for the simplest geometry, the procedure by which the requisite velocity of a blower or pump is determined does not readily inspire confidence. Empirical confirmation is wise. One method is to predict the surface temperature by calculating the exponent m from two successive measurements of flow rate in the actual equipment.[29]

It will be clear from the foregoing that a specifications sheet which states only the flow rate of the air blower or pump does not really provide adequate convection data. Nevertheless, some preliminary, albeit approximate, estimate of h_{cf} is often needed by the transformer designer. For large surfaces (1-ft square), the following empirical formula is sometimes used:

$$h_{cf} = 0.003 + \frac{0.008v}{60}$$

where h_{cf} is expressed in watts/(sq in) (°C). It is assumed that the air stream originates at least an inch from the surface and that it does not exceed 900 lin ft/min.

Another approach to the problem of estimating h_{cf} is to assume that the convective process is, or will be, fairly effective. Consequently, the rise ΔT in the fluid is small (say, 10°), and most of the power losses are absorbed by the fluid. We then employ the fluid rise equation, from calorimetry:

$$C_p \omega \, \Delta T = 3.143 W \tag{4.55}$$

to predict the permissible losses. The popular English set of units for this purpose would specify the specific heat C_p in Btu per pound per degree Fahrenheit [Btu/(lb)(°F)] (0.241 for air), the flow rate ω in pounds per hour (lb/hr), and ΔT in degrees Fahrenheit. The surface heat W will be in watts or, when multiplied by 3.143, in Btu per hour (Btu/hr).

---

* *Volume flow rate* is often expressed in cubic feet per minute (cu ft/min).

If we also assume that a good mix of inlet and ambient air occurs, then we can estimate surface rise to be:

$$\theta_{sf} \cong \frac{\Delta T}{2}$$

So simple an assumption permits us to make a rapid first approximation of the thermal design. To illustrate, suppose that $\Delta T = 18°F$, or $10°C$, and that five-sixths of the transformer surface is cooled by forced convection. Then $\theta_{sf} \cong 5°C$. If, without the air blower, the surface temperature rise is $30°C$ and the coil-compound rise is $10°C$, then forced convection has increased the natural coefficient h_{cv} by the factor $30/5 = 6$. This large factor relegates surface radiation and conduction to a minor role. The clear implication is that either a reduction in size or an increase in power rating has been made feasible.*

Another approach, which does not arbitrarily assume a good air mix, defines an overall temperature difference $\theta_{Mf} \cong \theta_{sf}$, which provides the thermal potential difference for the flow of heat over the entire configuration of the transformer and its neighbors.[8] The analysis equates the two expressions for forced convection:

$$W = C_p \omega \, \Delta T = h_{cf} A_{cv} \theta_{Mf} \tag{4.56}$$

where A_{cv} is the effective convection area of the transformer, typically $\frac{2}{3}A_s$ to $\frac{5}{6}A_s$. It can then be shown that:

$$\left. \begin{array}{l} \theta h_{cf} = \dfrac{\omega C_p}{A_{cv}} \ln \dfrac{T_s - T_\alpha}{T_s - T_\beta} \\[4mm] \theta_{Mf} = \dfrac{\Delta T}{\ln[(T_s - T_\alpha)/(T_s - T_\beta)]} \end{array} \right\} \tag{4.57}$$

In a rack-mounted chassis, as much as a threefold reduction in surface thermal resistance and temperature rise is obtainable with an air blower providing 50 cu ft/min of air.

The completely wound toroid would appear to be a favorable thermal geometry, since the entire coil surface is exposed to the ambient. Yet since the core is not exposed, *all* of the core power loss, manifested as heat, must flow through the coil. Usually, other than thermal considerations dictate the choice of a toroid. When the wound toroid *must* be small, as when its distributed capacitance must be kept to a minimum, the resulting high thermal flux density (watts per square inch) may result in an inordinately high coil temperature. (See Sec. 5.4.)

---

* We might then infer that the power rating may be increased by as much as $\sqrt{6}$.

Thus it may be desirable to employ forced liquid convection through tubes which are in contact with the surface of the transformer. In this case, h_{cf} depends on whether flow is laminar or turbulent, which in turn depends on the size of the Reynolds number (Re).*

For laminar flow:

$$h_{cf} \alpha \left(\frac{V}{D} \right)^{\frac{1}{3}} \qquad \mathrm{Re} < 2300$$

For turbulent flow:

$$h_{cf} \alpha \frac{V^{0.8}}{D^{0.2}} \qquad \mathrm{Re} > 2300$$

where D is the diameter of the tube. A small-diameter tube, or a narrow duct if the fluid is air, is desirable. Flagrant neglect of this observation leads to the Holland Tunnel effect, in which even high velocities can be ineffective because D is too large.

R. Lee has reported the dramatic thermal benefits of cooling a pulse saturable reactor with water by using a hollow circular collar in intimate contact with the toroidal core.[30] An average loss of 150 W, which resulted in a 225°C rise, produced only a 35°C rise when the water flowed at a rate of 0.5 gal/min. (The temperature of the water rose only 1°C.)

Forced convection has succeeded with fluids such as air, SF_6, water, transformer oil, and fluorochemical liquids. The liquids can increase the convection as much as tenfold when they are made to flow rapidly.

e. The Liquid-Filled Transformer

Occasionally, very high-voltage breakdown requirements necessitate that a liquid impregnant be used in a sealed metal can in preference to a solid-compound filling or to molding. Local convection currents and conduction in stable liquids, such as transformer-grade mineral oil and silicone oil, provide good heat transfer to the can.

Since the heat is transferred by both conduction *and* convection, however, it is difficult to predict accurately the oil temperature rise θ_i from coil to can. J. Meador's empirical equation is useful in making initial estimates.[31]

$$\psi = \frac{0.103}{\mu_1^{0.3}} \theta_i^{1.14} \qquad (4.58)$$

---

* The dimensionless Reynolds number is defined as DVd/μ, where D is the diameter of the tube, V is the velocity in feet per hour (ft/hr), d the fluid density in pounds per cubic feet (lb/cu ft), and μ the fluid viscosity in pounds per hour per foot [lb/(hr)(ft)].

where μ_1 is the viscosity in centipoises (cP). Typical values of $\theta_i/\psi = r_i$ (impregnant resistivity) are 10 to 15 at ordinary temperatures.

We may then base a computation of the equivalent film thickness Δa_i on the Fourier conduction equation:

$$\theta_i = \frac{\Delta a_i}{kA} W = \frac{\Delta a_i}{0.004}\psi$$

This yields a value of 40 to 60 mil for the equivalent thickness of the oil. Consequently, it is good thermal design practice to make margins, clearances, and ducts at least twice Δa_i, or about ⅛ in.

Unfortunately, coil-to-case temperature rise θ_i may become a substantial fraction of the total rise θ. This is most likely to occur when the exterior surface thermal resistivity $1/h_s$ is smaller than usual because ambient temperature and surface rise θ_s are high.

When thermal flux density is very great, as it is in miniaturized transformers, forced convection may prove to be inadequate or difficult to introduce, and heat transfer by change of phase may be indicated. A change of phase from solid to liquid has proved practical in very small components.[32]

f. Cooling by Vaporization

When coil operating temperatures are in the range of 150°C to 170°C, the designer may choose to use fluorochemicals, such as $(C_4F_9)_3N$ or $C_8F_{16}O$,* which vaporize in the vicinity of the coil's surface temperature. Under the best conditions, the fluorochemicals will yield a somewhat lower coil-to-case temperature differential θ_i than is obtainable with traditional mineral oils.

Alternatively, a change of phase from liquid to vapor may be employed. The use of this technique, called the *boiling mode of heat transfer*, has produced dramatically successful solutions to difficult thermal-dielectric problems. When a liquid is heated above its boiling point, two successive regions of behavior are distinguished as the surface temperature continues to rise: nucleate boiling and film boiling. In *nucleate boiling*, the superheated liquid at the surface forms mobile vapor bubbles, providing exceptionally great exterior conductivity h.[33]

At a still higher surface temperature, a film of vapor forms on the surface and *film boiling* results. However, since the film of vapor acts as an insulating blanket, nucleate boiling transfers markedly more heat than film boiling. It is customary to characterize both by the

---

* These liquids are also known by the trade designations FC-43 (boils at 177°C) and FC-75 (boils at 105°C).

more general term *vaporization cooling,* since it is difficult to distinguish the regions of nucleate and film boiling.

Condensation forms on the inside of enclosures containing power transformers and other heat-producing components. This surface is usually connected to a heat exchanger, and a continuous liquid-vapor-liquid cycle is maintained. Such a dynamic cycle has been called *autoconvection,* because the convection process is enhanced without the aid of a pump or fan.

If high voltages are present, a gas such as SF_6 may be added to produce a gas-vapor mixture of high electric strength. It is also desirable to increase the internal pressure above the atmospheric pressure, since the boiling temperature may then be brought closer to the point at which nucleate boiling occurs.

However, the film of SF_6 on the condensation surface slows the rate at which the heat is diffused by the vapor. This problem may, in turn, be substantially overcome by forcing diffusion to occur by using vane-axial fans, a procedure which tends to lower ambient pressure. When the ambient temperature varies widely, the high thermal coefficient of expansion will, in turn, create a large variation in the internal pressure, but this undesirable condition may be substantially reduced by using a gas (or a mixture of gases) which is soluble in the fluorochemical.

It is of interest to note that the nucleate boiling formula may be expressed in a form similar to that for natural convection Eq. (4.37):

$$\psi = C_v \theta_s{}^m = h_v \theta_s \qquad (4.59)$$

Representative values of the exponent m are 1.4 for the fluorochemical liquid FC-75 at 8.5 pounds per square inch gage (psig) and 1.75 psig for Freon 113. Typical values of h_v range from 0.5 to 6 W/(sq in.)(°C), two orders of magnitude greater than h_{cv} of natural convection.[34]

g. Total Thermal Design

There are various approaches to the thermal design of an electronic transformer.

The most practical, sensible, and popular approach is dictated by economic considerations. Thus we prefer existing manufacturing procedures and materials which are readily available. The desire for a transformer which performs reliably over a wide range of ambient temperatures is met by employing appropriate materials, such as suitable high-temperature insulation, high-temperature solder or welding, and encapsulants with small coefficients of thermal expansion. Exotic and difficult techniques are avoided.[35]

When cost is less important than miniaturization, it is sometimes possible to attempt an all-out thermal design of the transformer alone. Consider the instructive example, by Kilham and Ursch, of a thorough-going thermal design of a high-voltage transformer: copper heat-con-

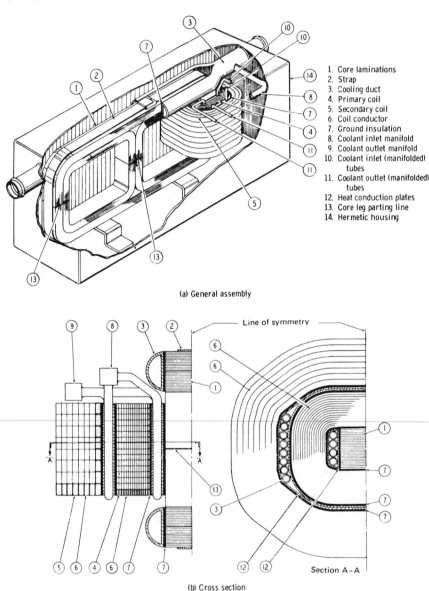

1. Core laminations
2. Strap
3. Cooling duct
4. Primary coil
5. Secondary coil
6. Coil conductor
7. Ground insulation
8. Coolant inlet manifold
9. Coolant outlet manifold
10. Coolant inlet (manifolded) tubes
11. Coolant outlet (manifolded) tubes
12. Heat conduction plates
13. Core leg parting line
14. Hermetic housing

(a) General assembly

(b) Cross section

Fig. 4.7 Conduction and fluid cooling. (*By permission, Westinghouse Electric Corp., Lima, Ohio.*)

ducting plates in the coil, a heat sink with an extended surface, a fluorochemical mixture under pressure used as a partial liquid filling, and a soluble gas which obviates the need for bellows. Coil insulation of woven glass promotes liquid penetration into the coil and permits nucleate boiling in the hot spots.[36] Another approach is to use conductive cooling in conjunction with forced fluid cooling by inserting ducts between the primary and secondary coils of the transformer.[37] (See Fig. 4.7.)

More total thermal design can be achieved by controlling the environment of the entire equipment. This arises from the requirement that equipment be of minimum size, which leads to compact packaging and high heat flux density. Such equipment is characteristically provided with a single heat exchanger (e.g., cold plate or thermoelectric cooler) for the *entire* package of components. The best field geometry for heat transfer is achieved by locating the greater heat producers in the center and the lesser heat producers radially, toward the periphery. Since the entire assembly is either encapsulated or hermetically sealed, the transformers are open, thus contributing to the reduction in volume and weight.

The need for total thermal design is surely indicated when an ultrahigh-power transformer (rated in megawatts) is to be airborne. In a system described by R. Lee, ebullient cooling is used with a fluorocarbon liquid and a heat exchanger.[38] The design procedure goes beyond conventional practice. An "optimum" design of the transformer must be assessed in the context of a larger matrix of considerations: the weight of the fuel attributable to the loss of energy (power dissipation X time) in the transformer, the weight of the liquid coolant, and the duration of the mission (which in this example is only a few minutes).

Perhaps the most impressive limit of the art of reduction of the size of the power transformer is to be achieved by extending temperatures downward. Refrigeration down to cryogenic temperatures, and ultralow temperatures ($4°K$ to $11°K$), where superconductivity occurs, virtually eliminates power losses. The drastic reduction in size and increase in rating which become feasible have not been fully assessed.

REFERENCES

1. W. Dersch, "Design of SIFERRIT Core Power Transformers with Sinusoidal Excitation," *Siemens Components Report*, Vol. 9, No. 2 (1974): 15–19.
2. F. Judd, D. Kressler, "Design Optimization of Small Low-Frequency Power Transformers," *IEEE Transactions on Magnetics*, Vol. MAG-13, No. 4 (July 1977): 1,058–69.

3. N. Grossner, "Transient Loading and Heating of the Electronic Transformer," *IEEE Transactions on Parts, Materials,* Vol. PMP-3, No. 2 (June 1967): 30–36.

4. P. Moon, D. Spencer, *Field Theory for Engineers* (Princeton, N.J.: D. Van Nostrand Company, 1960), pp. 389–409.

5. A. Rele, S. Palmer, "Cooling of Large Transformer Cores," *IEEE Transactions,* Vol. PAS-91, No. 4 (July 1972): 1,527–35.

6. J. Fourier, *The Analytical Theory of Heat,* trans. A. Freeman (New York: Dover Publications, 1955), p. 19.

7. A Lockie, "Thermal Classification of Insulating Materials," *Proceedings of the 9th Electrical Insulation Conference,* IEEE Publ. 69C 33-EI (September 1969): 77–79.

8. A. Kraus, "Heat Flow Theory," *Electro-Technology,* Vol. 63, No. 4 (April 1959): 123–42.

9. J. Holman, *Heat Transfer,* 4th ed. (New York: McGraw-Hill Book Company, 1976).

10. W. Rohsenow, J. Hartnett, *Handbook of Heat Transfer* (New York: McGraw-Hill Book Company, 1973).

11. L. Richardson, "The Technique of Transformer Design," *Electro-Technology,* Vol. 67, No. 1 (January 1961): 58–67.

12. Institute of Electrical and Electronics Engineers, *General Principles for Temperature Limits in the Rating of Electric Equipment,* IEEE Standard No. 1 (March 1969).

13. T. Konopinski, S. Szuba, "Limit the Heat in Ferrite Pot Cores for Reliable Switching Power Supplies," *Electronic Design* (June 1979): 86–89.

14. T. Higgins, "Formulas for Calculating the Temperature Distribution in Electrical Coils of General Rectangular Cross Section," *Transactions ASME,* Vol. 66 (1944): 665–70.

15. M. Carruthers, E. P. Norris, "Thermal Riting of Transformers," *IEE Proceedings,* Vol. 116(9), (London, 1969): 1,564–70.

16. J. Kaye, "Review of Industrial Applications of Heat Transfer to Electronics, *Proceedings IRE,* Vol. 44. No. 8 (August 1956): 977–91.

17. A. Scott, *Cooling of Electronic Equipment* (New York: John Wiley & Sons, 1974).

18. M. Mark, "Heat-Transfer Techniques for Magnetic-Core Components," *Electro-Technology,* Vol. 70, No. 2 (August 1962): 87–93.

19. G. Rezek, "Thermal Design and Analog Representation of a Thermoelectric Refrigerator," *IEEE International Convention Record,* Part 6 (1963), pp. 188–201.

20. H. Feder, *Miniature Power Transformers Having Wide Temperature Range: Final Report on Initial Development,"* Bell Laboratories Report 26,247G (April 1955). ASTIA Document 72–143.

21. Ray Lee, "Size Reduction of Airborne Transformers," *Transactions IRE,* Vol. CP-5, No. 3 (September 1958): 142–47.

22. A. Gilmore, "Development of High Temperature Transformers, *British IRE Journal,* Vol. 18 (April 1958): 254.

23. T. Irvine, J. Hartnett (eds.), *Advances in Heat Transfer*, Vol. 7 (New York: Academic Press, 1971), pp. 219–320g.

24. P. Dunn, D. Reay, *Heat Pipes* (New York: Pergamon Press, 1976).

25. J. Graves et al., "A Standard Load Center Converter Power Supply," *Power Electronics Specialists Conference*, Vol. 78CH 1,337-5 AES (1978): 325–30.

26. B. Sawyer, "Electrical Insulation System for 350°C Wafer-Coil Transformers," *Electro-Technology*, Vol. 67, No. 2 (February 1961): 131–34.

27. S. Jackson, *Optimization of Transformer Geometry by Method of Finite Increments*, AIEE Paper 61-789 (New York: AIEE, 1961).

28. E. Eckert, *Introduction to the Transfer of Heat and Mass* (New York: McGraw-Hill Book Company, 1950).

29. L. Fried, "Prediction of Temperatures in Forced-Convection Cooled Electronic Equipment," *Transactions IRE*, Vol. CP-5, No. 2 (June 1958): 102–107.

30. Reuben Lee, "Reducing Size of Radar Pulse Transformers," *Transactions IRE*, Vol. CP-9, No. 2 (June 1962): 58–61.

31. J. Meador, "Temporary Rise of Water Cooled Power Transformers," *Transactions AIEE*, Vol. 65 (1946): 19.

32. J. Gonzalez, D. Cawthon, "Phase Change Cooling of Thin Film Circuits," *IEEE International Convention Record*, Part 6 (1963), pp. 50–58.

33. E. Hahne, U. Grigull, *Heat Transfer in Boiling* (New York: Academic Press, 1977).

34. V. Asch, "A Study of Design Parameters in the Vaporization Cooling of Electronic Components," *Transactions IRE*, Vol. CP-9, No. 3 (September 1962): 105–14.

35. J. Rippin, H. Harms, G. Walters, "Ultrahigh Temperature Electronic Transformers: Design Optimization," *Transactions AIEE*, Vol. 80, Part 2 (July 1961): 302–309.

36. L. Kilham, R. Ursch, "Transformer Miniaturization Using Fluorochemical Liquids and Conduction Techniques," *Proceedings IRE*, Vol. 44, No. 4 (April 1956): 515–20.

37. P. Keuser, *Properties of Magnetic Materials for Use in High Temperature Space Power Systems,"* NASA Documents SP-3,043 (Springfield, Va.: National Aeronautics and Space Administration, 1967).

38. Ray Lee, "Ultra High Power Transformers for Airborne Applications," *NAECON 1977 Convention Record*, pp. 610–15.

CHAPTER **5**

The Power Transformer: Synthesis

Users of power transformers are particularly concerned with the transformer's size, especially when miniaturized circuitry is desired. For the transformer tends to dominate, physically and thermally, the layout and performance of the neighboring components in the equipment.

In this chapter, we shall concern ourselves with three important questions: (1) Given a set of specifications for a transformer, what will be its volt-ampere (VA) rating? (2) How big and how heavy will the transformer be? (3) How can we make the transformer smaller? The first question entails *analysis* of the relationships among the most pertinent variables. The second question entails *synthesis*—the determination of the size and design. The third involves problems of *optimum synthesis*—nonstandard design and transient loading.

5.1. VARIABLES WHICH AFFECT SIZE

Given a complete set of specifications, our objective is to identify the variables which will affect the size of the transformer. These variables were discussed in the three preceding chapters and are reviewed briefly now.

132

In Chap. 2, we noted such topics as the power factor, regulation, efficiency, voltage level, type of circuit (resistive, rectifier, inverter, or converter), type of winding connection (isolation or autoconnection), and dc polarization. We emphasized the fact that the VA rating of the transformer often exceeds that of the load.

In Chap. 3, we noted the effect of dielectric, thermal, and environmental stresses on reliability. These in turn determine the grade, internal operating temperature (or class), and life expectancy of the transformer. Directly or indirectly, these variables—grade, class, and life expectancy—influence size and VA rating.

In Chap. 4, temperature rise and thermal design were analyzed because of their direct effect on the size and rating of the power transformer.

In this chapter, we complete the discussion of rating and size initiated in Chap. 2. The relationships among regulation, line frequency, and geometry can be more adequately dealt with against the background of Chaps. 3 and 4. In the sections which follow, we first establish the design relationships among the variables that influence size. Next, a systematic procedure for estimating size is described. Standard and nonstandard design procedures are then discussed. Because magnetic components are widely used in regulated power supplies, we have chosen two examples for discussion: the ferroresonant transformer and the converter transformer. We conclude with transient loading— a topic which becomes important as the designer attempts to coordinate *all* variables to achieve minimum size.

5.2. DESIGN RELATIONSHIPS

Important properties of the transformer were inferred from a simple expression used in Chap. 2:

$$VA_T = P_o = E_1 I_1 = K_1 K_2 A_{Fe} A_{Cu} \qquad (2.5)$$

At that point we stressed the direct dependence of the VA rating (here called P_o) on the area product $A_{Fe} A_{Cu}$. The explicit introduction into this equation of the copper space factor K_{Cu}, iron space factor K_{Fe}, frequency f, current density J, and flux density B resulted in the basic formula:

$$P_o = k k_q K_{Cu} K_{Fe} 10^{-8} f J B A_{Fe} A_{Cu} \qquad (4.1)$$

where $k = \frac{1}{2}$, or the fraction of window occupied by the primary, and $k_q = 4.44$ (sine wave) or 4.0 (square wave).

Equation (4.1) makes it possible to assign ratings to transformers with great ease, provided only that J and B have been determined.

a. Temperature Rise

Another form of Eq. (4.1), one which takes the temperature rise into more direct account, is desirable. In Sec. 4.1 we developed expressions which equate the power rating P_o with θ (temperature rise), h (the heat transfer coefficient), and B:

$$P_o = C_2 fB(\text{vol})_{\text{Fe}} g_1 (h\theta)^{1/2} \qquad (4.12)$$

$$P_o = C_2 fB(\text{vol})_m g_0 (h\theta)^{1/2} \qquad (4.13)$$

and

$$\text{opt } P_o \cong C_5 f^a (\text{vol})_m g_4 (h\theta) \qquad (4.23)$$

Here C_2 and C_5 are constants; g_0, g_1, and g_4 are coefficients of the geometry; vol_{Fe} and vol_m denote the volume of the core and the geometric mean volume of core and coil, respectively; the exponent a depends on the choice of core material (see Sec. 6.1c); and h is a complicated function of ambient temperature, surface temperature rise, and other factors.

Note the inverse relation between the size and the temperature rise when the rating is fixed:

$$\text{vol} \propto \frac{1}{(h\theta)} \qquad (5.1)$$

Thus, miniaturization is facilitated by a large temperature rise.* We may also infer that if θ, rather than the rating, is kept fixed, then the product BJ will decrease when the size and rating are increased.

b. Losses and Efficiency

When θ is prescribed, the permissible total loss of power W may be computed from:

$$\psi = \frac{W}{A_s} = h\theta \qquad (4.5)$$

where ψ is the thermal density and A_s is the effective surface area.

The designer may compute ψ either by dividing the total power losses by the total equivalent area or by dividing the effective copper losses by the surface area of the coil. The two computations are not

---

* High temperature and a large rise in temperature are not synonymous and should not be confused. Only if the rise (i.e., the difference between the internal and external temperatures) is large can a large reduction in size be obtained.

identical, but the results, in design, should be similar. In Sec. 4.4, we derived the equation:

$$\theta = r\frac{W}{A_s} \tag{4.49}$$

Here r, thermal resistivity, is the reciprocal of h and has typical values in the range 830 to 950.

The heat transfer coefficient h is determined from such pertinent specifications as ambient temperature, altitude, and grade of construction, together with data such as we have recorded in Tables 4.3 to 4.7. The estimation and control of h, which are the major subjects of Chap. 4, are among the most difficult tasks in the design of transformers.

When flux density B can be set to an arbitrary value in a design, minimum total losses of power occur when losses in the iron and copper are equal. But this condition does not, in general, coincide with the condition of maximum VA.

In order to see this, we return to Sec. 4.1b and note that at frequencies above 200 (hertz) Hz it is possible to arrive at a maximum (optimum) VA rating by finding an optimum value for B. Thus, in the transformer designed with optimum values of B and J, power losses in the iron and copper can be expressed by:[1]

$$\left. \begin{aligned} W_{\text{Fe}} &= \frac{\theta}{r_{\text{Fe}}}\left(\frac{2}{2+m}\right) \\[2mm] W_{\text{Cu}} &= \frac{\theta}{r_{\text{Cu}}}\left(\frac{m}{2+m}\right) \end{aligned} \right\} \tag{5.2}$$

where m, an exponent of flux density in Eq. 4.14, depends on the magnetic characteristics of the core (see Secs. 5.3c and 6.1c). The ratio of power loss in the iron to loss in the copper will be:

$$\frac{W_{\text{Fe}}}{W_{\text{Cu}}} = \frac{r_{\text{Cu}}}{r_{\text{Fe}}}\frac{2}{m} \tag{5.3}$$

Note that power losses are equal only if $r_{\text{Cu}} = r_{\text{Fe}}$ and $m = 2$. In practice, however, this does not occur. The flow of heat in the transformer is not isotropic because its thermal resistivities are either unequal or nonuniform. Furthermore, the Steinmetz exponent m is less than 2 in metallic cores and usually greater than 2 in ferrites (see Sec. 6.1c). The implications are significant: the conditions for achieving minimum total power loss (i.e., maximum efficiency) and maximum VA are not always the same. Nevertheless it is expedient, in the initial phase program, to equate iron and copper losses and defer to a later stage any attempt to obtain minimal temperature rise or maximum VA.

c. Copper Regulation

The resistance R_w of a winding occuping half the window may be computed from the formula:

$$R_w = N^2 \frac{\rho\lambda}{K_{Cu}A_{Cu}/2} = 2N^2 A_R \tag{5.4}$$

where N is the number of turns, ρ is the resistivity of copper, λ is the mean length turn, and A_R is the resistance of a hypothetical bar of one turn which occupies the core window minus the copper space factor K_{Cu}. The use of the parameter A_R has the advantage, in the initial phase of design, of yielding copper resistance without recourse to a table of wire gauges.[2]

Copper regulation (reg) is computed from:

$$\text{reg} \cong \frac{R_{ab}}{R_L}$$

where R_{ab}, the total series dc resistance of the transformer, is the sum of primary resistance R_a and reflected secondary resistance $R_b = n^2 r_b$, n is the turns ratio, and R_L is the load resistance.

When wire sizes are chosen so as to provide uniform current density throughout the window, then $R_{ab} \cong 2R_w$ and the regulation may be computed from:

$$\text{reg} = \frac{W_{Cu}}{P_L} = \frac{2R_w}{R_L} \tag{5.5}$$

where P_L is the load power. Regulation may be computed from either a loss ratio or a resistance ratio.

The design for a specified regulation (reg) or copper loss ratio, Eq. (5.5), results in the formula for power rating:

$$P_o = k_r\,(\text{reg})B^2 f^2 (A_{Fe}{}^2 A_{Cu}/l_t) \tag{5.6}$$

where k_r is a constant which combines the core stacking factor, copper space factor, and resistivity.

d. Copper Space Factor

As we intimated in Sec. 2.4, some requirements increase the necessary insulation: electrostatic shielding, more than one secondary, taps, high operating voltages, and low capacitance. The consequent reduction in K_{Cu} from its nominal value results either in a reduction in the VA rating of the transformer or an increase in its size.

e. DC Polarization

In many electronic circuits (e.g., push-pull drive), the winding current includes an unbalanced dc component. This produces a magnetic bias which sets the core to flux density B_{dc}. The resulting polarization may be substantial, even in supposedly balanced circuits. Hence, to avoid magnetic saturation we must restrict the excursion of ac flux density B_{ac} as follows:

$$B_{ac} + B_{dc} = B < B_{sat}$$

The net result of this constraint is a polarized transformer whose size is larger than optimum by: $1 + B_{dc}/B_{ac}$.

5.3. THE DETERMINATION OF SIZE

In the practical exercise of the art, the transformer designer does not attempt the direct translation of basic equations and a set of specifications into a design. The first task is to estimate size, and the size of the transformer, as we have seen, depends on the equivalent VA rating, the permissible temperature rise, the regulation, and the operating frequency.

Having arrived at the equivalent VA rating, the designer selects a core size, usually from a table or graph. A nominal VA rating and regulation can be calculated for each core size with the formulas given in the preceding section. Because of the importance and great utility of such calculations, we shall outline the various considerations used in evolving a typical table (or graph) which relates the VA rating to size, temperature rise, regulation, and frequency.* The first step is to determine the losses which correspond to the temperature rise. The next step is to allocate the losses [see Eq. (5.3)].

The fraction of loss α to be assigned to the coil is large or small, depending on whether: (1) the frequency is low (25 to 60 Hz) or high (200+ Hz), (2) good regulation is preferred over the highest efficiency, and (3) the paths of heat flow from core and coil are assumed to be either uniform (isotropic) or nonuniform (anistropic), as discussed in Sec. 4.2.

---

*Detailed design procedures can be found in the references cited in the course of this discussion.

In most instances, good (low) regulation is of major importance, but it is difficult to achieve this in the small, 60-Hz transformer. Good regulation requires that the dc resistance of the transformer be lowered. To achieve this, we reduce the number of turns in the coil. The winding resistance becomes correspondingly smaller, the flux density is increased, and the exciting current must also increase (see Sec. 6.3b). However, the excitation bound of the core of the *small*, 60-Hz transformer is quickly reached. The upper bound of B_m is, therefore, set by the permissible exciting current, usually 20 to 40 percent of the full-load primary current.

It is also difficult to equalize the power losses in the core and coil of the small, 60-Hz transformer and thus promote efficiency. The number of turns in the coil must remain large; consequently, the losses in the coil remain very large in relation to those in the core. An α (i.e., W_{Cu}/W) of 0.8 is typical of this transformer.

Suppose we are freed from limitations resulting from constraints on frequency, regulation, and exciting current. We can then allocate losses in accordance with one of seveal strategems.

1. Maximum efficiency is assumed to occur when power losses in the coil and core are equal.

2. The procedure described in Sec. 4.1b, where the allocation of losses is not determined in advance, is used. The vital thermal coefficients of Eq. (4.15) are first determined by tests on a laboratory bench. This enables the designer to allocate losses by means of Eq. (5.3), solve equations for optimum values of B and J, and then proceed to determine the maximum power rating corresponding to a specific core shape, such as the scrapless E-I geometry or the standard E-C core.

3. The allocation of losses is based on the supposition of poor thermal coupling between the coil and core and of good coupling between the core and a heat sink. The designer then assigns a higher fraction of the power losses to the core than to the coil. Thus, if the loss in the copper is designed to be 40 percent, then that value of. B which results in a 60 percent loss in the core is selected.

4. The distribution of loss in power is based on the field-theory principle of generating uniform power loss *per unit volume* (see Sec. 4.5). To do this, losses are proportioned to comply with:

$$\frac{W_{Fe}}{W_{Cu}} = \frac{\text{iron volume}}{\text{copper volume}}$$

Uniform temperature throughout the transformer reduces the magnitude of the hot spot, although we are not assured of a design in which the temperature rise of the coil is minimum.

a. Constructing Tables of VA Ratings

We have, at several points, discussed the basic expressions relating VA, flux density B, current density J, iron-copper area product, and temperature rise. In principle, an optimum or suitable choice of B and J results in a maximum VA rating P_o. The synthesis of a rating from Eqs. (4.8) and (4.19) depends on our knowledge of many variables. Estimates of geometric and thermal factors, while readily arrived at, are difficult to state with an error of less than 5 percent. Computed VA ratings are confirmed, therefore, by temperature-rise tests.

Table 5.1 is a list of 60-Hz power ratings, at a 40°C temperature rise (class A insulation) for a range of small scrapless E-I laminations. In small, 60-Hz transformers, an optimum value of BJ is not readily achieved because B must not be allowed to reach saturation. In this table, the flux density of the grain-oriented core material has been restricted to 15.5 kilogauss (kG).

Practical considerations, such as inventory control and cost, dictate the use of scrapless laminations when possible and of class A, B, or F insulation. Table 5.2 is based on materials of moderate cost (i.e., grain-oriented laminations) and on the operation of the transformer so as to produce a rise of 75°C in the coil. If the ambient temperature is 75°C, the transformer will be in class F; that is, the operating temperature of the coil will be around 155°C.

When cost is the most important consideration, the designer often uses nonoriented laminations and class A (105°C) insulation.* When size is more important than cost, thin laminations and oriented C cores are used.

TABLE 5.1 Maximum VA Ratings: 60 HZ, 40°C Rise, 65°C Ambient Temperature

E-I lam-ination*	Tongue		Stack		Area product		VA	B_m, kG	J, A/sq cm	cmil/A	Effi-ciency, %
	in	cm	in	cm	sq in	sq cm					
625	⅝	1.59	⅝	1.59	0.114	0.735	11	15.5	600	328	75
75	¾	1.91	¾	1.91	0.237	1.53	20	15.5	450	438	82
100	1	2.54	⅞	2.22	0.656	4.23	45	15.5	315	625	89
100	1	2.54	1¾	4.45	1.313	8.47	88	15.5	300	656	92

* Table based on 0.014-in (0.36-mm), grain-oriented (M6) laminations.

SOURCE: Information for table from F. Judd, D. Kressler, "Design Optimization of Small Low-frequency Power Transformers," *IEEE Transactions on Magnetics*, Vol. MAG-13, No. 4 (July 1977): 1,058–69.

---

* Operation at higher values of flux density (e.g., 16.5 to 17 kG) is possible, provided adequate allowance is made by the designer for abnormally high line voltage and possible use at lower frequencies (e.g., 40 to 50 Hz).

TABLE 5.2 VA Ratings vs. Lamination Sizes: 60 Hz, 75°C Temperature Rise*

Lamination E-I type	Tongue width, in	Area product	VA average	Flux density, (kG)	Current density, A/sq in	Efficiency, %	Core loss, W	Copper loss, W	Copper regulation, %	Weight, lb Iron	Weight, lb Copper
625	⅝	0.114	9.1	14.0	4,060	63.5	0.5	4.7	52	0.37	0.98
75	¾	0.237	20.3	14.2	3,480	73.2	0.88	6.6	32.6	0.63	0.182
87	⅞	0.441	40.0	14.4	3,040	80.9	1.47	8.7	22	1.00	0.32
100	1	0.750	72.5	14.6	2,580	84.5	2.3	11.0	15.1	1.49	0.46
125	1¼	1.825	163	14.8	2,220	88.6	4.7	16.2	9.9	2.91	1.0
138	1⅜	2.66	229	14.8	2,130	90.2	6.2	19.0	8.3	3.88	1.44
150	1½	3.80	298	14.8	2,000	91.0	8.1	21.8	7.3	5.05	1.75
175	1¾	7.0	524	14.8	1,845	92.8	12.8	27.8	5.3	8.0	2.86
212	2⅛	15.3	1,050	14.8	1,550	94.7	22.6	37.4	3.6	14.2	5.18
250	2½	29.3	1,823	14.8	1,335	95.7	35.5	47.5	2.6	22.2	8.5
251	2½	68.8	3,551	14.8	935	96.5	49.8	79.2	2.23	31.1	26.6

*Table based on 29-gauge, grain-oriented (M6) silicon steel, square stack. Exciting VA/input VA is 23.5 percent (EI625) to 12.7 percent (EI251). Operating temperature = 75°C (amb) + 75°C (rise) = 150°C. Copper weight will ordinarily be less than values in the table.

SOURCE. I. Richardson, "The Technique of Transformer Design," *Electro-Technology*, Vol. 67, No. 1 (January 1961): 61.

THE POWER TRANSFORMER: SYNTHESIS 141

It is possible to construct other charts based on nominal values of the heat coefficients.[3] It should also be possible to arrive at a maximum VA rating for a given core by choosing optimum values of B and J. However, complications ensue when constraints or restrictions are placed on regulation, core material, and geometry. These topics are discussed in the three sections which follow.

b. Regulation and Size

In a power supply, the regulation of the transformer may account for a substantial fraction of the overall regulation. Good transformer regulation results from low winding resistance and, therefore, entails a small temperature rise.

Regulation is inversely proportional to size. The fact that many designs become resistance-limited before they become temperature-limited is of considerable practical importance. This condition is most common in the small, 60-Hz transformer (see Table 5.2). It is difficult to obtain better than 12 percent regulation of 60-Hz transformers weighing less than 1.25 pounds (lb) without reducing the load VA, the rating and, consequently, the temperature rise. When the transformer is miniaturized by allowing its temperature rise to increase, the regulation becomes poor.

The effect of leakage inductance on regulation is usually ignored in the small power transformer whose frequency is less than 600 Hz. In the rectifier circuit, for example, leakage inductance accounts for the commutation effect and entails but a small increase in regulation, less than 1 percent in single-phase rectifiers. In the dc converter circuit (see Sec. 2.10), small leakage inductance is required for the reduction of the amplitude of transient spikes. Such spikes can occur because of the discontinuity of the current during commutation.

c. Frequency and Size

We have already observed (in Sec. 4.1b) that a large increase in frequency permits a substantial reduction in size. To assay the extent to which size may be reduced by increasing the frequency, we must first consider the relations among flux density, frequency, and loss of power in the core.

When the frequency of the input to a transformer is increased, the flux density decreases. B varies inversely with f, as may be seen from the basic expressions for induced voltage, Eq. (4.1). The power loss in the core of the transformer also decreases as the frequency is increased. This may be inferred from the empirical expression, previously introduced as Eq. (4.14):

$$P_S = \frac{W_{Fe}}{M} = bf^n B^m \tag{4.14}$$

where M is the weight of the core in kilograms (kg) or pounds (lb), b is a constant, and the exponents n and m are determined from graphical plots of data (see Sec. 6.1c).

If the frequency applied to the transformer is changed from 60 to 400 Hz, the power rating can be increased in accordance with:

$$P_o \propto f^{1-n/m} \sqrt{h\theta} \tag{5.7}$$

For 14–milli-inch (mil) silicon steel, the rating and weight M should vary in accordance with the approximate proportions:

$$P_o \propto f^{0.2} \qquad M \propto f^{-0.15} \tag{5.8}$$

We have already noted that in most small, 60-Hz transformers the design is excitation limited; that is, the upper value assigned to B_m is restricted. This is done to limit the steady-state exciting current into the primary, or the *transient inrush current,* or the *stray flux,* or even the *audible noise* from the core. Consequently, the loss of power in the core is often substantially less than in the copper.

Increasing the line frequency to 400 Hz removes this drawback. It permits a reduction in α and an increase in the VA rating in excess of the amount predicted by Eq. (5.8). Thus, the 400-Hz transformer in which 14-mil steel is used has at least double (instead of 1.46 times) the rating obtainable at 60 Hz.

The VA rating of the 400-Hz transformer may be still further increased by employing thin (4- to 6-mil) laminations of oriented silicon or nickel-iron steel. The reduction in losses through eddy currents permit B_m to be further increased. The VA rating can be increased three- or fourfold when a higher-grade and thinner sheet material is used in the core.

(1) Core Materials.* The designer usually chooses core materials which can be operated at a high flux density. For applications *below a few kilocycles per second,* these include: (1) nonoriented and oriented silicon-iron; (2) 47 to 50 percent nickel-iron; and (3) 27 to 49 percent cobalt-iron alloys (e.g., Supermendur). Square-loop, oriented, 50 percent nickel-irons (e.g., Deltamax and Orthonol) are ordinarily reserved for saturable components such as the dc inverter transformer.

When cost is not a major consideration, premium materials are pre-

---

* See also Sec. 6.2.

ferred because their performance is generally superior. There are exceptions, however. Oriented silicon, for example, may be operated at a higher B_m than the more costly 50 percent nickel before saturation B_S ensues. But when the temperature rise is moderate (40°C to 55°C) and the power loss in the copper is equated with the loss in the iron (for maximum efficiency), nickel-iron may, at the same B_m, outperform the oriented silicon.

If, however, the specifications are such that the resulting transformer will have poor regulation or a large temperature rise, or will run very hot, the designer will unbalance the losses ($\alpha < 0.5$) and will specify a B_m higher than normal (which produces large losses of power in the core) in order to be able to use few turns and obtain less resistance in the coil. Then oriented silicon is indicated, or square-loop nickel, or even cobalt-iron. The expensive cobalt alloys, however, are ordinarily reserved for tape-wound toroids, in order best to exploit their extraordinarily high induction, in the 20- to 22-kG range.[4,5]

At frequencies *above about 2,000 Hz,* loses caused by eddy currents become so much greater than losses due to hysteresis that very thin steel strip (2- and 1-mil) must be used. But at such high frequencies, ferrite core material, which is much cheaper, has to be considered because of its high resistivity and its outstandingly small losses of power in eddy currents. It is not possible to establish, within the frequency spectrum, the exact point at which the ferrites become more advantageous than silicon, nickel, or cobalt sheet. But we can deal with this question by asking what the upper-frequency limit for metallic cores is and what the lower-frequency limit for ferrites is.*

Consider a ferrite transformer core (e.g., Ferroxcube 3 C8) in a power supply operating in the 20 to 50 kHz range. If we assume an operating flux density of 2 kG, the typical loss of power in the core is found to be about the same [6 to 20 watts (W)/lb] as that obtained with a thin-gauge [0.5-millimeter (mm)] nickel square-loop permalloy. Ferrite, far less costly than permalloy, is the preferred material in the commercial switching power supply.

Thus, economy indicates the use of ferrite when frequencies are high. What is a practical low-frequency limit for the use of ferrite? A practical answer can be obtained by comparing a metallic material of moderate cost, e.g., the 4-mil (or 0.1-mm) silicon C core (operated at high flux density) and a typical ferrite (operated at 3,000 G). Analysis

---

* The crossover frequency at which losses of ferrite and metallic glass are equal has been calculated to be in excess of 150 kHz.[6] Metallic glass is a useful alternative to ferrite at frequencies in the band of 40 to 225 kHz.[7,8] Since losses are low at flux densities in the range of 0.3 to 0.5 tesla (T), exceptionally small toroidal transformers are realizable in the switch-mode power supply.[9]

shows that the frequency at which equivalent VA ratings occur is a function of the size of the core.[10] This can be seen if equal performance is computed for two extremes of core volume, small and large:

Core size	Core volume, cu cm	VA	Frequency, Hz
Small	42–51	270–320	1,700–2,000
Large	408–480	1,000–1,100	600–700

Ferrite cores have certain disadvantages in power applications. If our design objective is to obtain maximum VA by choosing an optimum value of the BJ product, a low-frequency limit is encountered with ferrite because saturation occurs at low values of induction (2.4 to 4.7 kG). Moreover, their comparatively low Curie temperature (typically 150°C to 210°C) and their poor thermal conductivity [35 to 43 mW/(cm)(°C)] require that they be used with caution in high-temperature transformers. It is therefore customary in design work to place an upper bound on flux density. A maximum working density $B\mu$ is defined as that value of B at 115°C at which permeability μ drops to 50 percent of its value at 25°C.

The mechanical fragility of a ferrite core makes it susceptible to fracture and vulnerable to thermal shock.* Hence, if a ferrite transformer is to be used in aerospace, the designer has to guard against possible cracking by carefully compensating for factors such as the thermal coefficient of expansion of materials that are contiguous with the core. Regardless of their drawbacks, the low cost and the ease with which ferrite cores are assembled are strong incentives for their use even if minimum size or optimum performance is not quite achieved.

(2) Optimum Frequency. Mathematical analysis indicates that the higher the power frequency, the smaller the size and weight of the transformer, motor, or alternator. The upper limit for airborne equipment is around 4,000 Hz because of problems inherent in the design of the alternator. In ground-based equipment, solid-state ac-to-ac conversion is possible by use of the magnetic frequency multiplier, which can extend the 60-Hz power frequency to 800, 1,200, and 1,800 Hz and to frequencies in the low-kilohertz range.

Restrictions on the upper frequency are less severe with the dc converter (see Sec. 2.10). When the square-wave output of the inverter

---

* In high-voltage transformers, intense electric fields may occur in the region between a high-voltage terminal and the ferrite core (the ground). Large losses of power in the dielectric generate heat in the core (see Sec. 3.13a). Fracture of the core can be avoided by covering the ferrite with an electrostatic shield (screen). (See Snelling, ref. 10, p. 321).

circuit is rectified to obtain direct current, *the choice of operating frequency is within the discretion of the designer.* If the frequency chosen is equal to a tenth of the transistor's cut-off frequency,* values in the 20 to 50 kHz band are practicable. With the development of switching devices (e.g., MOSFET, the *metal-oxide semiconductor field-effect transistor*), which lose still less power during commutation (the transition interval between on and off), frequencies in the order of 100 to 500 kHz become feasible.

At frequencies above 100 kHz, other considerations must be taken into account, such as power losses due to eddy currents in the magnet wire. Copper resistance can increase substantially over its normal value at dc and low frequencies. And it is now useful to introduce the design factor:

$$F_R = \frac{\text{ac resistance}}{\text{dc resistance}} \tag{5.9}$$

When $F_R \gg 1$, the designer will reluctantly substitute stranded magnet wire (or braided strands: Litz) for solid magnet wire. Parameters such as distributed capacitance and leakage inductance now require close attention. Furthermore, when frequency is high and power is large, it may be necessary to take into account dielectric loss (and heat) in the insulation materials of the transformer. In the context of our discussion, we consider the operating frequency to be optimum when size, losses, *or* cost of the magnetic and switch components are minimal.

5.4. GEOMETRY AND SIZE

We use the term *geometry* to mean the dimensional attributes of the core and coil: the configuration or shape, the proportions of the window and core, and overall size. In the mathematical sense, we are concerned with the many ways that the geometric parameters can vary. In the development of the equations for power ratings (Chap. 4) and for inductance ratings (Chaps. 6 and 7) we have defined in text various geometric coefficients:

Function	g Coefficient†	Equation
Power	g_0, g_1, g_6	4.10
Optimum power	g_2, g_3, g_4	4.20, 4.21, 5.11, 5.12
Inductance	g_5, g_6, g_7	7.41, 7.74, 7.70
Inductance	g_2, g_1, g_4	6.67, 7.57, 7.58
Power or Inductance	g_8	5.13

---

* An alternative criterion for operating frequency is that the sum of rise, fall, and storage times of the switch not exceed 10 percent of the half-period $T/2$.

† For the values of g in the case of the scrapless E-I lamination, see Eq. (5.10).

Note that there are five explicit variables involved (A_s, A_{Fe}, A_{Cu}, l, l_t. We wish to investigate the conditions which result in an optimum geometry. We shall try to show that the optimum geometry depends on the application and that it changes with the design objective.

The two basic configurations of the magnetic assembly, the *core type* and the *shell type*, shown earlier in Fig. 2.4, are redrawn in Fig. 5.1. The core-type transformer has either one or two coils encircling its limbs. Note that the core type and the shell type are topological duals; that is, the two coils and one core of Fig. 5.1*b* are functionally interchangeable with the two cores and one coil of Fig. 5.1*c*. The designer can trade iron for copper through the medium of geometry.

a. Choosing the Shape of the Core

In most cases the designer has little doubt as to which core shape to choose for a specific design, although in principle it should be possible to justify the choice. A rational approach to the choice of core shape is described by Garbarino.[11] We summarize his procedures below.

First assume that we are to choose from one of the three configurations in Fig. 5.1: core type (one coil), core type (two coils), or shell

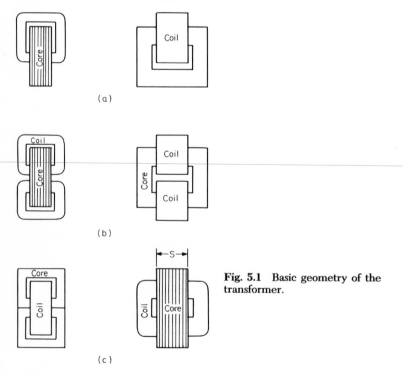

(a)

(b)

(c)

Fig. 5.1 Basic geometry of the transformer.

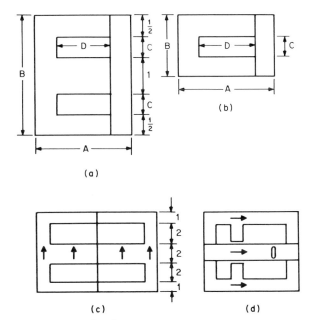

Fig. 5.2 Scrapless laminations.

type (one coil).   Now assemble the coil or coils onto either the E-I or U-I scrapless laminations depicted in Fig. 5.2.   Next, assume that the design specifications stipulate that we are to minimize one of the following four parameters, F:

F_1, volume ⎫
F_2, weight   ⎬   of coil (C) and iron core (I)
F_3, power loss ⎭
F_4, cost

Total F—that is, total volume, weight, power loss, and cost of coil and core—are to be expressed as a function of:

t, width of the core leg
s, stack ratio (stack height/leg width)
K, a weighting factor or design objective, here expressed as the ratio C_1/I_1, C_2/I_2, etc.

Now proceed to minimize $F = C + I$ with respect to the product of window area (A_{Cu}) and core area (A_{Fe}).   This area product [as shown in Eqs. (2.5) and (4.1)] is proportional to the VA rating of the transformer:

$$VA_T = K_3 A_{Fe} A_{Cu}$$

Estimates of the weighting factor K fall into three ranges of magnitude, for any of the four F parameters (volume, weight, loss, cost). Thus:

$K < 1$, minimum weight
$K = 1$, minimum volume
$K > 1$, minimum power loss; minimum cost

When the stack ratio is computed to yield minimum $C + I$, we arrive at two initial choices for the shape of the core:

1. For minimum volume or weight, choose the core type (Fig. 5.2b).
2. For minimum loss or cost, choose the shell type (Fig. 5.2a).

Further analysis indicates a preference for a stack whose area is larger than square; that is, $s > 1$.

The choice of core type cannot be regarded as fixed, especially when cost is the major consideration. From experience we find, all too often, that it is difficult to optimize more than one of the F parameters at a time.

b. Lamination Shapes

Scrapless lamination (Fig. 2.2) is stamped in such a way as to produce no waste material; hence, it has the virtue of minimum cost per pound. In Fig. 5.2c we see a scrapless design for an E-E lamination set punched so as to produce a large window. Power losses in the core are large because of the cross-grain characteristic of the limbs, but this is of little consequence since it is to be used in a welding transformer which operates only intermittently. The large losses are tolerable because the on time is much less than the thermal time constant of the transformer.

Other shapes, shown in Fig. 5.3, produce little waste when stamped. Of these, the L-L and L-T-L laminations are proportioned so that most of the flux path is with the grain, yielding an advantage when power losses must be held down.

When specifications call for a high output voltage, dielectric problems arise (see Chap. 3). Because it is imperative that corona be kept to a minimum, the wise designer chooses a simple core geometry (Fig. 5.2b) and selects a lamination whose window area is large enough to permit use of adequate margins, long creepage paths, and ample insulation. When voltage is so high as to indicate the need for a square window and L-L geometry, the cross-grain in one limb of the L can result in large power losses. One way to deal with this problem is to use L-L laminations stamped at an angle with respect to the grain (Fig. 5.3b). In doing so, one compromises optimum magnetic perfor-

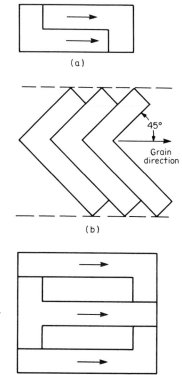

Fig. 5.3 Low-scrap laminations.

mance and settles for moderate power losses in exchange for a small amount of scrap and, therefore, low-cost laminations.

So far in this section, we have discussed the geometry of only a few transformer configurations. In truth, we have a wide choice of core shapes at our disposal (see Fig. 2.1) and can alter the relative proportions of iron and copper without changing the product of window and core areas. Our survey of prevailing practice indicates that designers favor the shell E-I shape (Fig. 5.2a).[12] Further, some designers will choose E-I and U-I structures over the more efficient C core when a vibration-proof mounting is desired at minimum cost.

Silicon-steel laminations, in the common 24, 26, and 29-gauge thicknesses, are less expensive than copper magnet wire. But when the laminations are big (as in the large, 60-Hz transformer) or numerous (as in the small, 400-Hz transformer in which 6-mil steel is used), their assembly becomes expensive. For this reason, designers are inclined to use a 4-mil C core in the 400-Hz transformer and a 12-mil C core in a shell configuration of the 60-Hz transformer. The higher cost of the core material is traded for the lower cost of assembly.

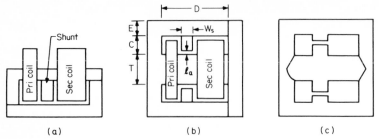

Fig. 5.4 High-reactance geometry with shunts.

The two-coil core construction is also employed when minimum winding resistance and copper regulation are desired. When power losses in the copper are larger than those in the core, as they are in the small 60-Hz transformer and in the magnetic amplifier, a large stack ratio and a large aspect ratio are desirable.[13] A stack ratio which is larger than square provides an additional advantage. The resulting transformer is rectangular, a shape which permits us to pack the most iron and copper into the rectangular can, which is the most popular shape for sealed transformers.

The greater cost of the core-type shape—whether a toroid, a wound C core, or L or U laminations—requires that it be selected over the shell-type transformer only for some good reason. For example, when the external magnetic field must be minimized without reducing the operating flux density or adding a surrounding enclosure, a toroid is indicated. Practical substitutes for the toroid permit normal winding techniques: they have an astatic construction of two coils on opposite legs of a D-U, U, L, or C core (as shown earlier in Fig. 2.4b).

In Fig. 5.4 we see examples of core-type and shell-type geometry used in high-reactance, ballast, and ferroresonant transformers. The magnetic shunt path between the input and output is obtained by means of short I laminations assembled into a stack. Also used in ballast transformers is the E-I-E geometry shown in Fig. 5.2d, which eliminates the need for the discrete shunts required by the L-T-L geometry of Fig. 5.3c.

An alternative geometry is the *cruciform* or *core-and-frame* geometry* of Fig. 5.4c. Here a moderate amount of window scrap is traded for saving in labor cost, which derives from the ease of assembly obtained by means of a hydraulic fixture that presses the core into the frame. Another high-reactance geometry (Fig. 5.5), dispenses with shunts and is discussed later in this chapter (see Sec. 5.6c).

---

* A U.S. patent for this geometry was assigned to J. G. Sola in 1939.

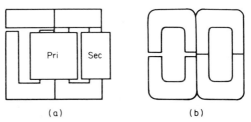

Fig. 5.5 High-reactance geometry without shunts.

c. Lamination Proportions

When the designer specifies window proportions (or an aspect ratio, i.e., window length/window width) which deviate from the normal, it is to achieve some special objective. Suppose thermal considerations are very important. If the convection currents are not favorable, the temperature drop in the internal coil must be kept within bounds. This is so, for example, when a transformer with a high temperature rise is to operate in a high ambient temperature. For such a transformer, the designer may prefer a long, narrow window, so that a long, thin coil can be used (see Fig. 4.5). A high aspect ratio is also used in the high-reactance transformer and in the constant-voltage transformer, where the long window permits the use of a magnetic shunt between adjacent primary and secondary coils.

A large window is indicated when voltages are high and clearances between the coil and the laminations must be ample. Sometimes it is necessary to split the secondary winding into several adjacent, pie-shaped segments; then a large, more nearly square window is better. Such a window is also used when the capacitance between the secondary and the primary windings must be kept very small. The area of these windows may substantially exceed the core area.

When the aspect ratio is small, a deep winding results and the large mean turn produces a high resistance in the winding. This resistance can sometimes be reduced significantly by dividing the original winding into two concentric windings, the inner one being of finer wire than the outer one.

Because the stack height of stamped laminations is easily changed, it is possible to alter the stack-tongue ratio while keeping the product of iron and copper areas constant. Hence, thermal and cost considerations aside, the same VA rating can be realized with a large lamination and small stack as with a small lamination and a large stack. This flexibility can be seen in Table 5.3, which shows the stack and tongue size of two assemblies of scrapless E-I laminations with the same area

TABLE 5.3 Iron and Coil Volume vs. Stack-Tongue Ratio* (in inches)

Lamination size	$A_{Fe}A_{Cu}$	Iron, I				Coil, C						$I + C$
		A_{Fe}			Gross volume, cu in	A_{Cu}			Volume (cu in)			Total volume
		Tongue t	Stack S	l		C†	D†	l_t	Gross $K_{Cu} = 1.0$	Net $K_{Cu} = 0.4$		
−100	1.125	1.0	1.5	6	9	0.5	1.5	8.0	6.0	2.4		11.4
−150	1.125	1.5	0.444	9	6	0.75	2.25	6.18	10.43	4.2		10.2

* Ratio (s) = stack (S)/tongue (t).

† C and D are window dimensions (see Fig. 5.2a).

product. Given the different proportions of iron and copper, however, the regulation and the cost of the two resulting designs will be different. We are saying, by means of this example, that the optimum proportions of a scrapless geometry are a function of the weighting factor K.

The geometric coefficients can be calculated as a function of the width of the center limb or tongue t (in cm) with a square stack ($s = 1$), as follows:

$$\left.\begin{aligned} g_1 &= 0.624t^{1/2} \\ g_5 &= 0.354t^{2.5} \\ g_6 &= 1.53t^{3.5} = V_{Fe}g_1 \\ g_7 &= 0.0589/t^{1/2} \end{aligned}\right\} \tag{5.10}$$

These coefficients, along with the corresponding values for the constants of our equations, are in a format which facilitates designing with the programmable calculator.

We now ask whether the proportions (aspect ratio) of the lamination window can be optimized so as to achieve some specified objective (i.e., weighting factor), such as minimum total cost, power loss, or weight.[14,15] The arbitrary alteration of window proportions is not easily fulfilled with either stamped laminations or ferrites without substantial tooling costs.

In the design of a ferrite pot core (see Fig. 6.6c) one may express the geometry as the product of a core factor and a coil factor:

$$\frac{1}{g_2} = \left(\frac{l}{A_{Fe}} \frac{l_t}{A_{Cu}}\right)^{1/2} \tag{5.11}$$

Both factors will vary as changes are made in dimensions of different sections of the core. With the aid of a computer program, g_2 is calculated and plotted as a function of the center pole diameter d. The optimum proportions are chosen to make g_2 a maximum, which also results in a design* that yields high Q.[16] (Q, or quality factor, can be defined as the quotient of stored energy and dissipated energy.)

However, the proportions of the wound C core can be readily altered by the vendor, with only a moderate increment in cost. For one study, a shell geometry comprising C cores (Fig. 5.1c) was chosen. The objective was to select those core proportions which result in maximum VA per unit volume of the envelope enclosing the transformer.[17]

In order to compare one geometry with another more readily, we have normalized all dimensions with respect to the center limb, as in Fig. 5.2a. The results of various studies of lamination geometry

---

* See Sec. 6.4b for further discussion of optimum geometry.

are summarized in Table 5.4. Perusal of the table reveals the significant fact that *the optimum proportions are different for each of the design objectives.*

If the thermal performance of the transformer or inductor is important, we focus on those geometric coefficients of which surface area is a component (g_0, g_1, g_3, g_4, g_6). Where the objective is a specified resistance or Q, the appropriate geometric factor (g_5, g_7) will not explicitly include the surface area. It is interesting to note, however, that coefficients such as:

$$g_4 = \frac{A_s}{l\lambda} \tag{5.12}$$

$$g_8 = \frac{A_s}{(A_{Fe}A_{Cu})^{1/2}} \tag{5.13}$$

are pure numbers. They may change, in a specific shape, when the internal proportions change. For example, in the E-I laminated transformer, when the stack is increased, g_8 changes in proportion (see also Sec. 7.7). The coefficients g_4 and g_8, therefore, serve as a useful figure of merit when comparing one type of geometry or shape with another. In this connection, W. McLyman's comparison of various types of cores in terms of the coefficient g_8 is of particular interest.[18]

Most studies conclude with the observation that large deviations from the optimum do not produce substantial improvements in performance. Hence, the pragmatic approach to design will often favor that geometry which yields equal weight, power losses, or cost of the iron and copper.

TABLE 5.4 Lamination Proportion: Shell Geometry*

Type	Center limb	Stack S	Window C	Window D	Overall A	Overall B	Objective	Reference†
1	1	0.5–2.0	0.5	1.5	2.5	3.0	Minimum scrap	Commercial practice
2	1	0.5–2.0	1.0	3.0	4.0	4.0	Minimum scrap	11
3	1	2.26	0.86	1.92	2.92	3.71	Minimum cost, $K=1$	15
4	1	2.27	0.70	1.58	2.58	3.39	Minimum cost, $K=2$	15
5	1	2.25	0.80	1.80	2.80	3.60	Minimum loss‡	15
6	1	1.5§	0.75	1.88	2.875	3.50	Minimum volume, $K=1$	14
7	1	1.5	1.6	4.0	5.0	5.2	Minimum weight, $K=0.2$	14
8	1	—	0.5	4.0	5.0	3.0	Maximum VA per unit volume	17

* Based on Fig. 5.2a.

† Numbers refer to references listed at end of chapter.

‡ Copper loss equals iron loss.

§ Stack ratio is chosen to comply with winding practice.

d. **Geometry and Cost**

The quest for the minimum combined cost of iron and copper has an important bearing on the choice of geometry. A consequence might be the observation that the geometry (e.g., window proportions or stack ratio) should change whenever the relative cost of copper and iron changes. In the practical world, if we ask that a lamination be priced at the lowest cost per pound and that it be readily available, we choose the scrapless E-I geometry. Yet there are studies which show that a scrapless geometry, examined in the context of *total cost* of the transformer, is not necessarily the best choice.

When the total cost of manufacture is the prime objective, our final choice of geometry is greatly influenced by other considerations, such as:

1. The decision to change from layer winding to bobbin winding of the coil
2. The decision to substitute aluminum for copper magnet wire*
3. The decision to assemble laminations by welds instead of by bolts or rivets
4. The need for the transformer to conform to a prescribed envelope (e.g., the housing of an oblong lighting fixture or an irregular pocket in a crowded electronic assembly)
5. The method of heat transfer

After a careful analysis of overall costs, enterprising designers may decide to live with practices that they might ordinarily shun. Thus, for example, they tolerate a cross-grain magnetic circuit or a butt-joint (rather than an interleaved) assembly, or may abandon the customary E-I shell geometry and choose a simple L-L shape because it can be assembled with fewer welds.

5.5. STANDARD POWER TRANSFORMER DESIGN PROCEDURES

The design of a power transformer, as we have already noted, begins with the computation of an equivalent VA rating. We showed, in Chap. 2, that the transformer rating may be substantially greater than the load VA, or power. The choice of circuit, voltage level, regulation, life expectancy, and duty cycle can significantly affect the rating and

---

* The substitution of aluminum for copper increases the resistance of a coil by 60 percent. However, a thorough redesign of the geometry can produce the same resistance in a transformer whose volume is larger than the original but which weighs less.

size of the transformer. Calculation of an equivalent rating is facilitated by use of a checklist and discrete multipliers to correct for each deviation from the nominal characteristics of a standard two-winding construction.

The principles discussed in this section are useful in the design of low-frequency (50 to 60 Hz, 400 to 2,500 Hz) power transformers. The design of ferroresonant and high-frequency converter transformers requires attention to additional considerations and is, therefore, discussed in separate sections later in this chapter.

a. Optimization Using Standard Materials

Our initial steps in the design of a power transformer are to make a careful study of the specifications and then to establish its equivalent VA rating. We proceed with a rational design procedure, making use of guidelines which will permit us to optimize one or more variables. But now let us be explicit about which of various objectives in a design program can be optimized. They are:

1. Maximum VA rating, given any of three constraints: a prescribed temperature rise, an arbitrary core geometry, or a specified exterior envelope
2. Minimize size: minimum iron-copper area product, volume, or weight
3. Minimum total power losses (i.e., maximum efficiency)
4. Minimum coil temperature rise
5. Minimum life expectancy or reliability
6. Minimum cost

Analysis may indicate the need to optimize one or more parameters, such as volume, weight, and power losses. But more often than not, a practical design approach prevails: the product is to be manufactured at *minimum cost* of materials and labor. Hence, initial choices of core material and geometric configuration are strongly influenced by considerations of cost, so much so that it is considered good practice for the designer to propose changes in the specifications which will help to reduce the cost. The goal of optimization, if strong enough, will result in trade-offs between one or more parameters or in some alteration of the arbitrary constraints of a set of specifications.

b. Empirical Design Formulas

Some designers are averse to using large quantities of data, tables, and curves. Analytic transformer designers prefer to make simplifying assumptions, especially in the early stages of a design. They may, by fitting a curve to a large amount of tabular and graphical data, derive an empirical relationship between the tongue t of a scrapless E-I lamination, the temperature rise, and the 60-Hz VA rating P_o:

$$P_o = C_4\left(\frac{\theta}{40}\right)^{0.6} st^\nu \qquad (5.14)$$

where C_4 is a constant which varies with the grade and gauge of the lamination and s is the stack-tongue ratio. The exponent ν is a value somewhere between 3.5 and 4, depending on the range over which the estimate is expected to be reasonably good.

We see that Eq. (5.14) is quite consistent with the theoretical expression derived earlier in Chap. 4:

$$P_o = C_2 f B g_6 (h\theta)^{1/2} \qquad (4.11)$$

and with the geometry of the scrapless lamination:

$$g_6 = 1.53 t^{3.5} \qquad (5.10)$$

A nonscrapless core is rated with the aid of a formula, derived from Eq. (2.8), which enables us to equate a nonstandard area product with the scrapless tongue:

$$st^4 = \frac{A_p}{0.75}$$

Another useful formula introduces the frequency variable and permits us to normalize all designs by referring them to a prototype rating P_R for a 60-Hz transformer with a 40°C rise:

$$\frac{P_o}{P_R} = \left(\frac{f}{60}\right)^{0.76}\left(\frac{\theta}{40}\right)^{0.63} \qquad (5.15)$$

Temperature-rise exponents such as 0.6 or 0.63, instead of the 0.5 used in Eq. (4.12), arise in empirical formulas because the nonlinear heat transfer coefficient h, which is a function of the surface temperature rise, is not explicitly used as a variable when the curve is fitted.

The design of the coil is facilitated by use of formulas which state winding parameters as a function of coil geometry.[19] A typical coil design, in programmable form, will express resistance as a function of wire gauge (or diameter) and *m*ean *l*ength of *t*urn (MLT) of the coil.

c. Data and Know-How

The design of a *reliable* power transformer is intricate and best left to the experienced transformer designer. Some notion of the large number of steps, iterations, and decisions involved in the design of even a simple power transformer can be gained from various works on the subject.[18,20,21] A large store of data is usually consulted in the process; moreover, a knowledge of standard production techniques is essential.

The construction of the coil must be sound, practical, and economical. A standardized body of tabular data about coil design is consulted to ensure mechanical strength and trouble-free dielectric performance. Adequate coil margins, creepage distances, and insulation thickness are essential to reliability. Data about magnet wire (e.g., the practical number of turns per inch) enable the designer to specify an integral number of winding layers so that end turns and taps may be easily spliced to the lead wires. Dielectric and thermal data about insulation, compounds, and oil are consulted, and derated to enhance reliability.

Basic magnetic data include information about the excitation properties—core power loss per pound and exciting volt-amperes per pound of the various core shapes (data about performance at 50, 60, and 400 Hz are sufficient for most purposes). The design of high-frequency transformers requires the use of data about ferrites.[10,22]

A given set of specifications may be satisfied by a multiplicity of designs, each employing a different geometry or core material. It is always good practice to fill the core window with copper and to keep the temperature rise close to the allowable maximum. An unfilled window and an abnormally small temperature rise are the signs of an uneconomical and amateurish design. On the other hand, a tight coil in the window and a rise in temperature which is exactly maximum reveal a lack of understanding of production tolerances, and the extra cost of labor or of rejects will outweigh the savings gained on material.

d. Computer-Aided Design

The process of designing a transformer may be regarded as a program containing the basic features of data storage, computation, and iteration (repeated trials). There are many tedious steps in the design procedure, and it is quite natural to consider using a digital computer.[23] The advantages of computer design—speed, uniformity, and freedom from computation errors—are impressive. In the field of electronic transformers, however, the large number of novel and changing specifications would require a large library of programs, many of which would need regular revision. It is of interest to note that the computer can be designed with the purpose of utilizing the designer's experience and judgment and hence provide for an easily modified program.[24-27]

An alternative system is used by many transformer firms. A large file of proven designs that have been thoroughly cross-indexed and cataloged may, when properly exploited, yield the flexibility which the stereotyped computer lacks. With the aid of similitude theory and judicious scaling of existing designs of similar rating, a new design can be rapidly achieved.

e. A Design Figure of Merit

We have commented on the fact that the design process, with or without the aid of a computer program, requires repeated computations in order to converge on an optimum value of a specified parameter. An algorithm associated with the *geometric mean* can facilitate the design process by providing an approximation of the optimum solution.

To illustrate, let us suppose that we have made an initial choice of geometry and now wish to assess that choice. With the lamination shape fixed, the design program proceeds as a succession of iterative calculations and, at some juncture, we will arrive at initial values of the weighting function F (weight, power losses, or cost of copper plus iron) and of the stack ratio. We now ask how far the design is from optimum; that is, how far is $F = C + I$ from its minimum?* To make the analysis more general, we will substitute symbols x, y, and k for C, I, and F. Thus, $C = x$, $I = y$, and $x + y = k$. Our task may now be stated as follows: Determine new values of x', y', and k' which make k a minimum. It can be shown that the answer is expressed in terms of the geometric mean:

$$k' = 2\sqrt{xy} < k$$
$$x' = y' = \sqrt{xy}$$

Note that the geometric mean is always equal to or less than the arithmetic mean, $(x + y)/2$. Calculation of the geometric and arithmetic mean can serve as a simple means of assessing whether the F parameter is close to a minimum. To do this, we will formalize the assessment by defining a figure of merit m as follows:

$$m = \frac{2\sqrt{xy}}{x + y} = \frac{\text{geometric mean}}{\text{arithmetic mean}} \tag{5.16}$$

As an illustration, suppose that a specific design has produced a coil and core which weigh 7 pounds (lb): $x = 2$ lb of copper and $y = 5$ lb of iron. It is useful to know that the minimum obtainable weight

---

* The symbols used here were introduced in Sec. 5.4a.

with these materials is $2\sqrt{2\times5} = 6.32$ lb. The merit factor is
6.32/7 = 0.90, or 90 percent. If our objective has been to obtain
minimum weight, then we will want to increase the ratio of copper
to iron. This can be done by altering the geometry. One procedure
is as follows: Select a lamination of similar proportions, but one whose
center leg (tongue) is larger and whose stack is smaller than the original.
This should result in less total net weight (see Table 5.3). Another
possibility is to choose a lamination with the same perimeter but having
a larger window area and a smaller center-leg width.

Now another example. Our objective is to minimize the combined
cost of copper and iron. To be specific, let us suppose, at the time
of manufacture, that the cost per pound of copper is three times the
cost of iron. We reason that minimum cost of core and coil requires
that the cost of copper and iron be equal, and this leads us to the
conclusion that the core should weigh three times as much as the
copper.

5.6. THE FERRORESONANT TRANSFORMER

As noted in Sec. 2.5c, transformers can be designed to regulate either
output voltage or current when there are large variations of input
voltage or load impedance. Ferroresonant transformers are designed
to stabilize output voltage. Ballast transformers are designed to stabilize output current. Both constant-voltage (CV) and constant-current
(CC) transformers are widely used in industrial products because of
their simplicity, reliability, and low cost. We will confine our discussion
to the CV ferroresonant transformer because of its utility in electronic
power circuits and because its performance is, in many respects, similar
to that of the CC ballast transformer. Designing a ferroresonant transformer raises interesting problems which have challenged the creative
engineer. Because of limitations of space, we will restrict our discussion of the design of these transformers to the topics of regulation,
design parameters, geometry, and core size.

a. Regulation

Models of magnetic regulating circuits utilize some combinations
of linear and nonlinear circuit elements. Figure 5.6a is a schematic
of the basic CV regulator circuit: a linear inductor in series with a
shunt combination of a saturable reactor with a resonant capacitor
connected *across* the load. In Fig. 5.6b, we depict the CC ballast
circuit, in which the capacitor is connected in *series* with the load.

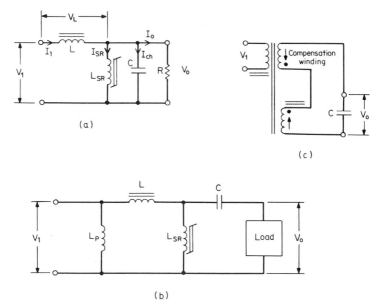

Fig. 5.6 Ferroresonant circuits.

The pi configuration of the circuit is useful in analysis, especially when the geometry of the core provides two separate flux paths for both input and output windings, as shown in Fig. 5.4. Typical loads of the ballast transformer have a nonlinear VA characteristic.

The analysis of a CV or CC magnetic circuit (Fig. 5.6) is facilitated by assuming quasi-linear behavior of the circuit elements.[28] In the ferroresonant transformer, the voltage V_o of the output winding, shown connected in parallel with the resonating capacitor, is proportional to flux density B_S:

$$V_o = 4fNA_eB_S\,10^{-8} \tag{5.17}$$

where V_o is peak volts (V) of the output winding, f is frequency, and A_e is the net area of the core on which N turns are wound. As the line input volts increase, shunt inductive reactance decreases and magnetizing current increases sharply. At resonance, flux density jumps to a region of saturation B_S. In the region of ferroresonance, the variation of flux density and volts at the output is much smaller than at the input. One way to study the variation, ΔB_S is to examine the slope of a *B-H* magnetization curve and to plot B vs. volt-amperes per pound. When this is done, we find that the percentage of variation decreases as the saturation density increases.

To improve regulation, we strive for maximum flux density, but this requires using a large capacitor. The magnitude of the capacitor's

VA has a decided effect on transformer size and affects the choice of other design parameters. We observe that good regulation incurs the penalty of high core volt-amperes, high core watts, and high copper power losses (due to high exciting current and high capacitor current). The ferroresonant transformer runs hot, even when lightly loaded.

Improvements in line regulation can be made by adding a subtractive *compensation winding* to the primary circuit (Fig. 5.6c), but the large phase angle between input and output windings requires a rather large compensation voltage and the allocation of substantial physical space for that winding.

Load regulation, that is, variation of output volts with changes in load current, is a function of winding resistance and is usually poorer than line regulation (the variation of output volts with changes in input voltage). When regulation requirements for variations in load, line, and frequency are all stringent, the use of negative feedback is indicated. Feedback techniques, magnetic or electronic, result in a *controlled ferroresonant transformer.*[29]

b. Design Parameters

Specifications for the CV transformer include, in addition to the basic requirements, parameters such as voltage range at the input, power factor, short-circuit current, and regulation. Additional constraints may be placed on such factors as stray flux and acoustic noise.

The design process begins with the selection of an appropriate circuit model containing the basic elements: L, leakage inductance; L_{SR}, shunt inductance (at saturation); and C, resonant capacitance. Following the choice of a circuit and its component elements, it is possible to designate the circuit variables:

V_1,	input voltage (at minimum line)
V_o,	output voltage
ω,	radian frequency
W,	load watts
$\cos\theta$,	input power factor
I_1,	input current
I_o,	output current
I_s,	short-circuit current
I_{ch},	capacitor current
I_{sr},	saturation current

Analysis of the basic circuit, Fig. 5.6a, results in a set of design equations and a preliminary estimate for the basic elements: L and C.[28]

The next phase of the design procedure consists of establishing nominal values for the various parameters. These include:

1. V_o/V_1, which determines the range over which regulation is effective.

2. The power factor.

3. I_s/I_o, which determines the value of leakage L and, therefore, the size of the shunt air gap.

4. Total cross-section area of the shunt stacks, which may be different from the core area of the primary winding.

5. Capacitor VA/load watts ratio, which determines the magnitude of regulation.

6. The resonance factor, which differs from unity* in a ferroresonant circuit. An approximate formula for the *ferroresonance factor* is:

$$\omega^2 LC = 0.76 \tag{5.18}$$

provides the designer with a direct means for computing C, once L has been established (see step 3 above).

The final phase is physical design: the selection of materials and geometry. Tentative choices of the following will be made:

7. Magnetic material (grain-oriented silicon is favored if a small regulation limit, e.g., 2 percent, is specified)

8. Saturation flux density B_S of the output winding

9. Input flux density, which is less than B_S so as to limit heating losses

10. Insulation material, chosen for its dielectric and thermal characteristics

11. Geometry of the core

12. Size of the core

In Table 5.5, the parameters given above are listed in a format suitable for a design program. The column at the right includes a typical range of values used in practical designs.

c. Integrated Geometry

There are many physical ways to realize the three basic elements of a ferroresonant circuit, although the integration of several elements into one magnetic structure is mechanically and economically desirable. The design of the integrated geometry of such a circuit can

---

* In the linear circuit, $\omega^2 LC = 1$.

TABLE 5.5   Design Parameters for a Ferroresonant Transformer

Item	Parameter	Symbol	Typical Range
1	Output-input voltage ratio	V_o/V_1	1.4–1.6
2	Power factor	$\text{Cos }\theta$	0.95–1
3	Short-circuit current ratio	I_s/I_o	1.1–3 (at low line)
4	Shunt area ratio	A_{sh}/A_e	1–1.53
5	Capacitor VA/ load watts	$\dfrac{\omega C V_o{}^2}{W}$	1.4–1.9; C tolerance, 6%
6	Ferroresonance factor	$\omega^2 LC$	0.61–0.76
7	Core material	—	grades 29M6, 26M22
8	Flux density, output	B_s	19.4 kG
9	Flux density, input	B_p	17–18 kG (at high line)
10	Insulation class	B, F, H	135°C–220°C
11	Geometry	E-I, core and frame	See Sec. 5.6c
12	Core size	$A_e A_c$	See Sec. 5.6d

be a challenging project in which the principle of duality serves as an aid to the development. This procedure permits the designer to equate elements of the equivalent circuit with physical elements in the integrated geometry.[30] There are many possible integrated configurations; we describe only those in common use.

The circuit behavior of a series inductor and transformer can be realized in one structure by placing a magnetic shunt between the input and output coils of a single-window transformer, as shown in Fig. 5.4. Figure 5.4c depicts a geometry in which the shunt is integral with the center limb of the lamination. The merits, in production, of this cruciform geometry were mentioned in Sec. 5.4b.

An alternative geometry (see Fig. 5.5) puts the leakage path in one outer limb of a shell structure and the saturated path in the other outer limb. The magnetic circuit can be constructed in several ways: (a) as a pair of U-I laminations or as a single E-I lamination with an air gap located in one limb, or (b) as a pair of C cores.[31] The rationale for the C-core geometry of Fig. 5.5*b* is based on two considerations: regulation is improved when the air gap resistance is reduced in the saturated path of the magnetic circuit, and losses will be minimal in a core structure whose laminations are totally with the grain.

In the cruciform geometry (see Fig. 5.4*c*) the gap reluctance of the saturated path is small but the power loss in the core is large in the cross-grain region of the frame. When the choice of geometry is strongly influenced by cost, the designer will choose either an E-I configuration which requires a separate shunt structure or its equivalent, the cruciform structure whose shunt is an integral part of the core and frame. The latter geometry is likely to be chosen when minimal interference from stray flux and low acoustic noise are required.

It is normal practice in design to provide uniform flux density throughout the magnetic circuit. However, there are occasions when it is beneficial to depart from normal practice and design for *nonuniform distribution of flux density*.[32]

Consider the geometry of Fig. 5.4. If uniform flux density is required in the shunt of an E-I transformer, then each shunt will have half the area of the center limb. Now suppose that the paths of shunt flux and saturation flux are altered in their cross section so as to produce regions of nonuniform flux density. When this is done, it is possible to obtain certain desirable characteristics. With magnetic saturation of the shunt path, we can obtain a choice of sharply defined contours for the short-circuit current. With different levels of flux density in the output region of the magnetic circuit, it is possible to effect a marked reduction of power losses, acoustic noise, and stray flux.

Table 5.6 contains a listing of a variety of E-I laminations designed for ferroresonant applications. In this table, a measure of nonuniformity of the distribution of flux in the output region of the lamination is given by the flux ratio:

$$\frac{\text{Outer leg area}}{\text{Center leg area}} = \frac{2E}{T} > 1 \tag{5.19}$$

where the dimensions E and T are as indicated in Fig. 5.4*b*.

d. Size of the Core

The final phase of physical design starts with the choice of a core shape and an estimate of its size. Size is related to the area product, which can be expressed in terms of the load volt-amperes (see

TABLE 5.6 Ferroresonant Power Ratings vs. Lamination Size, 60 Hz

E-I lamination	Output power W, dc W		Current density J, A/sq in	Core dimensions, in[t]				Flux ratio 2E/T
	Min	Max		Center leg T	Max stack	C	D	
750FR*	24	60	1,900	0.750	1.875	0.625	2.500	1.17
7	35	88	—	0.875	2.187	0.625	2.750	1.14
1,000FR	60	150	1,780	1.000	2.500	0.750	3.000	1.12
8	75	185	—	1.125	2.812	0.625	3.500	1.11
1,250FR	140	350	1,650	1.250	3.125	0.906	3.687	1.10
5,731FR	220	540	—	1.375	3.625	1.063	4.000	1.27
1,625FR	350	875	1,480	1.625	4.062	1.125	4.375	1.15
8,100	400	900	—	1.800	4.500	1.375	3.375	1.25
2,000	600	1,500	—	2.000	5.000	1.500	3.750	1.25
2,125FR	880	2,200	1,230	2.125	5.312	1.312	5.562	1.12
2,625FR	1,900	4,500	990	2.625	6.562	1.562	6.750	1.12
3,250FR	4,200	9,000	700	3.250	8.125	1.875	8.125	1.12

* FR type laminations are discussed in ref. 20.

[t] The output power ratings provide a starting point for FWCT capacitor input designs. These ranges yield a temperature rise of 55°C for the five smaller laminations and 80°C for the remainder, as well as good load regulation.

SOURCE: Based on data furnished by Optimized Program Services, Beria, Ohio.

Sec. 5.3a). When the output is rectified to furnish dc power W, we can recast Eq. (4.1) as follows:

$$W = 4fBJA_eA_c 10^{-8} \tag{5.20}$$

Here the subscripts e and c are used to denote the net area of the iron and copper. The coefficient 4 is used as the value of k_q because the trapezoidal waveform is functionally equivalent to that of a square wave.

The calculated value of the required resonating capacitance C is almost invariably too large to be practical, but it is a simple matter to use an auxiliary capacitor winding with N_c turns to obtain the desired capacitance. A small, practical capacitor C' is chosen to satisfy the formula:

$$\frac{C}{C'} = \left(\frac{N_c}{N_o}\right)^2 \tag{5.21}$$

where N_o is the number of turns in the output winding.

In Table 5.6 the power ratings (dc watts, W) are based on use of the standard FWCT (full wave rectifier circuit) and a ratio of capacitor VA to load watts of 1.9 (see item 5 in Table 5.5).[20]

The final phase of the design deals with the vital factors of dielectric, thermal, and cost performance. Calculations are, therefore, made of the following:

1. Voltage gradients of the insulation, in order to satisfy criteria of reliability
2. Temperature rise—a function of losses in the coil, core, shunts, and capacitor
3. Cost of materials and labor

There are many variables which must be dealt with, and assessed, in successive phases of an economical design. It may, therefore, be well worth the effort to translate the steps in the design into the format of a computer program.[25]

5.7. CONVERTER TRANSFORMERS[33]

Historically, our industrial society has transmitted power at low frequency. Yet there are industrial requirements for power transfer at high frequency, as in circuits for the induction heating of metals and the dielectric heating of plastics. In most situations, power is available only at the common line frequency of 50 or 60 Hz (400 Hz in aircraft). But in the design of some circuits, the designer is given the freedom to select the frequency at which power is transferred. Conversion from a low frequency to one much higher (e.g., 2.5, 25, and 500 kHz) is feasible and desirable (see Sec. 5.3c). We see immediate benefits in the magnetic components: reduced size, smaller power losses,[34] and improved regulation. These advantages are evident in the converter circuit of switch-mode power supplies and in the flyback transformer used in high-voltage circuits. We focus our discussion on the high-frequency converter transformer* because its design is explicitly motivated by the dual objective of reducing the size and the cost of the electronic power supply to a minimum.

As in designing the low-frequency transformer, our initial task is to estimate the size of the transformer, which will depend on the type of circuit, the permissible temperature rise, and the operating frequency.

Analysis of the specifications results in an equivalent VA rating. Now the designer will make an initial choice of the core material. Thin-gauge laminations operate at high flux density, are very efficient at moderately high frequencies, but are costly. Ferrite cores, although limited to lower values of flux density, are inexpensive.† Next, a deci-

---

* See Sec. 2.10 for a description of standard converter circuits.

† The choice of core material was discussed in Sec. 5.3c and is discussed further in Sec. 6.2.

sion about the geometry of the core is made. The choice among shapes, such as the toroid, cup core, U core, or E core, will be influenced by mechanical and thermal factors, i.e., the mounting layout and whether the heat is transferred to the chassis or a heat exchanger. A preliminary estimate of the size of the core can be gleaned from the basic equations of the converter circuit. In terms of the volt-second product ET and of EIT, the product of input energy EI (the VA product), and time T:

$$ET = 4NB_mA_eA_c\,10^{-8} \tag{5.22}$$

$$EIT = kJB_mA_eA_c\,10^{-8} \tag{5.23}$$

where the period T is inversely proportional to the switching frequency f, and the integer $k = 1$ or 2, depending on which converter circuit is chosen. The current density $J = 2NI/A_c$ when primary and secondary occupy equal space in the window. Earlier, in Sec. 4.1b, we saw that optimum values of J and B will yield a maximum VA rating for any given core material and temperature rise. An optimum value of flux density B_{opt}, albeit desirable and feasible with a metallic core, is not necessarily realizable with a ferrite core. When ferrite is used, an appropriate value of B_m is chosen after careful consideration of one or more of the following factors:[35]

1. Maximum temperature of the core (e.g., 100°C), which must be well below the Curie temperature of the material
2. Anticipated unbalance dc in the primary winding*
3. Magnitude and linearity of the magnetizing current (and inductance of the primary circuit), which affects the drive current and the magnitude of spikes (see Sec. 10.9a)
4. Magnitude of transient current during turn-on time (which may require using an air gap in the magnetic circuit to reduce retentivity)
5. Under conditions of transient loading, the need to limit B_m to a value which conforms to the ratio $\alpha \cong B_{sat}/B_m$ †

Having arrived at a conservative choice of B, a value of current density J is chosen. This will be based on a thermal estimate of allowable dissipation from the surface of the transformer (see Eqs. 4.17, 4.49).[36] Using assumed values of B, J, and the surface thermal coefficient, it is possible to compute power losses and temperature rise. Hence, for each type of geometry the area product can be assigned

---

* Without a suitable air gap (which markedly reduces retentivity B_r), the flux swing must be reduced to $\Delta B = B_m - B_r$.
† An alpha value equal to or greater than 1.5 is typical.

T A B L E 5 . 7 Forward Converter (Push-Pull) Transformer Power Ratings vs. Ferrite Core Sizes

E-C type no.	35.00	42.00	54.00	72.00
Weight, g	36.00	52.00	111.00	253.00
Centerpole diameter, cm	0.95	1.16	1.34	1.64
A_{Fe}, core area, sq cm	0.843	1.21	1.80	2.79
A_{Cu}, window area, sq cm	1.54	2.05	2.98	6.18
A_s, surface area, sq cm	43.50	59.00	91.00	150.00
λ, mean length turn, cm	5.30	6.20	7.40	9.70
l, magnetic path length, cm	7.74	8.93	10.50	14.40
V_{Fe}, core volume, cu cm	6.53	10.81	18.80	40.10
P_o, 20 kHz, 30°C rise*	80	130	220	470
P_o, 25 kHz, 40°C rise†	87	130	260	545
P_o, 30 kHz, 40°C rise‡	105	170	255	500

* Siemens catalog (1979), N27 core.

† Ferroxcube Division/Amperex Electronic Corporation, *SMPS Manual*, 1978. Center pole is at 100°C, with core clamped to a surface of constant temperature. Ratings are based on use of 3C8 core, with $K_{Cu} = 0.5$, $F_R = 1.25$, and $\alpha = 1.5$ to 1.72.

‡ Thomson CSF data, *Powercon Intl.*, Jan. 1981, p. 87.

a specific VA rating. It is possible, therefore, as it is in designing a low-frequency transformer, to construct a table of power ratings corresponding to each family of core types. Table 5.7 contains a list of VA ratings obtainable with four standard ferrite E cores. We chose the E-C core type for this example because of its circular leg results in a practical and economical coil geometry. Ratings of both the FWCT circuit of Fig. 2.14 and of the bridge circuit in which no center tap is required are listed. Note the marked increase in rating when the drive and load circuits are changed from the center-tap connection to the bridge type of winding. Tables of power ratings for other core shapes, such as the U, toroid, and open pot core operated at higher frequencies (e.g., 50 to 100 kHz), can be constructed by making use of curves and nomograms available from the core manufacturer.[37,38]

a. The Step-Up Flyback Transformer[39]

There are many applications in which the flyback circuit (Fig. 2.13*b*) is useful, especially when efficient step-up conversion is desired in portable equipment with a low-voltage battery.[40,41]

The power rating of the transformer is proportional to the energy stored in the primary circuit and to the air gap in the magnetic circuit:

$$P_o = c\frac{LI_p{}^2}{2}f \tag{5.24}$$

where c is a constant and I_p is the peak current flowing in the primary inductance.[42] The rating increases with the frequency f of the pulse repetition.

A basic disadvantage of the circuit stems from the presence of unbalanced direct current, which restricts flux excursion. When necessary, this handicap can be overcome by adding a reset winding and a diode whose polarity is chosen so as to produce a flow of current opposite to that of the load winding.[43]

An interesting and useful application of the flyback converter circuit occurs in the horizontal deflection circuit of the television receiver. We will describe the TV flyback transformer later, in Sec. 11.6b, because its design involves topics to be covered in later chapters of the book.

b. The Modulated Flyback Inverter

It is possible to obtain a large step of voltage by the conversion of direct current to a *sine wave* of low-frequency alternating current. In a circuit described by V. Brunstein, use is made of both *p*ulse *w*idth *m*odulation (PWM) and *a*mplitude *m*odulation (AM).[44] Current in the primary circuit of a flyback transformer is pulse width–modulated at a carrier frequency of 20 kHz. The bipolar secondary circuit produces a train of pulses whose *amplitude* is modulated by a 60-Hz driver. The result is an output voltage whose AM envelope is that of a 60-Hz sine wave, stepped up by the ratio given by Eq. 2.31. The transformer is small because it is transferring energy at a high frequency.[45]

5.8. TRANSIENT LOADING AND INTERMITTENT RATINGS

Under conditions of transient or intermittent loading of the power transformer, the duty cycle is less than 100 percent.[46] Hence, the life expectancy of the transformer will increase, the degree of increase depending on the thermal time constant of the transformer, the duty cycle, and the nature of the resultant thermal waveforms. As a consequence it is possible, in principle, to operate the transformer at a higher VA rating than normal, sustain a moderate increment of temperature, and yet retain the normal life expectancy of the insulation system.[47] This thermal strategem makes use of the concept of *relative aging* to either increase the VA rating or reduce the size of the transformer without recourse to exotic techniques of heat transfer.[48]

5.9. SUMMARY

The size and VA rating of the power transformer depend primarily on the reliability level, the temperature rise, the regulation, the voltage level, and the frequency. A reduction in size is more readily obtainable by increasing the operating frequency than by increasing the temperature rise, because a large rise in temperature adversely influences both regulation and reliability.

REFERENCES

1. F. Judd, D. Kressler, "Design Optimization of Small Low-Frequency Power Transformers," *IEEE Transactions on Magnetics*, Vol. MAG-13, No. 4 (July 1977): 1,058–69.
2. K. Macfadyen, *Small Transformers and Inductors* (London: Chapman & Hall, 1953), Chap. 7.
3. W. Muldoon, "Analytical Design Optimization of Electronic Power Transformers," *IEEE Power Electronics Specialists Conference Record*, 78CH 1,337-5AES (1978): 216–25.
4. J. Clark, J. Fritz, "Effects of Temperature on Magnetic Properties of Cobalt-Iron Alloys," *Electro-Technology*, Vol. 68, No. 5 (November 1961): 846–48.
5. R. Frost et al., *Evaluation of Magnetic Materials for Static Inverters and Converters*, NASA CR-1,226 (Washington, D.C.: National Aeronautics and Space Administration, 1969).
6. R. Boll, H. Warlimont, "Applications of Amorphous Magnetic Materials," *IEEE Transactions on Magnetics*, Vol. MAG-17, No. 6 (November 1981): 3053–3058.
7. W. Kunz, D. Grätzer, "Amorphous Alloys for Switched Mode Power Supplies," *Journal of Magnetism and Magnetic Materials*, Vol. 19 (1980): 183–184.
8. D. Chen, "High Frequency Core Loss Characteristics of Amorphous Magnetic Alloy," *Proceedings of IEEE*, Vol. 69, No. 7 (July 1981): 853–855.
9. J. Torre, C. Smith, M. Rosen, "Performance of Amorphous Metals in a a 1kW Switched Mode Power Supply," *Proceedings, International PCI Conference*, San Francisco (March 1982): 278–289.
10. E. Snelling, *Soft Ferrites* (London: Iliffe Books, 1969).
11. H. Garbarino, *Research and Development of a New Design Method for Power Transformers* (Chicago: Armour Research Foundation of I.I.T.), ASTIA Document 29,874 (February 1953): 78–108.
12. L. Giacoletto, "Proposal for Metric Standardized Scrapless E-I Laminations," *IEEE Transactions on Magnetics*, Vol. MAG-14, No. 5 (September 1978).
13. B. Bedford, C. Willis, G. Dodson, "An Analysis of Optimum Core

Configurations for the Magnetic Amplifier," *Transactions AIEE*, Vol. 74, Part 1 (March 1955): 62–70.

14. H. Garbarino, "Some Properties of the Optimum Power Transformer Design," *AIEE Transactions on Power Apparatus*, Vol. 74, Part 3A (June 1954): 675–83.

15. H. Hamaker, T. Hehenkamp, "Minimum Cost Transformers and Chokes," *Philips Research Reports*, Vol. 5 (1950): 357–94, Vol. 6 (1951): 105–34.

16. J. Ohita, T. Mitsui, M. Wasaki, "A New Ferrite Core for Use in HF Switched Mode Converters," *Power Conversion International* (September 1979): 57–66.

17. S. Jackson, *Optimization of Transformer Geometry by Method of Finite Increments*, AIEE Paper 61–799 (New York: AIEE, 1961).

18. W. McLyman, *Transformer and Inductor Design Handbook* (New York: Marcel Dekker, 1978), p. 63.

19. R. Seeley, "Transformer Winding Design Presented in Programmable Form," *IEEE Transactions on Parts, Hybrids and Packaging*, Vol. PHP-13, No. 1 (March 1977): 98–104.

20. O. Kiltie, *Design Shortcuts and Procedures for Electronics Power Transformers* (Cleveland: Harris Publishing Company, 1975).

21. L. Giacoletto, *Electronics Designers' Handbook* (New York: McGraw-Hill Book Company, 1977), Sec. 23.4g, pp. 94–109.

22. W. Dersch, "Design of SIFERRIT Core Power Transformers with Sinusoidal Excitation," *Siemens Components Report*, Vol. 9, No. 2 (1974): 15–19.

23. P. Odessey, "Transformer Design by Computer," *IEEE Transactions on Manufacturing Technology*, Vol. MFTT3, No. 1 (June 1974).

24. C. Jakielski, H. Tillinger, *Computer Design of Small Power Transformers and Inductors*, IEEE Conference Paper CP63-466 (New York, 1963).

25. P. Goethe, *TRANS: Transformer Design Program*, No. 25, 189 (Berea, Ohio: Optimized Program Service, 1976.)

26. H. Thode, "Designing Transformers and Inductors with a Programmable Calculator," *Solid State Power Conversion* (July 1977): 31–36.

27. F. Lilienstein, "A Universal Computer Program for the Design of Transformers, Inductors and Reactors," *Insulation/Circuits* (August 1981): 45–46.

28. H. Hart, R. Kakalec, "The Derivation and Application of Design Equations for Ferroresonant Voltage Regulators," *IEEE Transactions on Magnetics*, Vol. MAG-7, No. 1 (March 1971): 205–11.

29. ———, "A New Feedback Controlled Ferroresonant Regulator," *IEEE Transactions on Magnetics*, Vol. MAG-7 (September 1971): 571–76.

30. D. Peters, T. Maka, "An Analytical Procedure for Determining Equivalent Circuits of Static Electromagnetic Devices," *IEEE Transactions on Industry and General Applications*, Vol. IGA-2, No. 6 (November 1966): 456–60.

31. T. Wroblewski, A. Kusko, *Designing a New Low Distortion Ferroresonant Regulating Transformer*, Power Conversion Conference Paper, Powercon 7 (1980).

32. N. Grossner, "The Geometry of Regulating Transformers," *IEEE Transactions on Magnetics*, Vol. MAG-14, No. 2 (March 1978): 87–94.

33. A. Pressman, *Switching and Linear Power Supply: Power Converter Design* (Rochelle Park, N.J.: Hayden Book Co., 1977).

34. T. Wilson, Jr., et al., "DC to DC Converter Power Train Optimization for Maximum Efficiency," *IEEE Power Electronics Specialists Conference Record 1981*, Vol. 81CH 1,652–7.

35. F. Burgum, "Switched-mode Power Supply Design Nomograms," *Mullard Technical Communications*, Vol. 13, No. 129 (January 1976): 354–78.

36. T. Konopinski, S. Szuba, "Limit the Heat in Ferrite Pot Cores for Reliable Switching Power Supplies," *Electronic Design* (June 1979): 86–89.

37. G. Roespel, "Design of Power Transformers with SIFERRIT-Cores," *Siemens Components Report*, Vol. 10, No. 1 (1975): 5–11.

38. C. McLyman, *Magnetic Core Selection for Transformers and Inductors* (New York: Marcel Dekker, 1982).

39. D. Chen, H. Owen, Jr., T. Wilson, "Design of Two-winding Voltage Step-up/Current Step-up Constant Frequency DC-to-DC Converters," *IEEE Transactions on Magnetics*, Vol. MAG-9, No. 3 (September 1973): 252–56.

40. J. Geck, "Slash High-Voltage Power Supply Drain," *Electronics Design* (September 13, 1974): 158–64.

41. V. Brunstein, "Design Flyback Converters for Best Performance," *Electronic Design* (December 20, 1976): 66–69.

42. P. Cattermole, "Optimizing Flyback Transformer Design," *Power Conversion International, 1981*, pp. 74–79.

43. T. Gross, "Flyback-Inverter Efficiency Increases when the Transformer Is Loaded Properly," *Electronic Design* (October 25, 1977): 92.

44. V. Brunstein, "Modulating the Flyback Inverter Reduces Supply's Bulk," *Electronics* (August 2, 1979): 119.

45. B. Olschewski, "Unique Transformer Design Shrinks Hybrid Amplifier's Size and Cost," *Electronics* (July 20, 1978): 105–12.

46. Reuben Lee, *Electronic Transformers and Circuits*, 2d ed. (New York: John Wiley & Sons, 1955), pp. 56–60.

47. N. Grossner, "Transient Loading and Heating of the Electronic Transformer," *IEEE Transactions on Parts, Materials and Packaging*, Vol. PMP-3, No. 2 (June 1967): 30–36.

48. R. McNall, D. Lockwood, A. Gilmour, "Weight Algorithms for Adiabatic Transformers for Pulsed High Power Systems," *Proceedings IEEE International Pulsed Power Conference*, Vol. 76CH 1,147–8 (November 1976): IIE 5.1–5.4.

PART 2

The Magnetic
Circuit

CHAPTER **6**

The Magnetic Circuit:
Inductors without DC

The power which a transformer furnishes to a load accounts for the larger part of the volt-amperes (VA) furnished to the transformer. The exciting VA, the smaller part, operates the transformer itself and is partly lost as heat as the transformer functions. We now turn our attention to the excitation of the core, which is of central importance in the design and performance of transformers and inductors.

The exciting current, as we saw in Chap. 5, directly affects the size and rating of the low-frequency power transformer, limiting the flux density of the core and accounting, as heat is lost, for some of the temperature rise. The input inductance—our measure of excitation—has a direct bearing on the phase shift of the precision ratio transformer, the low-frequency cutoff and harmonic distortion of the wide-band transformer, and the top droop of the pulse transformer. And the quality factor Q which is specified for wave filters, inductors, and tuned transformers is, as we shall see, related both to their inductance and to their losses.

To determine the excitation characteristics of a transformer—the exciting VA, the exciting current, the losses in the core, and the inductance—we disconnect the transformer from its load. With its primary

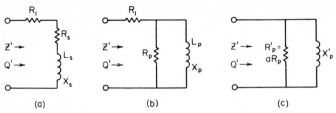

Fig. 6.1 Three-element LR network. (a) Series three-element RL circuit. R_1 = coil dc resistance. (b) Series-parallel equivalent of (a). (c) Two-element equivalent of (b) and (a). Total $Q' = aQ$, where $Q = R_p/X_p$, $a = R_s/(R_1 + R_s)$. $R_s = R_p/(1 + Q^2)$. $X'_p = X_s(1 + d^2/a^2)$.

coil no longer functionally connected to the secondary, the transformer behaves like an inductor with power losses in both the core and the coil. In network analysis, it may be regarded as a network with inductance L and a combination of dc and ac resistance R. That is, it behaves like an LR network with two terminals. The primary may be viewed as an inductor whose window area is but half filled with copper (the other half of the window being filled by the inactive secondary), and its excitation is well described by the theory bearing on the behavior of inductors. In this and the next chapter, when we refer to the properties of the *inductor,* we are also, by implication, referring to the excitation characteristics of the transformer.

In most circuits, the transformer or inductor is expected or required to behave like a linear component. We shall omit a review of the behavior of the linear LR network (Fig. 6.1). First we shall analyze the nonlinear ferromagnetic LR network and list the core materials most frequently used. Next we try to ascertain the conditions under which the iron-core inductor, at best a quasi-linear component, may be constrained to behave in reasonably linear fashion. This leads us to topics which are of great interest to the circuit designer—stability, high Q, and low distortion.

The special case of the transformer or inductor in which the flow of direct current results in a polarizing bias is treated in the next chapter. The discussion of ferromagnetic distortion is continued in Chap. 8.

6.1. THE FERROMAGNETIC LR NETWORK

In the linear inductor, typified by the low-frequency air-core coil, we assume that the inductance is linear:

$$L = N\frac{\phi}{i} = \text{constant} \tag{6.1}$$

where N is the number of turns, $\Phi = N\phi$ is the total flux linkage, and i is the magnetizing current.

The changes in inductance and ac resistance which are apparent at *high* frequencies are attributed to self-capacitance, losses of power in the eddy currents in the copper, and losses of power in the insulation and impregnant of the dielectric. In the iron-core inductor, however, the inductance and ac resistance are not characteristically constant, even at *low* frequencies.[1]

In this section we will describe a number of concepts which embrace the following topics: (a) definitions of permeability which have been derived from the *B-H* curve; (b) complex excitation and the calculation of core-loss; (c) the translation of core-loss data into simple formulas; (d) flux plots by means of the finite element method; (e) the $\mu'Q$ product; and (f) the magnetic skin effect.

a. The *B-H* Curve

The relation between ϕ and i is portrayed by the hysteresis curve, (Fig. 6.2) a plot of the relation between B and H, where:

$$B = \frac{\Phi}{A} = \mu H \tag{6.2}$$

and

$$H = 0.4\pi \frac{Ni}{l} \tag{6.3}$$

The flux density B is expressed in gauss (G) if the net core area A is in square centimeters (sq cm). The magnetizing force, or field strength

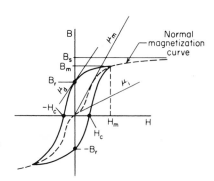

Fig. 6.2 *B-H* hysteresis curve.

H, is expressed in oersteds (Oe) or gilberts (Gb) per centimeter when the mean magnetic path l is in centimeters.*

The relationships between the vectors **B**, **H**, and intensity of magnetization **M** (boldface symbols indicate vectors) are:

$$\mathbf{B} = \mu\mathbf{H} = \mathbf{H} + 4\pi\mathbf{M} \text{ (emu)} = \mu_0\mathbf{H} + \mathbf{M} \text{ (mks)} \tag{6.4}$$

where μ_o, the permeability of free space, is $4\pi10^{-7}$ or $4\pi10^{-9}$, depending on whether the dimensions are in meters (m) or centimeters (cm).

Relative permeability u_r is:

$$\mu_r = \frac{\mu}{\mu_o} \tag{6.5}$$

and is related to susceptibility $\chi = M/H$:

$$\mu_r = 1 + 4\pi\chi \text{ (emu)} = 1 + \frac{\chi}{\mu_o} \text{ (mks)} \tag{6.6}$$

$\chi = 0$ *and* $\mu_r = 1$ in the absence of a ferromagnetic medium.

Inductance, when computed in terms of geometry, is seen to be proportional to relative permeability:

$$L = \frac{N\phi}{i} = \mu\frac{N^2A}{l} = \mu_0\mu_r\frac{N^2A}{l} \tag{6.7}$$

In this basic formula, N is the number of turns, A is the net area of the iron $K_{Fe}A_{Fe}$, and l is the effective length of the magnetic path.

Relative permeability or, more simply, permeability (without the subscript) is defined, for ferromagnetic materials, as a slope on some portion of the *B-H* curve. In Fig. 6.2, we see that it may have any one of several values, depending on where it is evaluated:

1. μ_i = initial permeability at the instep of the normal magnetization curve as H and B approach zero.

2. $\mu_d = dB/dH$, differential permeability, evaluated at any point on the hysteresis curve.

3. μ_{dc} = dc or static permeability, the quotient B/H evaluated anywhere on the normal dc magnetization curve.

4. μ_m = maximum value of μ_{dc} or μ_{ac}. Usually, $\mu_{dc} > \mu_{ac}$. Other measures of permeability may be inferred from the figure.

5. $\mu_z = B/H_z$, impedance permeability, where the apparent magne-

---

* Since small components have small dimensions, cgs units are used in order to avoid awkward numbers. Frequently used conversions are: 1 Wb (weber) = 10^9 maxwells; 1 Wb/sq m = 10^4 G = 1 T (tesla); 1 A (ampere)/m = $4\pi/10^3$ Oe; 1 Oe = 2.02 A-turns/in.

tizing force $H_z = 0.4\pi\sqrt{2} \ NI/l$, and root mean square (rms) current I is assumed to be sinusoidal. Impedance permeability is a pragmatic concept which permits the design engineer to calculate approximate values of core impedance and inductance from simple measurements of exciting current.

6. μ_L, inductance permeability associated with shunt inductance L_p of the parallel LR circuit (Fig. 6.1b).

7. $\mu_{pk} = B_m/H_m$, peak permeability, where the peak magnetizing force $H_m = 0.4\pi NI_p/l$. The peak value of the exciting current is, in general, not equal to $\sqrt{2} \ I$ because it is not sinusoidal at high values of flux density.

Depending on the magnitude of excitation, there are three ranges of permeability: initial permeability, high permeability, and low permeability. According to the domain theory of ferromagnetism, three physical processes account for these three regions of the normal magnetization curve. Initial permeability μ_i, at the instep of the curve, is physically associated with domain growth and reversible boundary displacement. High permeability at the knee is associated with irreversible boundary displacement. Decreasing and low permeability, above the knee, are associated with domain rotation and saturation.

Permeability is far from constant. It varies with (1) ac magnetic flux density, (2) frequency, (3) temperature, (4) mechanical strain, and (5) prior magnetic history. This last factor introduces time as a distinct variable, in addition to the remanent flux density B_r and the bias magnetizing force.

Ac power losses, represented by the loss resistance R_s or R_p (Fig. 6.1), are also nonlinear. They depend on (1) the area of the hysteresis loop; (2) the maximum flux density B_m; (3) the operating frequency; (4) certain constants of the core material, such as lamination or particle thickness and resistivity; and (5) the temperature.

Because L and R depend on such a multiplicity of variables, it is sometimes advantageous to assess the degree of nonlinearity and loss by viewing the B-H loop on the oscilloscope. This can be done using an Epstein frame or its equivalent, a transformer coil (Fig. 6.3). The magnitude of flux density B is, and should be, determined from the no-load voltage V_2 at the secondary. In the schematic Fig. 6.3, the voltage across R_1, which drives the horizontal amplifier, is proportional to H [Eq. (6.3)]. To obtain a voltage proportional to B to drive the vertical amplifier, we need a function which will be proportional to the integral of the secondary voltage. This can be understood if we first write the integral form of Faraday's law:

$$\int e \, dt = 10^{-8} \int dB$$

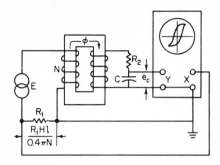

Fig. 6.3 Display of B-H loop.

Such a function is provided by the R_2C network across the output winding. The voltage e_c across the capacitor is proportional to the time integral of the current $i_L = V_2/R_2$ and the flux density:

$$e_c = \frac{1}{C}\int i_L\, dt = \frac{NA}{R_2 C 10^8}\int \frac{dB}{dt}\, dt \qquad (6.8)$$

$$= \frac{NA}{R_2 C 10^8} B$$

This equation is accurate if $R_2 \cong 250 X_c$, where X_c is the reactance of the capacitor. Typical values of its components are $R_2 = 5$ megohms (MΩ), $C = 0.1$ microfarads (μF), and $R_1 = 1$ ohm (Ω).

It is sometimes useful, in computer-aided design, to represent the magnetization curve by means of an exponential series.[2,3] This is done by attempting to fit data to a curve; the result is such expressions as:

$$\left.\begin{array}{l} B = k_1(1 - e^{-k_2 H}) \\ H = a_o + a_1 B + a_n B^n \\ H = (k_1 e^{k_2 B^2} + k_3)B \end{array}\right\} \qquad (6.9)$$

Here $e = 2.718$ (the base of natural logarithms), and coefficients such as a and k are derived from a program based on statistical procedures (e.g., regression analysis and the method of least squares).

It is difficult to obtain a good fit over a large excursion of the B-H curve. Hence, some designers simply store data in the computer to assure more accurate results, especially in the region of saturation.

b. Complex Excitation[4]

The intimate relationship between inductance and loss resistance (the ac resistance associated with power loss in the core), or between stored magnetic energy and dissipated magnetic energy, is further illuminated by the concept of complex permeability. This concept provides a formal link between the complex algebra of the linear LR network and the ferromagnetic LR network.

Complex permeability μ (also known as impedance permeability μ_2) is defined as:

$$\mu = \mu' - j\mu'' = \mu e^{-j\delta} = \mu \underline{/-\delta} \left.\right\}$$
$$\frac{\mu''}{\mu} = \tan \delta \qquad\qquad (6.10)$$

where μ' is the real, or in-phase, permeability and μ'' is the imaginary permeability.

Complex impedance $Z = R_s + jX_s$ is redefined as a vector in terms of a vector inductance $\mathbf{L}$:

$$\mathbf{Z} = j\omega\mathbf{L} = j\omega L_0 \boldsymbol{\mu} \qquad\qquad (6.11)$$

where $\omega = 2\pi nf$ and $L_0 = \mu_0 N^2 A/l$ is the air-core inductance.

We may now write:

$$\mathbf{Z} = j\omega L_0(\mu' - j\mu'') = \omega L_0 \mu'' + j\omega L_0 \mu' \qquad\qquad (6.12)$$

Ac loss resistance may be associated with μ'', and inductive reactance may be associated with μ':

$$\left. \begin{array}{l} R_s = \omega L_0 \mu'' \\ X_s = \omega L_0 \mu' \end{array} \right\} \qquad\qquad (6.13)$$

Real permeability μ' may now be identified with the familiar relative permeability. The computation of the ac resistance becomes simple if we measure $\mu''/\mu' = d = 1/Q$, since $R_s = X_s \mu''/\mu'$.

In a similar fashion we define a complex reluctivity $\boldsymbol{\nu}$:

$$\boldsymbol{\nu} = \frac{1}{\mu} = \frac{1}{\mu' - j\mu''} = \nu' + j\nu'' \qquad\qquad (6.14)$$

Conductance G and susceptance B may now be expressed in terms of imaginary and real reluctivities of the admittance $\mathbf{Y}$:

$$\left. \begin{array}{l} \mathbf{Y} = G - jB = \dfrac{1}{j\omega L_0}\dfrac{1}{\mu} \\[2mm] \mathbf{Y} = \dfrac{1}{j\omega L_0}\left(\dfrac{1}{\mu_p'} - \dfrac{1}{j\mu_p''}\right) = \dfrac{1}{R_p} + \dfrac{1}{j\omega L_p} \end{array} \right\} \qquad (6.15)$$

Imaginary shunt permeability μ_p'' and real shunt permeability μ_p' have a formal connection with parallel resistance $R_p = 1/G$ and with parallel reactance $X_p = 1/B$:

$$\left.\begin{array}{ll} \mu_p'' = \mu''(1 + Q^2) & \mu_p' = \mu'(1 + d^2) \\[2mm] R_p = \omega L_0 \mu_p'' & G = \dfrac{v''}{L_0} \\[2mm] X_p = \omega L_0 \mu_p' & B = \dfrac{v'}{L_0} \end{array}\right\} \tag{6.16}$$

Since

$$\frac{1}{\tan \delta} = Q = \frac{R_p}{X_p} = \frac{\mu_p''}{\mu_p'} = \frac{\mu'}{\mu''} = \frac{v'}{v''} \tag{6.17}$$

we can solve for imaginary reluctivity:

$$v'' = \frac{1}{\mu_p''} = \frac{1}{\mu_p' Q} = \frac{\tan \delta}{\mu_p'} \tag{6.18}$$

This expression identifies reluctivity with the inverse of the μQ product and with the *relative power-loss factor:*

$$\frac{1}{\mu_p' Q} = \frac{\tan \delta}{\mu_p'} \tag{6.19}$$

Both the μQ product and the relative power-loss factor are important concepts in magnetic circuit analysis. The μQ product serves as a figure of merit of core materials, and the power-loss factor is used by the core manufacturer in the graphical representation of tests of cores.

c. Graphical Representation: Core Tests

With the aid of graphs which relate the key variables—μ, B, f, and temperature—to each other, it is possible to calculate L and R_p for the ferromagnetic inductor and transformer.

The transformer designer needs much graphical data in order to design iron-core transformers and inductors properly. We illustrate the character of such data with but one example, 80 percent nickel-iron (Super Q 80 or Mumetal) ring laminations. The manner in which the loss or parallel resistance factor $\omega \mu_p''$ [see Eq. (6.16)] and the ac permeability (μ_p', μ_z, μ') vary with the flux density at a fixed frequency of 60 hertz (Hz) is shown in Fig. 6.4. The variation of $\omega \mu_p''$ and μ_p' with frequency, for several values of B, is shown in Fig. 6.5.

When the frequency is fixed, the permeability μ_p' rises from its initial value to a maximum at about 4 kG and then decreases sharply as saturation is approached. The shunt resistance factor decreases with

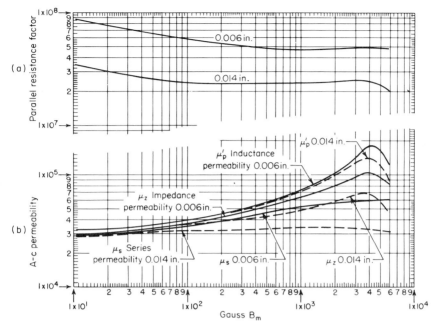

Fig. 6.4 Resistance factor and permeability vs. flux density at 60 Hz. Stamped-ring high-nickel (Super Q 80) laminations. (*a*) Parallel resistance factor vs. B_m. (*b*) Permeability (series, impedance, and shunt) vs. B_m. (*By permission, Magnetic Metals Co.*)

flux density. From Fig. 6.5 we can see that permeability decreases as frequency f increases. Although $\omega\mu_p''$ increases with f, division by ω results in the important conclusion that the $\mu_p'Q$ product decreases with frequency.

An alternative, more traditional, method of ascertaining the loss of power due to resistance R_p is to plot watts per pound (lb) P_s vs. B and f. We then calculate:

$$R_p = \frac{E^2}{P_s \times \text{core weight}}$$

where E is the applied voltage.

When ferrites are intended for power applications, graphs are constructed to depict losses of power at high flux densities: W/cu cm vs. B, f, and temperature. In low-power applications, the relative power-loss factor tan δ/μ [see Eq. (6.19)] is computed at low levels of flux density over a wide band of frequencies, and plotted as a family of curves.*

---

* The symbol μ (without subscript) is often substituted for μ_p'.

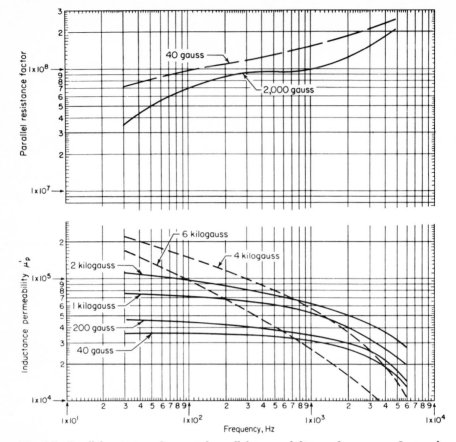

Fig. 6.5   Parallel resistance factor and parallel permeability vs. frequency.   Stamped-ring 6-mil high-nickel (Super Q 80) laminations.   Data obtained with modified Hay bridge.   (*By permission, Magnetic Metals Co.*)

Designers of an analytical bent find it convenient to use approximate formulas, derived from graphs, which are accurate over a wide range. The form

$$P_s = bB^m f^n \tag{6.20}$$

has long been in use for estimating the loss of power in the cores of transformers.   When silicon steel is used, and at moderate flux densities (2 to 10 kG), m is slightly higher than the Steinmetz exponent, 1.6. Eq. (6.20) was first introduced as Eq. (4.14) and was discussed in Sec. 5.3c.

Typical values of the exponents m and n in Eq. (6.20), obtained from graphical data, are as follows:[5,6]

Core Material	m	n	a	a_1
Nonoriented silicon	1.71	1.36	0.88	0.20
Oriented silicon	1.87	1.66	0.94	0.11
Mn-Zn ferrite	2.03–2.5	1.38	1.01–1.14	0.32–0.45

The columns a and a_1 at the right of the table require explanation. In Chap. 4, we described two solutions for the power rating of a transformer, depending on whether the product of B and J is optimum [see Eq. (4.19)] or nonoptimum. This results in two values of a which are related to the magnitude of exponents m and n as follows:

$$\text{optimum} \qquad a = 1 + \frac{n}{2} - \frac{n}{m}$$

$$\text{nonoptimum} \quad a_1 = 1 - \frac{n}{m}$$

When data or curves are not available, as often happens, the designer can perform tests in accordance with standard procedures.[7] Tests at commercial power frequencies are performed with a test circuit similar to that used with the Epstein frame.[8]

It is something of a problem to predict the loss of power in the core when the excitation waveform is distorted, especially at high induction.[9,10] When the applied waveform is rectangular or *square*, as in the dc inverter circuit, sine-wave charts no longer provide accurate estimates of the loss.[11,12] Many tests indicate that from 10 to 20 percent less power is dissipated by square-wave excitation (at the same flux density). Hence, designers are likely to use sine-wave data when square-wave data are not available.

d. Nonuniform Fields

When a coil is wound on one leg of a core-type transformer, there is sizeable stray leakage flux. Flux is not uniformly distributed in the core, and linear equations such as Eq. (6.19) become inaccurate.[13] The designer can, however, resort to a more general field approach to solve the problem.

In some components, the magnetic fields are in regions of irregular shape, having different permeabilities and conductivities. One approach to this type of problem is to use a *finite element model* (FEM) and a computer program to determine the field distribution.[14] Two-dimensional flux plots yield more accurate estimates of excitation and losses, especially when a local region of the magnetic circuit becomes saturated.

e. Analytic Representation

The design of iron-core inductors[15] has benefitted greatly from the analytic approach to the representation of the losses of ac power.

In the Legg equation

$$\frac{R_s}{\mu' f L_s} = hB_m + ef + c \tag{6.21}$$

each coefficient represents a basic factor contributing to the total loss.[16]

The *hysteresis* coefficient h of Eq. (6.21) can be shown to be:

$$h = \frac{8\pi W}{V_{Fe} B_m{}^3} = \frac{2\mathscr{A}_1}{B_m{}^3} \tag{6.22}$$

where

$$\frac{W}{V_{Fe}} = \frac{1}{4\pi} \oint H \, dB = \frac{\mathscr{A}_1}{4\pi} \tag{6.23}$$

W/V_{Fe} is the energy loss per cycle per unit volume (in ergs per cubic centimeter), and $\mathscr{A}_1$ is the enclosed area of the hysteresis loop (in gauss-oersteds). The connection between hysteresis and harmonic distortion is discussed in Sec. 6.5.

In laminated and powdered core structures, the *eddy-current* coefficient e of Eq. (6.21) is:

$$e = \frac{4\pi^3 a^2}{3\rho 10^9} \qquad e = \frac{2\pi^3 a^2}{5\rho r^{1/3} 10^9} \tag{6.24}$$

In the first expression, ρ is the resistivity (in ohm-centimeters) and a is the lamination thickness (in centimeters). In the second expression, a is the magnetic sphere diameter and r is the packing factor.

The *residual* coefficient c of Eq. (6.21) is associated with losses of power which cannot be attributed either to hysteresis or to eddy currents.

If we divide Eq. (6.21) by 2π, we obtain:

$$\frac{R_s}{2\pi f L_s \mu'} = \frac{1}{\mu' Q} > \frac{1}{\mu_p''} \tag{6.25}$$

which corresponds to the relative power-loss factor of Eq. (6.19) and the inverse product $\mu' Q$.

From Legg's equation, we can infer certain significant features in the behavior of $\mu' Q$ as a function of f and B_m. Since $hB_m/2\pi = \mathscr{A}_1/B_m^2$, we can infer that the $\mu' Q$ product can increase, or even decrease, with B_m, depending on how the area of the loop grows as B_m

increases. At some optimum value of flux density B_m, the $\mu'Q$ product may be at a maximum. The presence of the ef term in Eq. (6.21) indicates that $\mu'Q$ ultimately decreases with frequency.

f. Magnetic Skin Effect[15,17]

At high frequencies, eddy currents in the laminations may be great enough to limit the depth to which flux lines may flow unaltered. This phenomenon, called *eddy-current shielding* or *magnetic skin effect* is analogous to the more familiar copper skin effect.

The reduction in permeability from μ_i to μ because of reduced flux penetration is described by the function:

$$\frac{\mu}{\mu_i} = \frac{L}{L_i} = \frac{1}{\xi}\frac{\sinh\xi + \sin\xi}{\cosh\xi + \cos\xi}$$

(6.26)

where

$$\xi = \frac{a}{\delta} \quad \text{and} \quad \delta = \sqrt{\frac{2\rho}{\omega\mu_0\mu_i}}$$

ξ and δ define the depth to which flux penetrates a lamination sheet which is a meters thick and whose resistivity is ρ, in ohm-meters. There is a critical frequency f_c at which $\xi = 2$ and penetration is half the thickness of the sheet:

$$f_c = \frac{4}{\pi\mu_0}\frac{\rho}{\mu_i a^2}$$

(6.27)

When ρ is in microhm-centimeters and a is in centimeters, this equation becomes a useful formula:

$$f_c = 101.5\frac{\rho}{\mu_i a^2}$$

(6.28)

Thus, for 3 percent silicon, 50 percent nickel, and 80 percent nickel sheet, 14 milli-inches (mil) thick, f_c is 12,000, 1,300, and 200 Hz, respectively. Note that these are audio frequencies, much lower than the radio frequencies associated with the copper skin effect.

When the flux density is low and the frequency is low enough so that $\xi < 1.6$, the *series* eddy-current resistance increases with the square of the frequency. However, the *shunt* resistance R_p is independent of the frequency. Series, shunt, and initial μ are about equal. Core Q equals $6/\xi^2$ and decreases as the frequency increases.

Above the critical frequency, eddy-current shielding is substantial. In the region where $\xi > 2.5$, μ/μ_i is proportional to $1/\sqrt{f}$ and is approximately $1/\xi$ or $2/\xi$ depending on whether μ represents the series or the shunt permeability, respectively. In this region, the shunt μ_p' is twice the series μ', and the core Q equals unity. When we want inductance to be constant over a wide range of frequencies, we use thin laminations of high resistivity.

Ferrites have a resistivity about 10 orders of magnitude greater than sheet materials. This strikingly superior characteristic is intimately related to the fact that ferrites have an initial permeability which is uniform into the megahertz (MHz) region. To get comparable behavior from sheet material requires the use of costly ultrathin strip (½, ¼, or ⅛ mil thick).

We may conclude, for the present, that if we wish a ferromagnetic inductor to exhibit some degree of linearity, there is an advantage to operating the core at a low B_m and at frequencies at which $\xi <$ 1.6. Under such conditions, $R_p \cong R_e$, and the linear *shunt* circuit of Fig. 6.1b, whose elements are reasonably constant, is appropriate for circuit analysis. Linearity may be increased further by the judicious use of an air gap, which will be discussed later.

6.2. FERROMAGNETIC MATERIALS[18,19,20]

Many core materials are available, in a variety of shapes and sizes. The classes of material used by the transformer designer are:

1. Silicon-iron, nonoriented and grain-oriented[21]
2. Low carbon-iron (motor-lamination grade)
3. 50 percent nickel-iron, round and square loop
4. 80 percent nickel-iron, round and square loop
5. Cobalt-iron
6. Powdered iron
7. Powdered nickel-iron (powdered permalloy)
8. Ferrite, round and square loop[22]
9. Metallic glass (13 percent boron-iron)[23]

Representative low-cost sheet materials for use at commercial power frequencies are listed in Table 6.1. It is important to note the distinction between *fully processed* (FP) and *semiprocessed* (SP) materials.

The steel mill produces FP materials for use in laminations which, *after stamping* (AS), do not require further annealing by the lamination fabricator or by the customer, the manufacturer of a laminated transformer or motor.* The steel mill guarantees that FP grades meet specific standards of maximum power loss issued by the American Iron

---

* Some users, however, anneal to remove stresses caused by punching.

T A B L E 6.1   Epstein Properties of Silicon and Carbon Materials*

			Nonoriented silicon (NO) 0.025 in (0.64 mm)						Low-carbon iron 0.025 in (0.64 mm)
Grain-oriented silicon (GO) 0.0138 in (0.35 mm) fully processed AAS			Fully processed AS			Semiprocessed AAS			Semiprocessed AAS
	M6†								
Grade	Parallel	Transverse	M22	M36	M45	M36	M43	M47	2S
W/lb, max 60 Hz	0.66	2.1	2.18	2.40	3.60	2.13	2.30	3.50	4.90
W/kg, max 50 Hz	1.11	3.53	3.80	4.18	6.27	3.71	4.01	6.10	8.54
Peak permeability, 60 Hz‡	30,000	333	870	1,000	1,400	1,450	1,750	1,900	2,000
A-turns/cm, rms 60 Hz	0.3	18.0	8.0	6.5	5.0	4.5	4.0	3.5	3.2
B_{dc} (T) at 100 Oe	1.98	1.56	1.76	1.76	1.78	1.77	1.79	(1.80)	1.82
Specific gravity	7.65		7.65	7.70	7.75	7.70	7.70	7.80	7.85
% Silicon and aluminum	3.15		3.2	2.65	1.85	2.65	2.35	1.05	—

* All tests at 1.5 T. Tests on nonoriented and carbon are on Epstein strips with half-parallel, half-transverse grain.

† A highly oriented (HO) silicon, ASTM grade 30 HO83 (1.7 T), is used in wound cores.

‡ Conversion formula for nonoriented materials: (W/kg, 50 Hz) × 0.574 = W/lb, 60 Hz.

SOURCE: From AISI, *Flat Rolled Electrical Steel* (Washington, D.C., 1978) and ASTM, *Standards on Magnetic Properties.*

.nd Steel Institute (AISI) and the American Society for Testing and Materials (ASTM).   Foreign standards include Deutsche Industrie Norm (DIN), Japanese Industrial Standards (JIS), Norme Francaise (NF), and British Standard (BS).

In large power transformers, lamination surfaces are large, and a high volt (V) turns ratio can result in a significant flow of eddy currents between laminations.   The use of special high-resistance surface coatings (available on fully processed materials) will reduce total eddy-current losses to a minimum.

Semiprocessed materials are intended for use in laminations which will be annealed after stamping (AAS) by the fabricator or user. The specified magnetic properties are achieved only after a suitable "quality" anneal (QA) in an annealing furnace.   The heat treatment and gas atmosphere of such a furnace must be capable of relieving punching stresses in the laminations, decarburization (removal of carbon impurities), growth of grain, and adding a surface oxide.   An increase in the area of the grain (growth in size is viewed under the microscope) is desirable because it reduces power loss in the core.

The oxide film adds interlaminar resistance (to reduce loss of power in eddy currents and inhibit rust).

Thus the lamination producer with a suitable annealing and processing technology has the capability of transforming an SP material into an annealed lamination product which meets the loss of power in the core guaranteed of an unannealed FP material. This can be seen in Table 6.1, which shows, for example, that an M36 SP material, after a quality anneal (AAS), is capable of satisfying the core power loss specified for an M22-FP product. Since both products lose virtually the same amount of power in the core (at 1.5 T), they can each be labeled with the succinct designation M22. However, it is important to note that the SP product, after annealing, can yield higher levels of permeability (when flux density is high) than the FP product. This comes about because SP material usually contains less silicon than its FP counterpart. At low values of flux density, an FP material exhibits slightly greater permeability than its SP counterpart.

Unfortunately, the M-number designation for both FP and SP materials can create confusion when one material is substituted for the other. When more than one requirement (e.g., permeability and power loss) must be satisfied concurrently, problems can be avoided by the use of an unambiguous nomenclature. To assure interchangeability and to avoid misapplication of materials, both the ASTM and AISI have recommended the use of a six-digit symbol to replace the two-digit M number. The six-digit symbol is used to denote thickness, mill process, and loss (in watts per pound). Examples of this preferred type of designation are in the column at the right:

Thickness	M Number	Preferred Designation
0.64 mm (0.025 in)	M22-FP	64 F 218
0.64 mm (0.025 in)	M36-SP	64 S 213
0.35 mm (0.014 in)	M6 (1.5T)	35 GO 66
0.30 mm (0.012 in)	M5 (1.7T)	30 HO 83

The first two digits of the six-digit symbol denote thickness in millimeters (without the decimal point); the letters are abbreviations for FP, SP, GO, and HO materials; and the final digits denote specific power loss in the core material (decimal point omitted).

Many different trade names are used for materials which are either substantially the same or which serve essentially the same function. In Table 6.2 we have grouped the trade names into classes or categories of materials. Representative sheet materials are listed in the table, together with their distinguishing magnetic parameters. Note that materials with a high nickel content have very high initial permeability, but only when flux density is moderate. Metallic glass ribbon (about

1.1 mil thick), a comparative newcomer among magnetic materials, has characteristics roughly comparable with 50 percent nickel-iron alloy,* but more resistivity.[23,24]

Representative ferrite materials are listed in Table 6.3.† The large choice of grades makes it possible to obtain materials which are appropriate for the desired frequency range, flux density, permeability, Q, and operating temperature. Because thermal and mechanical stresses on ferrites can be crucial, the designer should have access to data about thermal conductivity and the coefficient of linear expansion. Core materials are constantly being improved and altered, hence the manufacturers' current catalogs (containing reams of data) are very useful to the transformer designer. We have restricted our tables to abbreviated generic data, partly to conserve space and partly to diminish the likelihood of their obsolescence.

6.3. THE SERIES GAP CIRCUIT

A variety of schemes has been designed to make the permeability of a ferromagnetic core behave like a constant over a large excursion of excitation. One is to straighten the magnetization curve with a superimposed high-frequency bias. Another is to use a premagnetization bias which constrains the useful flux excursion to the linear portion of the B-H curve.[25] The simplest and most common method of linearization is to add an air gap in series with the ferromagnetic path of the magnetic circuit.

The introduction of an air gap in series with the magnetic circuit has profound effects on the performance of the ferromagnetic inductor:

1. Exciting current increases and inductance decreases, in general.
2. Constancy, uniformity, and stability of permeability become feasible.
3. Quality factor Q is increased.
4. Harmonic distortion may be reduced.
5. In the polarized inductor and transformer, inductance may decrease, but it can be *increased* by certain optimum gaps.

a. The Gap Ratio

Figure 6.6 shows an air gap l_a in series with the magnetic path l of an F lamination, a toroid (annular core), and a pot core. In each case, the reluctance $\mathcal{R}$ of the magnetic circuit is increased, and ac

---

* At frequencies in the range of 20 to 150 kHz, some grades of metallic glass ribbon have lower core loss than either 50 percent nickel-iron tape or ferrite.

† Materials commonly used in pulse transformers are tabulated in Table 11.2.

TABLE 6.2 Typical Properties of Sheet Magnetic Materials

Type	Silicon		Low nickel content		High nickel content			Cobalt iron	Metallic glass
	Nonoriented	Grain-oriented	Round loop	Square loop	Round loop	Round loop	Square loop		Square loop
Typical composition	0.85–3.3% Si	3½% Si	50% Ni		80% Ni	Mo-Ni		49% Co	13% B 3% Si*
Specific gravity	7.65–7.80	7.65	8.15–8.25	8.25	8.6–8.74	8.74	8.74	8.15	7.32
μ_i	400	350	2,000–3,500	—	10,000–22,000	55,000	—	—	1,000–2,500
μ_m	10,000	50,000	20,000	—	40,000	500,000	—	66,000	100,000
$p \times 10^6$	23–52	45–48	50	40–50	58–60	58	58	26	125
B_m†‡	15–16	16.5–17.6	12	14.2	5.0	6.5	6.8	21.5	13–14
H_c	0.5–1.5	0.5	0.1	0.2–0.4	0.02	0.01	0.03	0.18	0.06
B_s, dc†	16.2	19.9–20	15.5	15.5–16	8.2	8.2	8.2–8.6	24	16.1
B_r	8–13	10.5–13	—	13–15	—	—	—	20	11.2
Curie temperature, °C	690	750	460	450–500	460	460	—	970	370

* METGLAS (Allied Corp.) grade 2605-SC. Other amorphous grades, including nickel-iron and cobalt-based alloys, are also available.

† In kilogauss.

‡ Values of B_m are typical maximum working values used in design work.

TABLE 6.3 Representative U.S. Soft Ferrite Materials

Material	Ferroxcube					Indiana General (d)					Magnetics-Spang			
	3C8	3B7	3D3	3E2A	4C4	8,100	06	H	TC9	Q1	P	G	A	R
(Notes)	(b)		(c)			(b)			(c)		(b)	(e)		(f)
f, MHz	—	0.3	0.2–2.5	—	1–20	—	0.5	1.0	0.5	10	—	0.15	2.0	—
μ_i (25°C)	2,700	2,300	750	5,000	125	2,200	4,700	850	2,000	125	2,700	2,300	750	2,300
B_s (25°C)	4,400 (H3)	3,800 (H2)	3,800 (H5)	3,600 (H2)	3,000 (H10)	4,600 (H10)	4,700	3,400	3,500	—	5,100 (H15)	4,000 (H10)	4,000 (H10)	4,000 (H10)
H_c, Oe	0.2	0.2	1.0	0.08	3.0	0.15	0.1	0.18	0.25	2.1	0.18	0.15	0.7	0.25
Loss factor, tan δ/μ														
10 kHz (g)	—	—	—	—	—	—	—	—	3	—	—	—	—	3
100 kHz	10	5	—	100	35	—	6	25	4.5	20	—	6	—	—
500 kHz	—	25	14	—	35	—	60	130	30	20	—	—	30	—
1 MHz	—	120	30	—	40	—	250	300	—	20	—	—	—	—
10 MHz	—	—	—	—	100	130	—	—	—	200	—	—	—	—
mW/cu cm (25 kHz)	115 1.6kG	—	—	—	—	130 2 kG	—	—	—	—	165 2kG	—	—	—
Curie °C	210	170	150	170	300	170	210	170	140	350	200	180	200	170

[a] Comparable imported materials are available from Siemens AG (West Germany) and TDK Corp (Japan).

[b] Similar materials are available from Stackpole Carbon Co., Allen Bradley Co., Fair-Rite Products Corp., and Ceramic Magnetics Inc. (MN 60).

[c] A linear μ vs. temperature characteristic is obtained with 3B9 and 3D3 material, TC (Indiana General) and Type J (Magnetics).

[d] Indiana General furnishes a series of cores rated for specified circuit performance.

[e] Higher μ materials are available: F (3,000); J (5,000); W (10,000).

[f] Material R is rated for pulse permeability.

[g] Loss factor in ppm (parts per million).

195

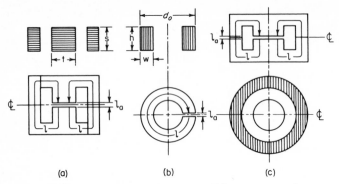

Fig. 6.6 The series air gap circuit. (*a*) F lamination with center gap. (*b*) Toroid (annular ring) $w' = w/d$, $h' = h/d$. (*c*) Pot core (gap in center post).

permeability μ (series or shunt) is lowered to an apparent, or effective, permeability μ_e:

$$\mathfrak{R} = \frac{l}{\mu A} + \frac{l_a}{1 \times A} = \frac{l}{\mu_e A} \tag{6.29}$$

The effective permeability is defined by either of the following equations:

$$\frac{1}{\mu_e} = \frac{1}{\mu} + \frac{l_a}{l} \tag{6.30}$$

$$\mu_e = \frac{\mu}{1 + \mu\beta} \tag{6.31}$$

where $\beta = l_a/l$ is the *gap ratio*.

The inductance formula [Eq. (6.7)] is now rewritten to reflect the reduction in permeability from μ_r to μ_e:

$$L = \mu_0 \mu_e \frac{N^2 A}{l} \tag{6.32}$$

And the gap ratio is more strictly defined:

$$\beta = \frac{l_a}{l} \frac{A_{\text{Fe}} K_{\text{Fe}}}{A_a} \tag{6.33}$$

where A_{Fe} is the gross area of the iron core, K_{Fe} is the stacking factor (typically 90 percent), and A_a is the effective cross-sectional area of the gap.

When the gap is small, $A_a = A_{\text{Fe}}$. When the gap is large, $A_a > A_{\text{Fe}}$ where fringing occurs. The net result of a stacking factor and of fringing is to increase the area and thus reduce the gap ratio. As a consequence, the effective gap is smaller than the physical gap.

When the magnetic circuit includes a large air gap, fringing flux in the region of the gap results in a marked increase in the power lost in local eddy currents. In the design of high-reactance and regulating transformers containing shunts (see Secs. 2.5c and 5.6), it is important to be able to compute the loss of power due to the gap, which can account for a significant fraction of the total loss in the iron.[26]

Fringing must be expected when the gap is a significant fraction of the tongue width t. For example, a gap is effectively reduced by 7 and 15 percent when $l_a/t = 0.01$ and 0.03, respectively. Thus, when the stack is a 1-in square, a gap of 30 mil is effectively reduced by 15 percent. In addition, the stray flux created by fringing has undesirable consequences when the gap is in the vicinity of a metal enclosure or in a circuit which is sensitive to flux. Since fringing is reduced by placing the coil snugly over a gap, the F lamination with an adjustable center gap is a preferred geometry.

It is extremely difficult to reduce the gap to some arbitrarily small value. A butt joint with no gap spacer still has a gap, and even interleaved laminations contain an equivalent air gap because the lines of flux must cross an air path. The gap ratios most frequently encountered are listed in Table 6.4. The stamped ring contains no air gap (see Fig. 2.1c), and this is the core used when the *intrinsic* μ of a sheet material is being measured.

When the excitation voltage is fixed, the flux remains constant, even if the presence of a gap has increased the reluctance. The magnetomo-

TABLE 6.4 Air-Gap Ratios

Core	Stacking method	Gap ratio $\beta = (l_a/l) \times 10^3$
Epstein sample $l = 37$ in (94 cm)	1×1	0.007
1 D-U, 0.014 in	1×1	0.010
150 E-I, 0.014 in	1×1	0.038
12 F, 0.014 in	1×1	0.04
75 E-I, 0.014 in	1×1	0.070
E-I lamination (average burr)	1×1	$0.6/l$
Center-leg F-laminations (gapped in center leg)	1×1	$1.5/l$
Small lamination (gapped in outer limbs)	Butt joint*	0.2–1.0
Small C core	Butt joint*	$1.25/l$
Large C core	Butt joint*	$2.5/l$

* Values of β in butt joints are susceptible to large variation. These values can be halved when surfaces are lapped.

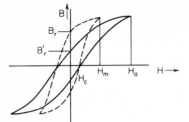

Fig. 6.7 Tilted hysteresis curve; the effect of an air gap. Note reduction of retentivity.

tive force (mmf or F) must also increase in accordance with the basic relationship: $F = \phi\mathcal{R}$.

Since the mmf (in A-turns or gilberts) must increase, the effect on the hysteresis curve (see Fig. 6.7) is to increase the field strength from H_m to H_a. This tilts the hysteresis curve and decreases the slope, the permeability, and the retentivity. The air gap, by increasing the series reluctance to R_a, has reduced the inductance L_1. In terms of an equivalent circuit (Fig. 6.8), an air-core inductor L_a, whose $\mu_r = 1$ has been placed in *shunt* with L_1:

$$L_a = \frac{L_1}{\mu\beta}$$

b. The Exciting Current

In most cores, the unavoidable* air gap increases the *reactive volt-amperes* (var) which must be furnished to the winding.

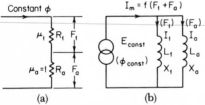

Fig. 6.8 Magnetic and electric equivalent circuits for series flux. (*a*) Series reluctances and magnetomotive forces F_1, F_a. (*b*) Shunt reactances are x_1 and x_a. I_m is the magnetizing component of the exciting current I_x. $L_1 = \mu_1 L$, $L_a = (l/l_a) L$; where $L =$ inductance without ferromagnetic core.

---

* As we have noted, the stamped ring is free of an air gap. So is the homogeneous toroid. The spiral wound-strip toroid and the D-U lamination structure are almost free of a gap.

The exciting VA is the vector sum of power loss in the core and the var:

$$VA = \sqrt{W_{Fe} + (var)^2}$$

The exciting current I_x may be determined in several ways. It may be computed from a chart showing VA/kg vs. B_m or, in the absence of such a chart, we may compute it from:

$$I_x = \sqrt{I_{Fe}^2 + I_m^2}$$

Rms magnetizing current I_m is computed with either of two formulas:

$$I_m = \frac{10}{4\pi \sqrt{2}} \frac{B_m l}{\mu N} \quad \text{or} \quad I_m = \frac{E}{\omega L_e} \tag{6.34}$$

C cores have a discrete gap (typically 1 mil), and gap current I_a must be determined and added to I_m to obtain the total magnetizing current. To determine I_a:

$$I_a = \frac{10}{4\pi \sqrt{2}} \frac{B_m l_a}{N} \tag{6.35}$$

The exciting rms VA of Epstein, ring, or toroid samples can be computed from:

$$\frac{VA}{kg} = \frac{10^4 F_f}{\pi F_c} \frac{B^2 f}{\delta \mu_{pk}} \tag{6.36}$$

where the form factor $F_f = 1.11$ when the applied voltage is sinusoidal. F_c, the crest factor (peak-to-rms ratio) of the exciting current, depends on the magnitude of B (in tesla, T). When the laminations are silicon, we will assume that $F_c = \sqrt{2}$ if B is less than 1.0 T (10 kG), and increases to about 1.8 when B is 1.5 T. The density of core material δ is in g/cu cm, and μ_{pk} is peak permeability.

When the laminations are stamped, the exciting rms VA can be computed as follows:

$$\frac{VA}{kg} = 3,183 \frac{F_f B^2 f}{F_c \delta \mu} (1 + \mu\beta) \tag{6.37}$$

Here the excitation is seen to be the sum of the VA of the core material and the air gap, the latter a function of the gap ratio β.

In the laminated transformer, it is not unusual for the gap VA to exceed the VA of the core material as determined by Epstein tests. Hence only a modest improvement in excitation can be expected from the use of highly permeable (so-called high-mu) material. Note also that the flux density is not always uniform, especially in the region of mounting holes and the joints between the E and I laminations. In cores composed of scrapless laminations, flux paths are partly in the direction of the grain and partly transverse to it. Thus, reliable estimates of the exciting VA of such cores are obtained from tests of the actual cores.[5,7]

Example.   Assume that we wish to calculate the exciting current of a small transformer and that only Epstein curves are available. The excitation of the primary is sinusoidal, and the basic data are:

120 V, 60-Hz sine wave
$B = 1.5$ T, E-I 100 stack weighs 0.68 kg (1.5 lb)
$\mu = 2{,}000$ (from B vs. μ_{pk} or B vs. H_p curve)
$F_f = 1.11$, $F_c = 1.8$, $\delta = 7.70$
Interleave $= 1 \times 1$, in the laboratory, so we assume that:
   $\beta = 0.1 \,(10)^{-3}$ (see Table 6.4).

Our initial estimate of the excitation [see Eq. (6.37)] is:

$$\frac{\text{VA}}{\text{kg}} = 17.2 \,(1 + 2000 \times 0.1 \times 10^{-3}) = 20.6 \text{ (or 9.36 VA/lb)}$$

The quantity 17.2 VA/kg in front of the parenthesis is attributed to the Epstein core sample (3-cm $\times$ 30.5-cm strips). But if less care is taken in the assembly of laminations, the gap ratio can easily be increased. In production, an increase in β to $0.2 \,(10)^{-3}$ results in 17.2 (1.4) = 24.1 VA/kg or 10.92 VA/lb, which is 40 percent greater than the Epstein value. A conservative estimate for exciting current would therefore be that $I_x = 24.1 \,(0.68)/120 = 0.137$ A, maximum.

Power loss in the core may be computed from a W/kg chart or from the formula $I_{\text{Fe}} = E/R_p$.

Shunt resistance R_p is determined from Eq. (6.16), and the parallel factor $\omega\mu_p''$ is obtained from a curve such as Fig. 6.5.

Although an air gap increases the steady-state exciting current, it reduces the transient inrush of current to the transformer.[27] This can be attributed to the reduction in retentivity from B_r to B_r', shown in Fig. 6.7. Since the change $B_s - B_r'$ is greater than $B_s - B_r$, the transient inductance is increased, and the inrush current is therefore lower than that in the core without a gap.

c. Composite Cores

Providing a core material which possesses an arbitrary combination of characteristics can be difficult and sometimes impossible. A simple, but perhaps costly, solution is to combine two cores into a single assembly, each core with one of the desired characteristics. In Fig. 6.9a, composite (mixed) cores are encircled by a single winding. In one application, core A (e.g., Supermalloy) will have very large initial permeability μ_i but small saturation B_s. Core B (e.g., Silectron, Deltamax, Orthonol), however, will have small or moderate μ_i and large B_s. In another application, a transformer must operate over a very wide band of frequencies. Core A (e.g., permalloy) is chosen for its high μQ (see Sec. 6.1e) in the kilohertz band and core B (e.g., ferrite or powdered iron) for its high μQ in the megahertz band.

In Fig. 6.9b, we depict a swinging dc choke structure (see Sec. 7.3) in which the cores are of the same material but have different air gaps. Core A (small gap) has a high value of μ_e when dc current is small, while core B (large gap) has been designed to obtain optimum μ_e when dc current is large.

For purposes of analysis, the composite core structure is viewed as a parallel magnetic circuit in Fig. 6.9c and as a series electrical circuit in Fig. 6.9d. Note that these circuits are duals of the series magnetic circuit (Fig. 6.8a) and the parallel electrical circuit (Fig. 6.8b)

d. Stabilization of Permeability

As we noted in Sec. 6.1, the permeability of the core is anything but constant. It is a function of many variables, such as the input voltage (and therefore the flux level), the temperature, and even the production lot.

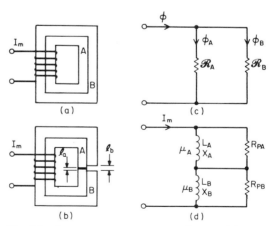

Fig. 6.9 Composite cores. (a) A and B are different materials. (b) A and B have different air gaps. (c) Shunt reluctances (magnetic circuit). (d) Series reactances (electrical circuit).

Suppose any one of the numerous factors has caused a change $\Delta\mu$ in the material's intrinsic permeability μ, producing the fractional change $\Delta\mu/\mu$. The insertion of a series air gap (see Fig. 6.6) results in the fractional change $\Delta\mu_e/\mu_e$. If we take the derivative $d\mu_e/d\mu$ of the basic Eq. (6.31), we arrive at the following important expression:

$$\frac{\Delta\mu_e}{\mu_e} = \frac{\Delta\mu}{\mu}\frac{\mu_e}{\mu}(1-\beta) \cong \frac{\Delta\mu/\mu}{\mu/\mu_e} \tag{6.38}$$

This formula has far-reaching consequences on the design procedure for the ferromagnetic LR network. We infer that linearity and stability, as measured by $\Delta\mu_e/\mu_e$, depend on the extent to which we reduce the permeability, by adding a gap. Suppose that for any of a host of reasons, such as variation of flux or temperature, μ changes by 10 percent. If we wish to reduce this change to 1 percent, we simply add a gap sufficient to reduce μ by a factor of 10.

This factor, which we call the stabilizing ratio, is defined as:

$$\frac{\mu}{\mu_e} = 1 + \mu\beta \tag{6.39}$$

The choice of gap is seen to depend on the μ of the material and on the length of the magnetic path, since $\beta = l_a/l$.

Magnetic aging, the change of magnetic characteristics with the passage of time, is virtually nil in laminations made of silicon alloys. Aging in the order of 10 percent is possible, however, in carbon motor-lamination steel.* Tests of aging are usually made before and after subjecting samples to a temperature of 100°C for 600 hours (hr) or of 150°C for 100 hr.

The permeability of ferrites changes slightly in the time which elapses between manufacture and testing. This variation is defined in terms of a disaccommodation factor (DF):

$$\frac{\Delta\mu_e}{\mu_e} = DF\left(\log\frac{t_2}{t_1}\right)\mu_e \tag{6.40}$$

The manufacturer usually makes t_1 equal to 2 hr, and t_2 equal to 20 hr in ascertaining DF. Suppose, for example, that 100 hr after a core is manufactured we introduce a gap which reduces μ_i to a μ_e of 125. If DF = 4 ppm (parts per million), then 10,000 hr later, about 1.1 years, μ_e will have logarithmically decreased by 1,000 ppm, or 0.1 percent. The same reduction, or drift, will occur in 11 years if we provide the gap 1,000 hr after the core is manufactured.

In stable, resonant LC networks, it is desirable that no change in

* Minimal aging is obtained with a carbon content below 0.005 percent.

the resonant frequency occur as the ambient temperature varies. We can equate the negative temperature coefficient of the capacitor with the positive temperature coefficient of the inductor by selecting that β which results in:

$$\frac{\mu_e}{\mu}\left|\frac{\Delta\mu/\mu}{\Delta T}\right| = \left|\frac{\Delta C/C}{\Delta T}\right| \tag{6.41}$$

where the quantities in brackets stand for the temperature coefficients of the core material and the capacitor, respectively, and ΔT is the variation in the ambient temperature.

We should not overlook the fact that the size of the air gap itself varies, usually because of the temperature coefficient of the linear expansion of the core material:

$$\frac{\Delta\mu_e}{\mu_e} = \frac{\Delta l_a}{l_a}\left(\frac{\mu_e}{\mu}-1\right) \tag{6.42}$$

where the variation in the size of the gap is Δl_a.

We may infer that stability is enhanced as the air gap is made larger. The most stable inductors will be those which have small inductance.

The effect on the core Q of a series air gap is determined from:

$$\frac{Q_e}{Q} = \frac{R_p/\omega L_e}{R_p/\omega L} = \frac{\mu}{\mu_e} \tag{6.43}$$

Thus, Q is increased to Q_e when μ is decreased to μ_e. Note that the merit factor is unchanged:

$$\mu Q = \mu_e Q_e \tag{6.44}$$

Since the μQ of a core does not depend on the presence or absence of a gap, μQ is a practical concept and a merit factor for evaluating core materials.

6.4. DESIGNING FOR HIGH Q

High Q is most needed when the inductor or transformer is part of an LC wave filter. In low-pass and high-pass filters, inductors with moderate Q's (10 or less) are adequate. A high Q (greater than 10) is most necessary in filters with a low insertion loss and a narrow band pass. The following relation is a direct illustration of the fact that a narrow bandwidth requires a high Q:

$$\frac{2\,\Delta f}{f_r} = \frac{1}{Q} \tag{6.45}$$

where $2 \Delta f$ is the half-power bandwidth and f_r is the resonant frequency.

a. Optimum Gap and Maximum Q

In the analysis of the ferromagnetic high-Q inductor which follows, we make two assumptions: (1) that the frequency range is small enough so that performance is not complicated either by the magnetic skin effect (see Sec. 6.1f) or by self-resonance and (2) that the voltage level, and therefore the flux level, is sufficiently low so that shunt loss R_p (see Fig. 6.1b) is reasonably independent of frequency.

The appropriate equivalent circuit is the parallel representation of L_p and R_p for the core, in series with the copper resistance R_1.

Since the μQ product is constant, we can expect that an air gap which reduces μ will increase the Q of the core. Q is maximum when the copper and core losses are equal. Analysis shows that there is an optimum gap and an optimum frequency at which maximum Q occurs.[28]

The Q of the inductor is given by:

$$\frac{1}{Q} = \frac{R_1}{X_p} + \frac{X_p}{R_e} \tag{6.46}$$

where $X_p = 2\pi f L_p$. The symbol R_e, the resistance due to the eddy currents, is used in place of R_p to remind us that the flux level is low (typically 40 G or less).

Q_o, maximum or optimum Q, occurs at an optimum frequency f_o:

$$Q_o = \frac{1}{2}\sqrt{\frac{R_e}{R_1}} \qquad f_o = \frac{\sqrt{R_1 R_e}}{2\pi L_p} \tag{6.47}$$

When the frequency is not optimum, Q can be computed from:

$$\frac{Q}{Q_o} = \frac{2}{\Omega + 1/\Omega} \tag{6.48}$$

where $\Omega = f/f_o$. When values are plotted on logarithmic paper, a symmetrical curve is obtained (Fig. 6.10).

As the gap is increased, f_o increases inversely with inductance and, in accordance with Eq. (6.45), Q_o remains essentially constant. Its invariance is shown in Fig. 6.11.

This concept simplifies our design problems. For each core structure (i.e., grade and thickness of material), the maximum Q obtainable will occur at an optimum frequency. At any other frequency, Q will be less by an amount which can be determined from Eq. (6.48).

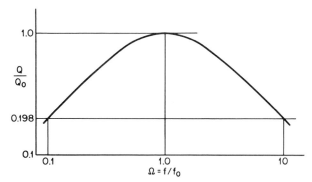

Fig. 6.10 Optimum frequency and maximum Q. The gap ratios l_a/l and μ_e/μ are fixed. Curve is drawn to half scale for log-log graph paper 2½ in per decade (Keuffel and Esser No. 359–120).

Note that when our objective is to maximize Q, inductance becomes a dependent rather than an independent variable. If high Q is mandatory in an LC circuit, we fix C and the circuit impedance *after* we have ascertained the optimum value of L.

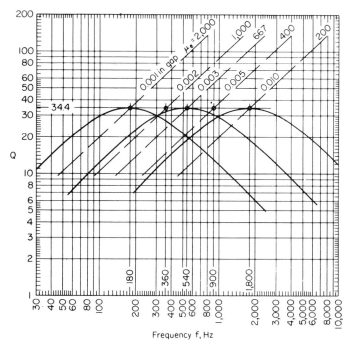

Fig. 6.11 Maximum Q, optimum f, and optimum μ_e. The entire family of curves is based on a constant Q of 34.4 at 1,800 Hz.

The utility of the design technique can be extended by expressing the $\mu_e f_o$ product and Q_o in terms of the copper and core material constants and of the geometry. McElroy shows that the shunt loss resistance can be expressed as:[29]

$$R_p = \frac{12\rho_{Fe}}{a^2} \frac{N^2 A_{Fe} K_{Fe}}{l} = \frac{12\rho_{Fe}}{a^2} \frac{10^9 L}{4\pi\mu} = \frac{4\pi^2 L_p}{e\mu} \tag{6.49}$$

A comparison of *shunt* R_e with the *series* eddy-current resistance $ef^2 L_s$ [derived from Eq. (6.21)] confirms that the shunt resistance is independent of frequency.

Optimum Q and optimum frequency are shown to be:

$$Q_o = \frac{\sqrt{3}}{a} \sqrt{\frac{A_p'}{l\lambda} \frac{\rho_{Fe}}{\rho_{Cu}}} \tag{6.50}$$

$$\mu_e f_o = \frac{10^9 \sqrt{3}}{4\pi^2 a} \sqrt{\frac{l\lambda}{A_p'}} \rho_{Fe}\rho_{Cu} \tag{6.51}$$

where a is the lamination thickness, A_p' is the *net* area product $A_{Fe} A_{Cu} K_{Fe} K_{Cu}$, ρ is resistivity in ohm-centimeters, λ is the coil mean-turn, and all dimensions are in centimeters.

One very important conclusion to be drawn from Eqs. (6.50) and (6.51) is that maximum Q varies directly with the cube root of the core volume and that optimum frequency varies inversely with the cube root of the core volume. To double Q, we must increase volume eightfold. This is not an observation to gladden the hearts of those of us who are interested in miniaturization.

If we multiply Eqs. (6.50) and (6.51), we obtain:

$$\mu_e Q_o f_o = \frac{3(10)^9}{4\pi^2} \frac{\rho_{Fe}}{a^2} \tag{6.52}$$

Now we see that the product of the three key variables is a constant whose magnitude depends on the core's properties alone. Just as the product μQ serves as a merit factor for the core, the triple product $\mu_e Q_o f_o$ can serve as a merit factor for the high-Q inductor. To continue, suppose we again regard Q_o as invariant and that we rearrange Eq. (6.52) as follows:

$$\mu_e f_o = \text{const} \frac{1}{Q_o} \tag{6.53}$$

This form makes it clear that for a specified combination of high Q and frequency, there is an optimum permeability $\mu_{opt} = \mu_e$ which occurs at f_o. Recognition of this principle permits us to express Eq. (6.48) in a more general form:

$$\frac{Q}{Q_o} = \frac{2}{v\Omega + \dfrac{1}{v\Omega}} \tag{6.54}$$

where $v = \mu_e/\mu_{opt}$ and μ_{opt} is the optimum permeability at the optimum Q and optimum frequency, and $\Omega = f/f_o$. Figure 6.10, in which $v\Omega$ can be used in place of Ω in the abscissa, is a graphic representation of this equation.

Example. We can perhaps best illustrate the use of the foregoing ideas by considering some typical problems encountered in the design of a 1-in cube inductor. We select a square stack ($\frac{1}{4}$ in) of 6-mil 24–25FG or Mumetal (high-nickel) laminations. Structures such as Super Q 80, the F lamination, and the pot core, which allow for an adjustment of the gap to maximize Q as well as to stabilize L, are particularly attractive.

We begin with three questions:

1. What is the maximum quality factor Q_o when the exciting frequency is low?
2. If we want a stability ratio of about 100:1, how big is the gap, and what should be the value of μ_e?
3. What is the optimum frequency f_o at which we obtain Q_o?

In our answers, we employ the constants in Eq. (6.50): $K_{Fe} = 0.9$, $K_{Cu} = 0.25$ [based on a test coil of 4,000 turns of size 40 American Wire Gauge (AWG)], $\rho F_e = 55(10)^{-6}$, $\rho_{Cu} = 1.274(10)^{-6}$, and $\lambda = 2$ in. We compute a Q_o of 34.4.

Since $\mu_i \cong 20,000$ and $\mu_i/\mu_e = 100$, then $\mu_e = 200 \cong 1/\beta$, and $l_a = 2/200$, or 10 mil. For this core, then:

$$\mu_e f_o Q_o = 1.8(10)^7 \tag{6.55}$$

and f_o occurs at 2,620 Hz.

Now, what would Q be a decade lower, at 262 Hz? If $\Omega = 0.1$, then $Q = 34.4(2)/(0.1 + 10) = 6.8$.

Suppose we need a Q of 10, rather than 6.8, at 262 Hz? Then we can increase the volume of the core by $(10/6.8)^3$, or 3.2.

If this increase in size is too great, what alternative do we have? We can increase μ_e and decrease the gap, although this makes the behavior of the inductor less linear. Solving,

$$\frac{10}{30} = \frac{2}{0.1v + 10/v}$$

and $v = 1.7$. Since μ_e must be increased to $1.7(200) = 340$, the gap must be reduced by about the same factor—to $10/1.7$, or 6 mil.

Suppose we want the Q *to be 34.4 at 262 Hz and we may not increase the size? How do we proceed?* μ_e must be drastically increased, from 200 to 2,000. This results in a gap ratio of about $1/2,000$, less than the minimum of $0.0015/2$ (which we obtain from line 8 of Table 6.4). We can meet the specifications by abandoning the preferred F lamination and using the "gapless" EE24–25 lamination, but our inductor will become subject to fluctuations arising from variations in material, from mechanical strain, from thermal stress, and from variations in voltage. We end up with an unstable, nonlinear inductor.

(1) **High Operating Frequencies and Voltages.** At the beginning of this discussion, we made two assumptions: that operating frequency is low and that the voltage level is low. When this is not the case, the performance of the inductor will be different. A discussion of the differences is in order.

As the exciting voltage is increased, the flux density increases, and so does hysteresis. In consequence, the shunt R_p decreases as L_p increases. The result is that Q_o and f_o decrease somewhat, the extent depending on the increments $\Delta\mu_e$ and ΔR_p. The Q curve shown in Fig. 6.10 is flattened at the top and shifted to the left.

In the frequency region where eddy-current shielding takes place ($\xi > 2$), if the flux level is moderate, we may infer from the analyses of Welsby[15] and Legg[16,17] that the series hysteresis resistance may actually decrease as L_s and L_p increase slightly. The net result is that Q_o can increase slightly, instead of remaining invariant with frequency. When there is a specific gap, the slope of the Q curve to the right of f_o may be less than to its left. That is, the curve becomes skewed.

At still higher frequencies, self-capacitance, dielectric losses, and copper eddy-current losses tend to reduce the maximum Q. The area to the right of f_o decreases, and the slope increases, resulting in a further departure from symmetry.

(2) **Skin and Proximity Effects.** At high frequencies, well above the audio band, components of the magnetic field which flow in the coil, as well as in the core, result in losses of power due to eddy currents in the conductor. Studies have indicated two aspects of this loss: a skin effect factor and a proximity effect factor. The *skin effect* arises from the magnetic field concentric with current flow originating in the conductor, which restricts the flow of current to the skin of the conductor. But there is also a *proximity effect,* which arises from current flowing in adjacent conductors, and produces a transverse magnetic field perpendicular to the axis of the winding.* Field plots using

---

* The transverse field is not uniform throughout the window area. In the transformer, ac resistance depends on the number of layers as well as on the ratio d/Δ.

the *finite* *e*lement *m*ethod (FEM) show that the skin effect in a local region of the winding is magnified by the current flowing in a neighboring region of the coil.[30] Snelling, however, attributes the major power losses due to eddy currents in the inductor winding to the proximity effect.[22] Whatever the cause, the ac resistance of the coil exceeds dc resistance by a factor F_R, described earlier in Eq. (5.9):

$$F_R = \frac{\text{ac resistance}}{\text{dc resistance}} \tag{6.56}$$

The frequency f and wire diameter d (in millimeters) at which the skin effect becomes significant can be useful information when the coil is being designed. Using an analysis similar to that in Sec. 6.1f (with $a = d$ and $\delta = \Delta$), we can arrive at the expression:

$$f = \frac{k^2}{\Delta^2} \tag{6.57}$$

where k is a constant, equal to 65.5 in copper at 20°C, and Δ is the penetration depth (in mm), the depth at which the magnetic flux is attenuated to $1/\epsilon$ of its value at the surface.

We regard the skin effect as negligible when $F_R = 1.02$ (a 2 percent increment) and $d/\Delta = 2$. With these values substituted into Eq 6.57, we can define a critical frequency f_c as follows:

$$f_c = \frac{4k^2}{d^2} = \frac{17.16(10)^3}{d^2} \tag{6.58}$$

By way of example, if an inductor is to operate at 40 kHz, our initial estimate is that the diameter of the conductor should not exceed 0.655 mm (about 22 AWG). However, if the proximity effect is taken into account, we can estimate the increment of resistance to be an order of magnitude higher, say 20 percent. The substitution of stranded conductors for solid conductors is made only after comparative tests indicate that the improvement is worth the trouble. Some designers use conductors made of bunched insulated strands. In many instances a reduction in strand diameter d is also accompanied by a decrease in the copper space factor K_{Cu}, and therefore produces no significant improvement in Q.

There are applications, usually in the megahertz band, when power losses due to eddy currents in the conductor (as well as losses in the dielectric) can become limiting factors in the design. When this occurs, it is wise to forego an arbitrary specification of inductance in favor of a specification of maximum Q. Design of the inductor would then begin with the geometry of the coil; the winding is deliberately restricted either to one layer of solid circular wire or to a few layers of thin foil. The coil former is made of a dielectric with low power losses or is dispensed with entirely.

It should be remembered that the air gap, as measured by calipers, is not the effective gap. The absence of a measurable gap does not mean zero l_a (see Table 6.4); furthermore, when a gap is large, fringing reduces its effective area. When accurate predictions of Q are required, we construct graphs similar to Fig. 6.11, and obtain the actual Q_o and f_o for several gaps. A template whose shape is similar to that predicted by Eq. (6.48) expedites the design process.

The principal advantage of a theoretical, rather than an empirical, approach is that it enables the designer, with the aid of similitude theory, to predict with facility all the major trends in the variation of Q and L which result from a variation in any of the many (more than 11) parameters involved.

b. Geometry

We have seen that core material can be assessed on the basis of the relative power-loss factor $(\tan \delta)/\mu$ or its reciprocal μQ. There is an alternative, perhaps more direct, means of relating the Q of the inductor to its geometry. First note that Q is maximum when power losses in the core are nil. Hence, minimum dissipation can be expressed as:

$$\frac{1}{Q_{max}} = \frac{R_1}{\omega L} = \frac{N^2 R_{Cu}}{\omega N^2 A_L} = \frac{R_{Cu}}{\omega} \frac{1}{A_L} \tag{6.59}$$

Here we have made use of Eq. (6.7), in which $\omega = 2\pi f$, inductance L is defined as a function of turns N, and the dc resistance $R_1 = R_{Cu}$. We have also introduced A_L, the normalized inductance which is also called the *inductance factor*:

$$A_L = \frac{L}{N^2} = \frac{\mu_0 \mu_e A_e}{l} \tag{6.60}$$

where A_e is the net area of the core. The parameters of A_L will be recognized as the permeance of the magnetic circuit.

The inductance factor is analogous to the dc resistance factor introduced in Eq. (5.4) during our discussion of regulation in Sec. 5.2c:

$$A_R = \frac{\rho \lambda}{K_{Cu} A_{Cu}} \tag{6.61}$$

We recall that Q is the reciprocal of the dissipation factor—the quotient of resistance and reactance. By analogy, we can obtain a geometric expression for dissipation by forming the quotient A_R/A_L:

$$\frac{A_R}{A_L} = \frac{\rho\lambda}{K_{Cu}A_{Cu}K_{Fe}} \frac{l}{\mu_0\mu A_{Fe}} = \frac{\rho}{\mu_0\mu} \frac{\lambda l}{A_p'} \qquad (6.62)$$

where $\mu_0 = 4\pi 10^{-9}$

   μ = core permeability
   ρ = copper resistivity
   $\lambda = l_t$, the mean-length turn of the coil
   l = mean length of the magnetic path
   A_p' = the product of the net core area and the net copper area

The reciprocal of Eq. (6.62) corresponds to the time constant τ of the LR network of Fig 6.1:

$$\tau = \frac{A_L}{A_R} \qquad (6.63)$$

When copper and core power losses are equal, the optimum Q_o equals one-half the maximum Q_{max}. The reciprocal of Eq. (6.59) will enable us to express Q_o in terms of the material constants and the geometry of the core:

$$\frac{Q_o}{\omega} = \frac{A_L}{2A_R} = \frac{\mu_0\mu}{2\rho} \frac{A_p'}{\lambda l} \qquad (6.64)$$

An alternative expression can be developed in terms of the shunt ac resistance R_p in Fig. 6.1. To do this, we define a normalized ac resistance factor A_{Rp} and a reactance factor A_X:

$$A_{Rp} = \frac{R_p}{N^2} \qquad A_X = \frac{X_p}{N^2} \qquad (6.65)$$

where $X_p = \omega L_p$ is the reactance of the shunt inductance L_p. Starting with Eq. (6.45) and replacing R_e with R_p, we obtain $R_{Fe} = 4Q_o^2 R_{Cu}$. Then:

$$\omega Q_o = \frac{A_{Rp}}{2A_L} \qquad (6.66)$$

and it is quite possible to obtain, analogously to Eq. (6.63), an expression in terms of ac resistivity and the geometry. Curves which plot R_p/N^2 and X_p/N^2 vs. frequency are useful in design work, especially since we can equate these parameters with the power-loss factor, $\tan \delta = X_p/R_p$ [see Eq. (6.17)].

A glance at Eq. (6.63) indicates that the magnitude of Q is a function of the geometric coefficient g_2:

$$g_2 = \left(\frac{A_{Cu}A_{Fe}}{\lambda l}\right)^{1/2} \tag{6.67}$$

which was introduced in earlier chapters as Eqs. (4.20) and (5.11). Hence g_2 can be regarded as a merit factor in the selection of a core by the designer.

Our next question is: How can we assess whether a specific core structure has been proportioned so that its dimensions will produce

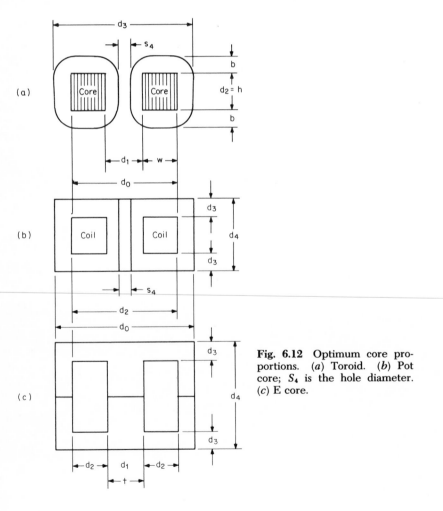

Fig. 6.12 Optimum core proportions. (a) Toroid. (b) Pot core; S_4 is the hole diameter. (c) E core.

maximum Q?[21] The design task might be phrased as follows: Assume that we have chosen a specific core shape and that there are arbitrary constraints on the outer dimensions. Now choose an optimum combination of internal dimensions which results in a maximum for the geometric parameter g_2.

In the study by B. Astle, the basic core shapes (toroid, pot core, and E core), shown in Fig. 6.12 are analyzed with the objective of achieving maximum Q given a choice of two constraints on the volume: minimum volume of the envelope containing the inductor or minimum volume of the core material.[31] The best proportions result in a minimum value of the shape factor ψ when Q is maximum.

In Table 6.5, we summarize the results obtained under the constraint most often encountered in design work, i.e., minimum volume of the envelope. The shape factor ψ assigned to a core is based on a number of conditions:

1. Whether the coil has multiple layers or a single layer of magnet wire
2. The fraction of window y occupied by the coil former (or bobbin)
3. The exponent m in Eq. (6.20)
4. The proportions of the inductor, normalized with respect to the dimension d_o (the outside diameter of the toroid, or the outside width of the E geometry)

As expected, the toroid and pot core are the best shapes for obtaining maximum Q. It is reasonable to conclude that the same geometry which yields an inductor of maximum Q also produces the most effi-

TABLE 6.5 Optimum Inductor Geometry

Shape	Notes	ψ	d_1/d_o	d_2/d_o	d_3/d_o	d_4/d_o
Toroid	Low-frequency multilayer	154	0.595	0.337		
	Low-frequency single layer	26	0.53	0.47		
	High-frequency single layer	14	0.46	0.54		
Pot core	$y = 0,\ a = 0$	145	0.390	0.841	0.113	0.649
	$y = 0.02,\ a = 40°$	224	0.411	0.866	0.106	0.685
E core	$y = 0$	208	0.233	0.214	0.136	0.815
	$y = 0.02$	273	0.215	0.252	0.143	0.917
E-C core	$y = 0$	223	0.374	0.152	0.123	0.802
	$y = 0.02$	311	0.362	0.189	0.134	0.920

Table is based on constant overall volume, with $m = 2$ (exponent of B). $K_{Cu} = 0.5$ in a toroid and 1.0 in the other shapes. a = angle of the cut-out for the exit of leads from a pot core.

cient transformer.[32] (For further discussion of lamination proportions, see Sec. 5.4c).

c. Powder Permalloy

The F lamination and the pot core are particularly useful when our objective is to achieve a specific degree of stability μ/μ_e by adjusting the gap. However, the very presence of a discrete physical gap often creates costly mechanical difficulties. One solution is to use the pot core with an adjustable threaded slug which passes through the center of the core gap and varies the reluctance of the gap.

Another and excellent engineering solution to the problems of stabilizing the gap and the permeability is the compressed powder toroid developed at Bell Telephone Laboratories.[33] The compressed powder toroid has a distributed, rather than a discrete, gap. A high ratio of μ to μ_e is obtained by using an alloy with great permeability, such as molybdenum-nickel permalloy, which has inherently low power losses due to hysteresis. Grinding such materials into small particles cuts the power losses due to eddy currents greatly, and the introduction of a nonmagnetic binding substance in a controlled proportion (the packing factor) produces a large effective gap ratio. The possibility of a change in the gap through mechanical vibration or shock is virtually eliminated.

Legg has derived the following equations for the wound powdered toroid, neglecting hysteresis and residual losses:

$$\left.\begin{aligned} Q_o &= \frac{420d}{\sqrt{10^9 e}} \\ \mu_e f_o &= \frac{7.4(10)^6}{d\sqrt{10^9 e}} \end{aligned}\right\} \tag{6.68}$$

where d is the mean diameter of the toroid and e is the eddy-current coefficient. The coefficients are based on $K_{Cu} = 0.5$.

Multiplying these equations, we obtain:

$$\mu_e f_o Q_o = \frac{\pi}{e} \tag{6.69}$$

Using $\mu_e f_o Q_o$ as a merit factor for core structures, we can compare the permalloy F lamination structure with the powdered permalloy toroid. First we readjust the copper space of the F core from 0.25 to 0.5 by multiplying Eq. (6.55) by $\sqrt{2}$. The merit factor, using the

6-mil F lamination, is $\mu_e f_o Q_o = 2.54(10)^7$. For the 125-$\mu$ toroid, where $e = 19(10)^{-9}$, $\mu_e f_o Q_o = 16.4(10)^7$.

That the merit factor of the toroid is roughly six times that of the laminated core simply reflects the increase in resistivity effected through the use of 120-mesh powder. Even smaller values of e can be selected if powdered cores with lower μ_e are used. Manufacturers maintain a reasonably consistent set of ratios of dimensions, making possible a large range of possible values of f_o and Q_o obtainable with but six standard diameters. Note that although the use of cores with low μ_e increases Q, f_o also increases. If a high value of Q is specified, it is easier to achieve a small inductance at a high frequency than a large inductance at a low frequency.

Above 30 kHz, it may be necessary to use stranded wire to obtain maximum Q. Note that the copper space factor is reduced from 0.5 to about 0.25 or less when we employ stranded wire and low-capacitance winding techniques such as pi windings (see Sec. 9.4c).

The utility of the design approach initiated by Arguimbau is demonstrated by Legg in a universal curve (Fig. 6.13). It is possible to choose among five levels of μ_e and eight diameters and to obtain, over a span ranging from 2.1 kHz to 1.16 MHz, a peak Q which ranges from 68 to 1,020.

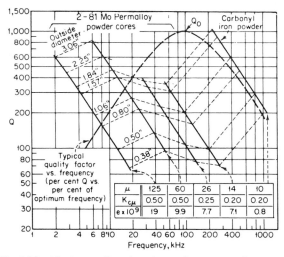

Fig. 6.13 Maximum Q and optimum frequency of annular cores. Hysteresis and residual losses are assumed to be negligible. Dimension ratios: $w' = 1/3$, $h' = 3/8$, $d' = 1/6$. (*From V. E. Legg, "Analysis of Quality Factor of Annular Core Inductors,"* Bell System Technical Journal, *Vol. 39, No. 1, January 1960, Fig. 2.*)

d. Ferrites[22]

Although ferrites have much lower permeability than the nickel permalloys, their resistivity is extremely great. Their merit compares interestingly with that of permalloy powder, especially when operating frequencies are above 50 kHz.

For the 125-μ powdered permalloy toroid, we compute a core $\mu' Q$ product (at 10 kHz) of $2\pi / ef = 32,800$, and for the 60-μ core a product of 62,800. At the same frequency, a $\mu' Q$ product of 50,000 can be obtained with nickel-zinc (Ni-Zn) ferrite and a $\mu' Q$ of 300,000 can be obtained with manganese-zinc (Mn-Zn) ferrite.

The power lost in the eddy currents becomes so small that the residual losses predominate. The addition of a higher order term gf^2 to Legg's equation (6.21) takes into account the more rapid increase of losses in ferrites at very high frequencies.

It now follows that the product of optimum μ and optimum Q of coil and core is no longer independent of frequency. And, although the $\mu' Q$ of these ferrite cores generally decreases with frequency, the inverse relationship is not linear.[34] Our equations can be modified, but they become awkward, and we rely on graphical data furnished by the manufacturer when reasonably accurate predictions are desired.

At upwards of 100 kHz, the permeability-quality factor products of some ferrites exceed those of virtually all other materials. This makes them particularly suitable in the 0.1 to 50 MHz region and in transformers which pass narrow pulses. It has been found that exceptionally high inductor Q (between 300 and 1,000) is more readily achieved with ferrite cores than with powdered cores when the operating frequency is very high.

The trend in circuit design toward the use of higher frequencies, in the megahertz (and eventually gigahertz) band, naturally raises the question of the useful bandwidth of ferrite materials. At a sufficiently high band of frequencies, all ferrites exhibit the properties of a complex dielectric, having high permittivity ϵ and resistivity ρ.[23] Hence, from the point of view of circuit analysis, a coil with a ferrite core ultimately behaves more like an RC network than an LR network.

In the analysis of H. Schlicke, μQ appears to be proportional to $1/f^{0.75}$ for manganese-zinc, nickel-zinc, and lithium ferrites. The envelope of maximum μQ products of temperature stable ferrites (from 10 kHz to 100 MHz) appears to follow the relation:

$$(\mu Q)_{\text{max}} = \frac{44,000}{f^{0.6}} \tag{6.70}$$

J. Watson has analyzed the experimental evidence which shows the dependence of initial permeability on frequency and has noted that the envelope of curves for Mn-Zn and Ni-Zn ferrites can be matched to an empirical formula:[35]

$$\mu_i f = 3,000 \tag{6.71}$$

where f is in megahertz.

As a possible guideline for the design of inductors, the formula appears to set an upper bound on the frequency usable with ferrite cores. It is consistent with what has been learned from experience: When high frequencies are used, we must accept materials with low permeability. If this equation is taken at face value, then at the useful limiting frequency of 3,000 MHz, where μ_i equals unity, the core can be entirely dispensed with. However, such a limit appears optimistic when we note loss of power in the core increases sharply with frequency. This, then, sets a practical upper bound on frequency that is lower by an order of magnitude, in the neighborhood of 300 MHz.

The comparatively high temperature coefficient of the permeability of ferrites makes it necessary to use an air gap for stability. The ferrite inductor can provide the same degree of stability μ'/μ_e as the powdered toroid only if the designer can accept small inductance in exchange for the very high inductor Q_o that is feasible.

The initial permeability of the ferrites appears to be inversely proportional to the Curie temperature. The designer of a high-temperature coil is therefore likely to choose a ferrite with a small μ_i in order to avoid a low Curie temperature. Researchers are working intensively to develop ferrites which will have high permeability in a high ambient temperature.

Occasionally a coil is accidentally subjected to an excessive polarizing dc bias or to an abnormally high temperature. These can permanently alter the permeability of some ferrites. In precision circuits, or when great reliability is required, the cores are thoroughly tested; when such tests are not practical, the coil designer will be inclined to favor the permalloy or iron-powder toroid, whose ability to recover its permeability is excellent.

Perhaps the most impressive characteristics of the ferrites, from the point of view of the designer, are that they are low in cost and available in a large variety of shapes. Impressive, also, is the almost continuous improvement in the magnetic properties of ferrite cores, which appears to diminish progressively the few disadvantages we have noted.

e. Powdered Iron

An alternative to ferrites, when frequencies are in the 10 to 300 MHz band, is powdered iron. Iron power was recommended for practical use by Heaviside in 1887 and was introduced for commercial use in the United States about 1921.   Since 1933, it has been used for the core material of high-Q radio-frequency (RF) coils.*   Iron powder is available with permeability values in the range of 3 to 75 and, in most RF applications, its permeability remains fairly stable as the temperature changes (35 to 170 ppm per °C).[36]   Because of its high dc saturation density (typically 7 to 9 kG, about double that of ferrite), iron powder can also be used effectively in high-frequency storage inductors which operate at low ac flux density (see Sec. 7.6).   Iron powder shares with ferrite the advantage of low cost compared with permalloy powder.

6.5. FERROMAGNETIC HARMONIC DISTORTION

The flow of harmonic currents in a winding or the presence of harmonic voltages across a winding is usually undesirable.   One consequence of such harmonics is the flow of undesirable residual currents in null and resonant circuits; another is harmonic distortion and cross-modulation, which either mask the intelligibility or reduce the fidelity of communications.   In certain nonlinear circuits, however, distortion is deliberately accentuated for some special purpose.   The frequency multiplier is an example.

In this section, we shall stress those relationships among the core parameters—the shape of the hysteresis loop, the flux density, the volume of the core, and the size of the air gap—which influence the magnitude of harmonic currents in the iron-core coil.

a. Wave Shapes

Harmonic currents or voltages result from the nonlinear relationship between flux and current which is exemplified in the hysteresis curve.

The wave shapes of the current or voltage will depend on the circuit (see Fig. 6.14) as well as on the shape of the hysteresis loop.   If the source resistance is small, we assume that the generator provides a constant voltage and that flux is sinusoidal.   If source resistance is great, we assume that the current is constant and sinusoidal.

If the source resistance is small and the flux ϕ is a sine function of time, then the induced voltage (the derivative of the sine function)

---

* Experimenters such as radio amateurs will sometimes replace ferrite with iron powder simply because then the accidental application of direct current to a coil does not permanently alter its permeability.

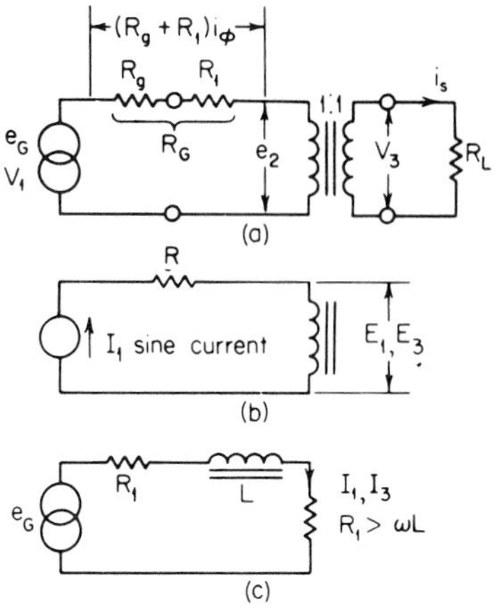

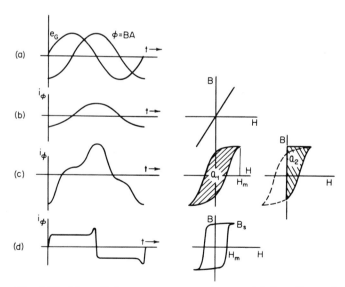

Fig. 6.14 Distortion circuits. (*a*) Constant voltage, shunt coil or transformer. R_G is small; sine-wave flux. (*b*) Constant current source. R is large; sine-wave current. (*c*) Series circuit. $R_1 >$ ωL.

has, because it is a cosine function, the same wave shape as the flux. The wave shape of the current, however, will depend on the shape of the hysteresis curve, as shown in Fig. 6.15. And the shape of the hysteresis curve depends on the properties of the core material. Metallurgists and physicists have developed materials whose hysteresis curves do not have the classical form.

Fig. 6.15 Magnetizing current wave shapes. (*a*) Sine flux and cosine induced voltage. (*b*) Sine current; linear *B-H* curve. (See Fig. 6.16 for effect on voltage of nonlinear *B-H* curve.) (*c*) Peaked current; round *B-H* loop. (*d*) Square-wave current; square *B-H* loop.

If the hysteresis curve, then, is linear, the current will be sinusoidal and in phase with the flux. If the hysteresis curve is the classical rounded loop, the magnetizing current is distorted. The curve has a characteristic peak when the flux density is high and, since it is lopsided, it is almost, but not quite, in phase with the sinusoidal flux. If the hysteresis curve is a rectangular loop, the shape of the wave of current is deformed into a rectangle. It is now out of phase with ϕ and in phase with the voltage or cosine function. What would have been a peak (on the rounded loop) becomes a spike if the excitation is great enough to cause the flux density to reach saturation (see Fig. 6.14d). The impedance e/i, now distinctly nonlinear, is established by the magnitude of the coercive force H_m. The transient response also differs from that of the linear or quasi-linear inductor. If the applied function is a step of voltage, the current is *not* exponential but is a rectangular staircase function.[37] Such characteristics will ordinarily make a core material with a rectangular hysteresis loop unsuitable for use in a linear or quasi-linear inductor.

If the source resistance is large and the current is a sine function of time, the flux is not sinusoidal. It has a flattened top and steep sides (see Fig. 6.16). The voltage wave (instead of the current) peaks as the slope of the flux steepens.

Since it is either the current or the voltage which is distorted, we can expect odd harmonics of current or voltage to appear in the circuit. The consequences are of practical importance. Peaked current flowing through the series resistance produces a peaked voltage across that resistance. Although the input cosine voltage e_G is undistorted,

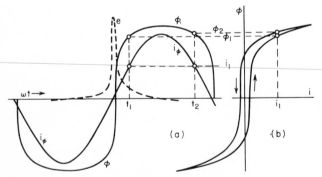

Fig. 6.16   Sine current excitation. Graphical construction for the determination of the wave forms of flux and induced voltage when the exciting current is a known sinusoidal function of time. (*From M.I.T., Dept. of Electrical Engineering*, Magnetic Circuits and Transformers, *Fig. 6, p. 165. Copyright 1943, The MIT Press, Cambridge, Mass. Reproduced by permission of The MIT Press.*)

there must appear, across the primary winding, a distorted voltage e_2. In this way the distorted current i_ϕ produces, at the output winding, a distorted voltage whose shape is a mirror reflection of the current. This may be seen in Fig. 6.17. Note how the two distorted shapes, when added, produce the undistorted sine wave of the generator.

b. Distortion at Low Voltage Levels

The magnitude of distortion is related to the width of the hysteresis loop. If the excitation level is low and the loop area small, there is little distortion. For the purpose of analysis, it is desirable to have available an expression for the hysteresis resistance R_h. This may be obtained from the ratio of dissipated hysteresis energy to stored energy:

$$d_h = \frac{R_h}{2\pi f L_s} = \frac{h\mu' B_m}{2\pi} = \frac{\mathscr{A}_1 \mu'}{\pi B_m{}^2} = \frac{1}{2\pi}\frac{\mathscr{A}_1}{\mathscr{A}_2} \qquad (6.72)$$

where h is the hysteresis coefficient of Eq. (6.22) and $\mathscr{A}_1$ is the enclosed area of the hysteresis loop.

An alternative expression used in International Electrotechnical Commission (IEC) standards is:

$$\tan \delta_h = \eta_B \mu' B_m \qquad \eta_B = h/2\pi \qquad (6.73)$$

where η_B is a hysteresis material constant.

The hatched area in Fig. 6.15c, $\mathscr{A}_2$, represents the energy absorbed by the core as the flux moves from $-B_r$ to B_s along the right flank. If $\mathscr{A}_1$ varies with the cube of B_m and $\mathscr{A}_2$ with the square of B_m, then d_h will be directly proportional to B_m.

Next we consider how d_h is related to the magnitude of distortion. The analytic solution for harmonic distortion requires a suitable approximation of the hysteresis curve. Rayleigh used two parabolic segments

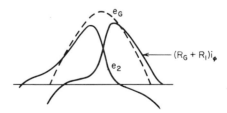

Fig. 6.17 Sum of distorted shapes equals a sine wave. Waveforms correspond to voltages in Fig. 6.14a. *(From M.I.T., Dept. of Electrical Engineering,* Magnetic Circuits and Transformers, *Fig. 5, p. 164. Copyright 1943, The MIT Press, Cambridge, Mass. Reproduced by permission of The MIT Press.)*

to approximate the hysteresis curve (Fig. 6.15c) when flux density is small.

In Rayleigh's representation, the flux density is given by:

$$\frac{B}{\mu_o} = (\mu_i + \alpha H_m)H \pm \frac{\alpha}{2}(H_m{}^2 - H^2) \tag{6.74}$$

where α is a measure of the width of the hysteresis curve.

First assume that the current is sinusoidal, as in Fig. 6.14b.   Using either Rayleigh's or Peterson's approximation (described later) of the hysteresis curve, it can be shown that the peak third-harmonic voltage E_3 is proportional to the hysteresis resistance and to the peak magnetizing current I_m:

$$E_3 = \tfrac{3}{5}R_h I_m \tag{6.75}$$

In the analysis of distortion in toroidal cores, a modulation factor m_h, the ratio of third-harmonic to fundamental voltages,[3] is defined as:[38]

$$m_h = \frac{V_3}{V_1} = \frac{3h\mu B_m}{10\pi} \tag{6.76}$$

If we now substitute $h = 2\mathscr{A}/B_m{}^3$ [derived from Eq. (6.72)] and compare the result with Eq. (6.74), we find that:

$$m_h = \tfrac{3}{5}d_h \tag{6.77}$$

Thus, when the input current is sinusoidal, voltage distortion is 60 percent of the hysteresis dissipation factor.   Analysis shows that a somewhat higher coefficient than $\tfrac{3}{5}$ can be expected when there is intermodular distortion, which occurs when the applied excitation consists of two frequencies, closely spaced.

In the course of the equivalent circuit analysis by Welsby,[15] it is shown that distortion may be viewed as due to the slight but continuous variation of both the ac resistance and inductance over a cycle of excitation.   The hysteresis resistance is:

$$R_h = F_h I L_i{}^{3/2} f \tag{6.78}$$

where

$$h = F_h = \frac{8\sqrt{2}}{3}\frac{\alpha}{\mu_i{}^{3/2}\sqrt{\mu_o V_{Fe}}} = 2\pi\eta_B \tag{6.79}$$

Here F_h is defined as the hysteresis factor of the core; I is the rms value of I_m; L_i is the inductance corresponding to the initial permeability; V_{Fe} is the volume of the core, and $\alpha = d\mu/dH$, a measure of the curvature of the normal magnetization curve of the B-H loop.

Measurements on a core of fixed size, and with standard values of μ and f, provide the designer with a convenient means for estimating distortion due to hysteresis.* To facilitate such computations, it is useful to relate the hysteresis material constant η_B [see Eq. (6.73)] to a hysteresis core constant η_i:

$$\eta_i = \eta_B \sqrt{\mu_0 \mu_i{}^3 / V_{\text{Fe}}} \qquad (6.80)$$

Combining this equation with Eq. (6.79) produces the formula:

$$\alpha = \frac{3\pi}{4\sqrt{2}} \eta_i V_{\text{Fe}} \qquad (6.81)$$

Now consider the series circuit (Fig. 6.14c) and assume that its matched source and load resistance R_1 are greater than the coil reactance. This approximates the conditions for constant current. Welsby has derived an expression for the ratio of third-harmonic current to fundamental current:

$$\frac{I_3}{I_1} = \frac{0.6}{4\pi} F_h \frac{\omega L^{3/2} I_1{}^2}{R_1} \qquad (6.82)$$

The distortion of the current is small if the resistance is large and the driving current is small.

The shunt coil, or the transformer fed from a low-resistance source (Fig. 6.14a), is of great importance in electronic circuits. For matched source and load resistance R_1, when $R_1 \ll \omega L$, Welsby derives the following formula for fractional voltage distortion:

$$\frac{V_3}{V_1} = \frac{0.3}{6\pi} F_h \frac{V_1 R_1}{\omega^2 L^{3/2}} \qquad (6.83)$$

We infer that small voltage distortion requires small resistance, low voltage, and high shunt inductance. A large core of high initial permeability is advantageous.

Peterson's more general analysis is especially interesting, for he does not restrict his investigation, as did Rayleigh, to a very low level of excitation and flux density. Peterson has used two power-series expansions to represent each branch of the hysteresis curve.

It will be of interest to some readers to see that it is possible to make a mathematical connection between remanence, hysteresis energy loss, and the magnetizing force H. Peterson shows that the derived power series for remanence starts with the second power of H, while the power series for hysteresis loss starts with the third power of H.

---

* Snelling observes that direct measurement of harmonics is a more reliable guide to distortion than the use of hysteresis coefficients.[22]

Of the various cases analyzed, we summarize two which seem most pertinent to linear circuits—the inductor and the voltage-fed transformer. The analysis shows that the degree of distortion depends on which of the several parameters the coil designer elects to vary while holding the others fixed.

For an inductor with fixed L, Peterson shows, the third-harmonic voltage is directly proportional to α. The coefficient $\alpha = d\mu/dH$, as noted earlier, is a measure of the curvature of the normal B-H magnetization curve. The third-harmonic voltage is inversely proportional: (1) to $\mu^{3/2}\sqrt{V_{\text{Fe}}}$, if the number of turns N is the variable; (2) to μ, if the core area A_{Fe} is the variable; and (3) to μ^2 if the magnetic length (πd) is the variable.

In this solution for distortion of voltage in the transformer, the total source resistance $R_G = R_g + R_a$ and the load resistance R_L need not be matched.

Secondary current i_s is:

$$i_s = \frac{\gamma V_1^2}{3X_p} = \frac{R_G}{R_L(1 + R_G/R_L)^3} \tag{6.84}$$

where $\gamma = (\text{const}) \times d\alpha/N^3 A_{\text{Fe}}{}^2\mu^3$ if L is a variable.

Substitute into Eq. (6.84) the following: $L = N^2 A_{\text{Fe}}\mu/d$, $b = R_G/R_L$, $X_p = 2\pi fL$, and $N = V_1/BA_{\text{Fe}}f \times 1/(1 + b)$. Now the equation may be rewritten in terms of the distortion quotient:

$$\frac{V_3}{V_1} = (\text{const})\frac{\alpha B_f}{\mu^2 L}\frac{R_G}{X_p}\frac{1}{(1 + b)^2} = \frac{(\text{const})\,\alpha B}{2\pi}\frac{R_G}{\mu^2\,L^2(1 + b)^2} \tag{6.85}$$

Distortion is minimum when b is near either zero or infinity—that is, when *either* the source resistance or the load resistance is very small. Distortion is maximum when $R_G = R_L/2$. Since α and μ are nonlinear functions of B, the solution of the equation involves the measurement of specific cores at several discrete levels of flux density.

c. Distortion at High Voltage Levels

When the level of excitation is high, the power series has many terms and converges so slowly that its use is not practical. Nevertheless, the analysis at low levels reveals major relationships which are useful in the design of the wide-band transformer (see Sec. 8.4).

In Partridge's analysis,[39] the harmonic component I_h of the magnetizing current can be determined by measurements at various levels of B_m. It is shown that the harmonic voltage across the primary V_h is related to the fundamental input voltage V_f as follows:

$$D = \frac{V_h}{V_f} = \frac{I_h}{I_f} \frac{R}{Z_c} < \frac{R_G}{X_p} \tag{6.86}$$

where I_f is the fundamental component of the magnetizing current, R is the parallel impedance of the reflected load resistance R_L, and the source resistance $R_G = R_g + R_a$. Z_c is the open-circuit impedance of the primary at the voltage level producing B_m.

Macfadyen shows that Eq. (6.86) can be confirmed by fundamental physical considerations and that it can be stated as a useful theorem: "The distortion factor due to a particular material at a given fundamental frequency is a unique function of the ratio R/Z_c and the fundamental flux density only."[40]

Equation (6.81) may be plotted as a simple family of curves with the fundamental component of flux density B_f as a parameter. From such curves for silicon steel and nickel steel at 50 Hz, we find that D does not exceed $0.6R/Z_c$ when B_f is less than 10 kG.[41]

An important question is: What is the upper bound of ferromagnetic distortion as a function of the variables at our disposal? One answer might be inferred from a consideration of a square-loop core material which will produce a rectangular magnetizing current rich in harmonics. In this case $d_h = 1/\pi$ and, according to Eq. (6.77), $m_h = 19.1$ percent. Alternatively, we might reason that the harmonics of the magnetizing current I_m cannot exceed I_m itself in magnitude. Hence, when flux density is large, the maximum distortion factor is asymptotic to:

$$D_{max} = \frac{R}{X_p} > D \tag{6.87}$$

Circuit designers find that distortion may be more economically reduced by decreasing the source resistance than by increasing the size of the core in order to reduce flux density.

d. Distortion and an Air Gap

The introduction of a series air gap tilts the hysteresis curve (see Fig. 6.7), but although the slope and μ_e decrease, the enclosed area does not change.

The presence of an air gap increases the total magnetizing current and reduces the inductance. The hysteresis factor F_h drops to F_{hg}, and the inductance L drops to L_e:

$$\frac{F_h}{F_{hg}} = \left(\frac{L}{L_e}\right)^{3/2} = \left(\frac{\mu}{\mu_e}\right)^{3/2} = (1 + \mu\beta)^{3/2} \tag{6.88}$$

They both change by the same amount; consequently, the voltage distortion, which is proportional to $F_h/L_e{}^{3/2}$, is *unchanged* when only the air gap is varied. In terms of Eq. (6.86), although the percentage of harmonics in the magnetizing current drops, the reactance also drops, and by the same amount.

To reduce distortion without disturbing inductance, we must increase the number of turns as well as the air-gap ratio β. The decrease in distortion will be inversely proportional to the three-halves power of the stability ratio μ/μ_e:

$$m_{hg} = m_h \left(\frac{\mu_e}{\mu}\right)^{3/2} \tag{6.89}$$

In the inductor, the gap which stabilizes inductance and optimizes Q is also therefore instrumental in reducing distortion. In the transformer, however, such an increase in the number of turns reduces the efficiency of the copper. Consequently, distortion is minimized by choosing a large core and a large window, a structure which provides the conditions necessary for small flux density, large inductance, and little copper loss. The drawback is that the weight is increased.

If the air gap is made arbitrarily large, the core no longer serves a useful purpose, and can simply be eliminated. Size must be increased, but we have the satisfaction of knowing that our large air coil produces no distortion due to hysteresis.[42,43]

Harmonic distortion also can arise in electronic circuits for causes other than hysteresis. For this reason, the discussion of distortion is continued in Secs. 8.4 and 8.5d.

REFERENCES

1. S. Charap, "Magnetic Hysteresis Model," *IEEE Transactions on Magnetics*, Vol. MAG-10, No. 4 (December 1974): 1,091–96.
2. W. Macfadyen, R. Simpson, "Representation of Magnetisation Curves by Exponential Series," *Proceedings IEE*, Vol. 130, No. 8 (August 1973): 902–904.
3. J. Brauer, "Simple Equations for the Magnetization and Reluctivity Curves of Steel," *IEEE Transactions on Magnetics*, Vol. MAG-11, No. 1 (January 1975): 81.
4. K. Macfadyen, "Vector Permeability," *Journal IEE*, Vol. 94, Part 3 (1947): 407–14.
5. S. Charap, F. Judd, "A Core Loss Model for Laminated Transformers," *IEEE Transactions on Magnetics*, Vol. MAG-10, No. 3 (September 1974): 678–81.
6. C. McLyman, *Magnetic Core Selection for Transformers and Inductors* (New York: Marcel Dekker, 1981).

7. American Society for Testing and Materials: *ASTM Standards on Magnetic Properties,* A 697–74 (Philadelphia, 1976).
8. *IEEE Standard Test Procedure for Magnetic Cores,* Standard No. 393–77 (New York: Institute of Electrical and Electronic Engineers, 1977).
9. J. Lavers, P. Biringer, H. Hollitscher, "Estimation of Core Losses when the Flux Waveform Contains the Fundamental Plus Odd Harmonic Component," *IEEE Transactions on Magnetics,* Vol. MAG-13, No. 5 (September 1977): 1,128–30.
10. R. Newbury, "Prediction of Loss in Silicon Steel from Distorted Waveforms," *IEEE Transactions on Magnetics,* Vol. MAG-14, No. 4 (July 1978): 263–68.
11. D. Chen, "Comparisons of High Frequency Core Losses: A Sinusoidal Voltage and a Square Wave Voltage," *IEEE Power Electronic Specialists Conference,* PESC Record 78CH 1337-5 AES (1978), pp. 237–41.
12. J. Triner, "Analyze Magnetic Loss Characteristics Easily Using High Power Wideband Operational Amplifiers," *Powerconversion International 1981,* pp. 55–63.
13. C. Uhlig, N. Shah, *Magnetic Flux Distribution in Single Phase Core Transformers with Primary Confined to One Leg,* IEEE Power Conference, Paper No. C72 244–7 (February 1972).
14. J. Brauer, *Finite Element Analysis of Electromagnetic Induction in Transformers,* IEEE PES, Paper No. A77 122–5 (February 1977).
15. V. Welsby, *The Theory and Design of Inductance Coils,* 2d ed. (London: Macdonald Company, 1960).
16. V. Legg, "Magnetic Measurements at Low Flux Densities Using the Alternating Current Bridge," *Bell System Technical Journal,* Vol. 15 (January 1936): 39–62.
17. V. Legg, "Survey of Magnetic Materials and Applications in the Telephone System," *Bell System Technical Journal,* Vol. 18 (July 1939): 438–64.
18. C. Heck, *Magnetic Materials and Their Applications* (New York: Crane, Russak & Co., 1974).
19. C. Chen, *Magnetism and Metallurgy of Soft Magnetic Materials,* (Amsterdam: Elsevier North-Holland, 1977).
20. R. Boll, *Soft Magnetic Materials* (London: Heyden & Son, 1979).
21. M. Littmann, "Iron and Silicon-Iron Alloys," *IEEE Transactions on Magnetics,* Vol. MAG-7, No. 1 (March 1971): 48–60.
22. E. Snelling. *Soft Ferrites* (London: Iliffe Books, 1969).
23. D. Nathasingh, C. Smith, "A New High-Flux, Low-Loss Magnetic Material for High Frequency Operation," *Proceedings of Powercon 7, 1980,* B2/1–12.
24. D. Raskin, L. Davis, "Metallic Glasses: A Magnetic Alternative," *IEEE Spectrum* (November, 1981): 28–33.
25. T. Bose, "Theory and Design of Premagnetized Current Transformers," *Transactions AIEE,* Vol. 79, Part 3 (December 1960): 1,029–33.
26. Reuben Lee, D. Stephens, "Influence of Core Gap in Design of Current-Limiting Transformers," *IEEE Transactions on Magnetics,* Vol. MAG-9, No. 3 (September 1973): 408–10.

27. T. Specht, "Transformer Inrush and Rectifier Transient Currents," *IEEE Transactions: Power Apparatus,* Vol. PAS-68, No. 4 (April 1969): 269–76.

28. L. Arguimbau, "Losses in Audio Frequency Coils," *General Radio Experimenter,* Vol. 11, No. 6 (Concord, Mass., November 1936).

29. P. McElroy, "How Good is an Iron-Cored Coil?" *General Radio Experimenter,* Vol. 16, No. 10 (Concord, Mass., March 1942) and "Those Iron-Cored Coils Again," *General Radio Experimenter,* Vol. 21, No. 7 (December 1946).

30. R. Rodriguez, J. Dishman, "Proximity Effect Calculations for Rectangular Planar Conductors," *Proceedings IEEE Intermag Conference* (Grenoble, France, 1981).

31. B. Astle, "Optimum Shapes for Inductors," *IEEE Transactions on Parts,* Vol. PMP-5, No. 1 (March 1969): 3–15.

32. J. Ohta et al., "A New Ferrite Core for Use in High Frequency Switch Mode Converters," *Power Conversion International, 1979,* pp. 57–66.

33. V. Legg, E. Given, "Compressed Powdered Molybdenum Permalloy," *Bell System Technical Journal,* Vol. 19 (July 1940): 385–406.

34. H. Schlicke, *The Essentials of Dielectromagnetic Engineering* (New York: John Wiley & Sons, 1961), pp. 61, 89.

35. J. Watson, *Applications of Magnetism* (New York: John Wiley & Sons, 1980).

36. M. DeMaw, *Ferromagnetic Core Design and Application Handbook* (Englewood Cliffs, N.J.: Prentice-Hall, 1981).

37. G. Attura, *Magnetic Amplifier Engineering* (New York: McGraw-Hill Book Company, 1959), p. 28.

38. E. Peterson, "Harmonic Production in Ferromagnetic Materials at Low Frequencies and Low Flux Densities," *Bell System Technical Journal,* Vol. 7, No. 4 (October 1928): 762–96.

39. F. Connelly, *Transformers* (London: Sir Isaac Pitman & Sons, 1950), p. 386.

40. K. Macfadyen, "The Calculation of Waveform Distortion in Iron-Cored Audio Frequency Transformers," *Proceedings IEE,* Vol. 97, Part 2 (1950): 800–11.

41. ———, *Small Transformers and Inductors* (London: Chapman & Hall, 1953), p. 203.

42. A. Thiele, "Air-Cored Inductors for Audio," *Journal of Audio Engineering Society,* Vol. 24, No. 5 (June 1976): 374–78.

43. T. Gross, "Multilayer Coil Design Is Made Easy with a Programmable Calculator," *Electronic Design* (March 15, 1978): 102–106; (September 13, 1978): 11, 20.

CHAPTER 7

The Magnetic Circuit: Inductors with DC

When direct current is introduced into a transformer or inductor winding, the hysteresis loop (see Fig. 6.2) becomes asymmetrical about the *B-H* axes and the core is said to be polarized or to have dc bias.

In linear circuits, the consequences are undesirable because a dc bias reduces inductance. Smoothing chokes, storage inductors, single-ended transformers, and push-pull transformers with unbalanced direct current are all adversely affected in some way. Size must often be substantially increased.

In nonlinear circuits, however, the effect of polarization is often put to specific use. The nonuniform variation of inductance as the bias is changed is used to vary the series or shunt impedance of the saturable reactor and saturable transformer and to control ac power. Remanence is essential to storage and memory functions of logic and switching transformers in digital circuits. Saturation is exploited in the coil pulser, which can perform as a magnetic harmonic generator, as a magnetic timer, and (in the magnetic pulse modulator) as a generator of narrow, steep pulses.

In this chapter, we are concerned with the quasi-linear, steady-state performance of the biased inductor. It should be understood, when

229

we speak of an inductor or choke, that the discussion is directly applicable to a transformer winding. For example, an analysis of the behavior of a single-ended output transformer carrying direct current involves the same considerations used in analysing a choke.

7.1. MINOR HYSTERESIS LOOPS

One conspicuous feature of the minor hysteresis loop of the biased inductor is its asymmetry with respect to the *B-H* axes. This results in an absence of half-wave symmetry in the exciting current. Harmonic currents, according to Fourier analysis, will therefore contain even as well as odd multiples of the excitation frequency.

Another distinctive feature is the markedly reduced slope of the loop. The size and slope of the loop shown in Fig. 7.1 depends on the magnitude of the excitation voltage as well as on the bias H_{dc}. The large ac signal of Fig. 7.1a results in a large increment in the flux density ΔB and also produces a large increment in the magnetizing force ΔH. Note, however, that the slope is less during the positive half-cycle than during the negative half-cycle of ΔB.

Incremental permeability, μ_Δ, defined as:

$$\mu_\Delta = \frac{\Delta B}{\Delta H} \tag{7.1}$$

will therefore vary during the cycle. The small ac signal and small ΔB of Fig. 7.1b are characteristic of the behavior of the core of the capacitor-input dc smoothing choke and of the storage inductor in switching converter circuits.

If the excitation is a unidirectional pulse train with a low repetition

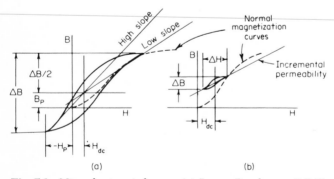

Fig. 7.1 Minor hysteresis loops. (*a*) Large B_{ac} change ($\Delta B/2$); constant polarizing bias H_{dc}. (*b*) Small B_{ac} change; constant polarizing bias H_{dc}.

rate, then the lower tip of the loop coincides with the retentive flux density B_r.* The adverse effect of dc bias, $H_{dc} = 0.4\pi NI_{dc}/l$, on $\mu\Delta$ is quite conspicuous in the typical family of curves for oriented silicon steel laminations (Fig. 7.2). Note that it is the dc ampere (A)-turn *product*, not the current alone, which determines the extent of the bias. One ampere flowing through one turn produces the same bias as 1 miliampere (mA) flowing through 1,000 turns. However, in the latter case, B_m will have one-thousandth the value, which may result in a smaller inductance even though the bias is the same.

7.2. THE SERIES AIR GAP

When a small air gap is introduced in series with the magnetic path, the bias H_{Fe} in the core becomes less than the bias H_{dc} in the winding. The incremental permeability is increased. The effective permeability μ_e, which may increase *or* decrease, is determined from the reluctivity expression:

$$\frac{1}{\mu_e} = \frac{1}{\mu_\Delta} + \frac{l_a}{l} \tag{7.2}$$

where l_a is the length of the air gap in the magnetic path of the length l (see Fig. 6.6).

Although this expression is identical in form to Eq. (6.30), there is an important difference. Effective permeability depends both on $\mu\Delta$ and on the gap ratio $\beta = l_a/l$. If we increase β, then $\mu\Delta$ also increases. But if the gap is made too large, $\mu\Delta$ *decreases*. Functionally:

$$\mu_\Delta = f(\beta, H_{Fe}, B_{ac}) \tag{7.3}$$

and our problem is to find that gap which minimizes the effective reluctivity of Eq. (7.2) and which therefore makes effective permeability a *maximum*.

The core bias H_{Fe} (which determines $\mu\Delta$) is found as a function of the air gap l_a by solving the following simultaneous equations for the series magnetic circuit:

$$F = H_{Fe}l + H_a l_a \tag{7.4}$$

$$B_{dc} = f(H_{dc}) \tag{7.5}$$

The first equation states that the mmf of the coil ($4\pi NI_{dc}/10l$) is the sum of the drop in magnetic potential in the core and in the gap. The second nonlinear equation states the functional dependence

---

* Sections 10.8 and 11.4 contain discussions of permeability during unidirectional pulse excitation.

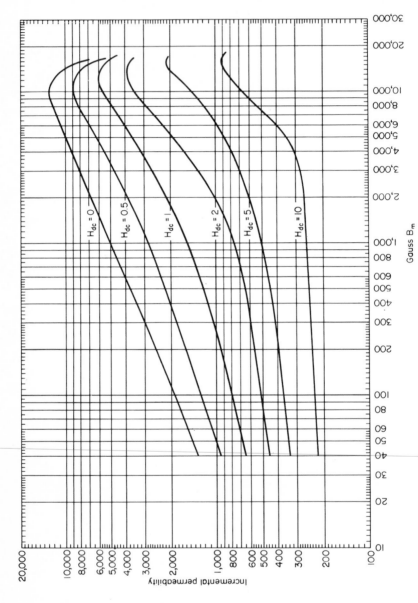

Fig. 7.2 60-Hz incremental permeability μ_Δ curves. 1-D-U laminations, 100 percent interleaved, 0.014-in thick, grade M6 oriented silicon steel. (*By permission, Magnetic Metals Co.*)

232

of dc flux density on the magnetizing force as depicted by t.
magnetization curve.

In the graphical solution, H_{Fe} is found from the intersection of
"negative air-gap line" and the normal magnetization curve, as sho...i
in Fig. 7.3.[1] The intercepts with the B and H axes are $F/l_a = H_a =$
B_a and $F/l = H_{dc}$. The two curves intersect at (H_{Fe}, B_o), B_o being that value of dc flux density which exists both in the core and in
the gap. The negative air-gap line is so called because its slope is
$-1/\beta = -B_a/H_{dc}$.

It is clear that the intersection (and therefore the bias in the core)
depends on the size of the gap:

$$l_a = \frac{H_{dc}l}{B_a} = 0.4\pi \frac{NI_{dc}}{B_a} \qquad (7.6)$$

A small gap increases both B_o and H_{Fe}, and a large gap decreases
them (see Fig. 7.3b).

To avoid saturation, the flux density B_o in the core should be chosen
to satisfy:

$$B_o + \frac{\Delta B}{2} = B_m < B_s \qquad (7.7)$$

Determination of maximum μ_e now depends on finding values for
pairs of β and $\mu\Delta$ (the latter from H_{Fe} and B_{ac} on a curve such as
Fig. 7.2) which will reduce Eq. (7.2) to a minimum. Inductance can
then be computed from:

$$L = \frac{0.4\pi N^2 A_{Fe}K_{Fe}\mu_e 10^{-8}}{l} \qquad (7.8)$$

where $A_{Fe}l$ is V_{Fe}, the volume of the core.

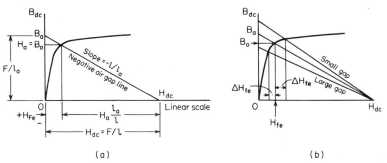

(a)   (b)

Fig. 7.3 Graphical solution: nonlinear core in series with an air gap. (a)
$B_o/H_o = -l/l_a = -1/\beta$. (b) Note that H_{Fe} varies less if the gap is large.

The foregoing procedure for finding the optimum gap for maximum inductance, while instructive, can be time consuming.* Hanna has shown that it is possible to relate LI_{dc}^2/V_{Fe} to NI_{dc}/l within a range of gap ratios.[2] When an envelope is then drawn tangent to a family of β curves, a smooth curve is obtained. The ordinates of such a curve permit us to determine the maximum inductance obtainable with an optimum gap. The accurate use of Hanna's method requires that we have curves available for several levels of ac induction.[3]

7.3. THE QUASI-LINEAR INDUCTOR

It is not always wise to adjust an iron-cored choke for maximum inductance by using an optimum gap. There are occasions when a smaller or larger gap is preferable.

In rectifier and converter circuits, which have large fluctuations in direct current (I_{max} to I_{min}), it is important to exceed the critical inductance I_c (see Secs. 2.8 and 2.10). A gap which satisfies this requirement at I_{max} may result in too little inductance at I_{min}. There are several solutions. One is to make the choke large enough to allow for the necessary L_{min} and L_{max} over the range of current; another is to design a "swinging" choke, whose gap will be smaller than the so-called optimum.[4] A third solution is to use a core with a divided gap, that is, a small gap (e.g., 30 percent of the stack) for the condition of I_{min} and a large gap (70 percent) for the condition of I_{max} (see Fig. 6.9b). Any of these techniques results in a somewhat smaller component.[5,6]

Normally, in the construction of chokes, there is an almost unavoidable variation in the size of the air gap. There is usually a variation in the incremental permeability of different production lots of laminations, also. The conservative designer anticipates these variations by specifying a gap ratio greater than optimum. The slight increase in size is a small price to pay for avoiding painstaking, and therefore costly, mechanical adjustments during production.

The designer may deliberately operate a biased inductor below the knee of the normal magnetization curve (note the lower air-gap line in Fig. 7.3b). Variations in the gap will then produce smaller changes in H_{Fe} than they would if the load line were steeper—that is, higher on the knee. The gap ratio β is:

$$\beta = \frac{l_a}{l} = \frac{H_{dc}}{B_0} - \frac{H_{Fe}}{B_0} = \frac{H_{dc}}{B_0} - \frac{1}{\mu_{dc}} \qquad (7.9)$$

---

* See Sec. 11.4c for a discussion of the optimum air gap when pulses are unidirectional.

Let us go back to the equations in Hanna's classic paper:

$$\frac{LI^2}{V_{Fe}} = \frac{B_o{}^2\,[(1/\mu_{dc}) + (l_a/l)]^2\,10^{-8}}{0.4\pi\,[(1/\mu_\Delta) + (l_a/l)]} \qquad (7.10)$$

$$\frac{NI}{l} = \frac{B_o}{0.4\pi}\left(\frac{1}{\mu_{dc}} + \frac{l_a}{l}\right) \qquad (7.11)$$

Note that the gap ratio $l_a/l = \beta$ appears in both numerator and denominator of Eq. (7.10). We can, therefore, infer two features of a solution to the equation: Maximum energy density LI^2/V_{Fe} occurs at small values of β and, when β is adjusted to produce a maximum value for inductance, effective permeability μ_e will be a function of three variables. That is:

$$\mu_e = f(\mu_{dc}, \mu_\Delta, B_{ac}) \qquad (7.12)$$

where μ_{dc} and μ_Δ are each a function of the dc bias:

$$H_{dc} = 0.4\pi\,\frac{NI}{l} \qquad (7.13)$$

The above equations, in which several of the variables are nonlinear and interdependent, is more easily solved in the laboratory than on paper. In the laboratory, we can do as Hanna did. We treat B_o as fixed and vary the air gap, and hence β, to arrive at the maximum value of LI^2. Tests are performed over a range of values of β and NI/l. A plot of the optimum results for the specific core material is obtained.

In the design approach adopted in this chapter, the author has several assumptions about the magnitude of β in relation to μ_{dc} and B_{ac}. For the low-frequency inductor, we make the following arbitrary assumptions:*

$$\beta \gg 1/\mu_{dc} \qquad B_{ac} \ll B_o \qquad (7.14)$$

We can now proceed with an analytical solution to the Hanna equations.

If we square Eq. (7.11) and substitute $B_o{}^2$ into Eq. (7.10), we obtain a cumbersome expression. But the assumptions of Eq. (7.14) permit a simpler expression for energy density to emerge:

$$\frac{LI^2}{V_{Fe}} = \frac{10^{-8}}{0.4\pi}\left(\frac{\mu_\Delta}{1 + \beta\mu_\Delta}\right) H_{dc}{}^2 \qquad (7.15)$$

---

* In Sec. 7.6, where we discuss the high-frequency inductor, these assumptions become reasonable rather than arbitrary.

The magnetizing force H_{dc} corresponds to the dc bias in the coil, and the effective permeability (stated in parentheses in the equation) is:

$$\mu_e = \frac{\mu_\Delta}{1 + \beta\mu_\Delta} \tag{7.16}$$

Approximations (which will be explained later) can be made of μ_e in relation to β, in both the low-frequency and high-frequency inductor:

$$\mu_e \cong 0.85/\beta \quad \text{(low frequency)} \tag{7.17}$$

$$\mu_e \cong 1/\beta \quad \text{(high frequency)} \tag{7.18}$$

In the latter case, the energy density in Eq. (7.15) can be expressed by:

$$\frac{LI^2}{V_{Fe}} = \frac{10^{-8}}{0.4\pi} \mu_e H_{dc}^2 \tag{7.19}$$

Or, alternatively, when $H_{dc} = \beta B_o$:

$$\frac{LI^2}{V_{Fe}} = \frac{10^{-8}}{0.4\pi} \beta B_o^2 \tag{7.20}$$

When Eq. (7.20) is recast in terms of the volume of the air gap V_{air}, which equals $A_{Fe} l_a$, then:

$$\frac{LI^2}{V_{air}} = \frac{10^{-8}}{0.4\pi} B_o^2 \tag{7.21}$$

We can infer, from this simple expression, that the inductive energy we have been discussing is stored *within the air gap* of the magnetic circuit. The elimination of the nonlinear parameters permits us to obtain a simple, direct solution to the design of a high-frequency inductor with dc bias.

The design of the low-frequency inductor with dc bias is simplified only when the assumptions which we make about β, μ_{dc}, and B_{ac} are reasonable. Certain of the assumptions we make about an equivalent inductance with an air gap imply, in effect, the concept of a time-invariant inductance. When β is not large (so that no longer is $\mu_e \ll \mu_\Delta$), *inductance varies with time*. In that case our analysis is subject to serious criticism on theoretical grounds. An examination of this subject is, however, outside the scope of this book.[7]

7.4. LOW-FREQUENCY INDUCTORS

In this section, and the next (Sec. 7.5), we are concerned with biased inductors used in low-frequency [50 to 800 hertz (Hz)] power circuits.

We begin by investigating the conditions which favor quasi-linear performance of the biased inductor. When the coil bias H_{dc} is much greater than the core bias H_{Fe}, as it most frequently is, then $\beta \gg 1/\mu_{dc}$, and B_a is not much larger than B_o. The gap is then:

$$l_a = \frac{F_a}{B_a} = \frac{0.4\pi N I_{dc}}{B_a} \qquad (7.22)$$

In grain-oriented (M6) silicon steel, typical values are:

H_{Fe}, Oe	B_o, kG	μ_{dc}
0.25	10	40,000
0.5	14	28,000
1.0	15.5	15,500

From Fig. 7.2 we see that when the bias H_{dc} is 1 Oe, μ_Δ may vary between 600 and 6,000. If our smallest gap is a butt joint without a spacer, then minimum β will range from 10^{-3} to 0.2×10^{-3}, depending on the size of the lamination (see Table 6.4). If the bias is arbitrarily assumed to be 1 Oe (2.02 = A turns/in) and the ac flux density is 100 gauss (G) (a typical value), μ_Δ will be 800. When β is 1/800, μ_e will be 400.

The extensive tables of linear-choke data compiled by I. Richardson show that μ_e ranges from 350 in E-I 625 laminations down to 200 in E-I 150 laminations when grain-oriented (M6) steel is used and that B_{ac} ranges between 7,850 and 1,360 G.[4]

Values of μ_e less than 500 may be regarded as evidence that gap reluctance is dominant. Some measure of linearity is assured if we see to it that $\beta > 1/\mu_e$; moreover, behavior becomes linear when $\beta \gg 1/\mu_\Delta$. In that situation, effective permeability is simply:

$$\mu_e \cong \frac{1}{\beta} \qquad (7.18)$$

and is reasonably independent of the core material.

Inductance L, when the air gap is large, is now found with the *linear* formula:

$$L = \frac{0.4\pi N^2 A_{Fe} K_{Fe} 10^{-8}}{l_a} \qquad (7.23)$$

This formula will hold, approximately, when $l_a/l > 1/400$, or 0.25 percent. If, for example, $l = 4$ in [102 millimeters (mm)], the gap should be larger than 10 milli-inches (mil) (0.25 mm) if linearity is desired.

Rather large gaps are sometimes necessary. In the dc charging inductor, linearity is assured by keeping the flux density low (e.g., 8,500

G) and by employing a large gap ratio (e.g., 0.6 percent). When the gap is large, however, the possibility of fringing and its adverse effects must be weighed [see discussion following Eq. (6.33)].

It should be noted that the costly nickel-iron alloys rapidly lose their permeability, even when the bias is small.*   When the gap is large, therefore, they offer no advantage over silicon steel.   For that matter, whenever minimum size is not mandatory, the thicker gauge silicon grades, such as 0.025 in (0.64 mm) M19, M45, or even lamination grade 2S, are economic and satisfactory substitutes for 14-mil M15 nonoriented or M6 grain-oriented material.

7.5. SIZE AND TEMPERATURE RISE

An important conclusion may be inferred from an empirical form of Hanna's relationship:

$$\frac{LI^2}{V_{Fe}} = \text{const} \left(\frac{NI}{l}\right)^{3/2} \qquad (7.24)$$

The LI^2 product increases directly with the core's volume if the core path l is kept constant.   If we double the stack, we can either double the inductance or increase the current by 41 percent.   We may assume that the LI^2 product is proportional to the product A_p [defined in Chap. 2, Eq. (2.6)].

In practice, dc resistance is restricted to some maximum value, along with a specified inductance and direct current.   This is clearly necessary in smoothing chokes for dc power supplies, which must not introduce an excessive drop in voltage if they are to meet a required level of regulation.

The addition of this third requirement, resistance, complicates the design process and introduces the problem of temperature rise as well.†   Hanna curves, widely used for low-level chokes, now serve only as an initial step in a cumbersome series of trial solutions.

Ideally, one should be able, given a set of specifications including the six factors—L, R, I_{dc}, E_{ac}, f, and θ (temperature rise)—to proceed directly to the solution (i.e., synthesis) for the four major design factors:

---

* Lower design values of B_0 and H_{Fe} are used: typically, 9,000 G and 0.1 Oe for 50 percent nickel alloy; 4,000 G and 0.05 Oe for 80 percent nickel alloy.

† If ac excitation is small, the power loss in the core will be negligible.   Since the heat source is now in the coil alone, the maximum operating temperature will be midway in the coil and the total temperature rise of a low-level choke will be less than in a transformer dissipating the same losses.

size of the core (volume or area product A_p), turns N, wire size, and air gap.

The problem is of great practical importance, and a variety of solutions has been recorded in the literature.[8-11]   An exact solution to the equations of the low-frequency inductor is made easier by appropriate computer programs. We will describe briefly three examples of computer-*a*ided-*d*esign (CAD).

1. In the first example, solving for μ_e begins by first performing empirical tests on the core material in an Epstein frame, or better still, in the shape or configuration it will assume in the physical inductor.[12]   We then plot two curves: μ_Δ versus H_{dc} (similar to Fig. 7.2) and the normal magnetization curve, B_{dc} versus H_{dc} (Fig. 7.3). From the latter curve, we can compute and plot the slope, $\mu_{dc} = B_{dc}/H_{dc}$. H_{Fe}, the magnetizing force in the iron portion of the circuit, is related to the gap ratio by the expression:

$$H_{Fe} = \frac{H_{dc}}{1 + \beta \mu_{dc}} \tag{7.25}$$

Hence it is not necessary (although desirable) to make the many laborious tests with physical gaps l_a in the laboratory, as Hanna did. The designer can use a curve-fitting routine and a power series to represent $\mu_\Delta = F(H_{dc})$ and $\mu_{dc} = F(H_{dc})$. Next, a program is written which performs the operations described in Eq. (7.25) for various values of β. One then calculates H_{Fe}, determines μ_Δ, and finally solves for μ_e [Eq. (7.16)].   A computer printout provides a comprehensive summary of the capabilities of a given core to satisfy specifications for L, I, and R.

2. In the second example, the analysis is amplified to include large ac excursions of flux.[13]   Three algorithms are developed: (1) a program for computing turns and air gap; (2) an optimum program for obtaining the *minimum number of turns;* and (3) a nonoptimum but simpler program (suitable for a pocket calculator), in which a specified, but arbitrary, quiescent point on the dc magnetization curve is assumed.

3. In the third example, a solution is obtained for the more general case which includes prescribed values of both resistance *and ac flux density.*[14]   A sequence of algorithms allows us to: (1) select a core and a magnetic bias point, (2) find the optimum time constant L/R, and (3) obtain the optimum (minimum) number of turns corresponding to the maximum L/R.

I have based the procedure to be described in this chapter on a pragmatic point of view. I have avoided some of the difficulties of a strict analysis by making assumptions that enable the designer to translate laboratory specimens into production units. In effect, I am trading an optimum magnetic solution for an optimum solution on the production line.

I begin by assuming an arbitrary quiescent bias point, and then introduce dc resistance and temperature rise as explicit requirements of the design. The outcome of this approach is not optimum in the sense of providing a core of minimum size and turns. However, the production department finds the design optimum: minimum size has been traded for reproducibility and productivity by a designer who has deliberately chosen an air-gap ratio that is larger than the optimum.

To the user of chokes and power inductors, a graphical solution such as Fig. 7.4, which relates LI^2, size, temperature rise, and time constant L/R to each other, can be of considerable value. Such a chart makes it clear that size must be substantially increased when we insist on a small resistance. The code number in the figure designates (to the transformer designer) a specific core size which fits comfortably into a standard can listed in the MIL-T-27 specification.

I begin with the following assumptions:

1. Gap reluctance > 10 X core reluctance, and therefore $H_{Fe} \ll H_{dc}$
2. $\beta \cong H_{dc}/B_0$ [from Eq. (7.9)]
3. A bias point (H_{Fe}, B_0), appropriate for the core material, is selected
4. $B_{ac} < B_0$

It is axiomatic in transformer design that the most efficient use of core and coil entails allowing the maximum safe temperature rise to occur. In keeping with this principle, it is of value to construct a table in which each size of coil is assigned a nominal copper power loss W_{Cu} which is proportional to the surface area. This approach is consistent with the concepts discussed in Chap. 4 , and it will be shown later (see Sec. 7.6c) that LI^2 or energy density LI^2/V_{Fe} can be expressed directly in terms of W_{Cu}:

$$\frac{LI^2}{V_{Fe}} = 10^{-8} \frac{B_0}{l} \left(\frac{W_{Cu}}{A_R} \right)^{1/2} \tag{7.26}$$

Here A_R, the normalized resistance, [Eq. (6.61)], is defined as the resistance of a single-turn copper bar which fills the window. That is:

$$A_R = \frac{R}{N^2} = \frac{\rho\lambda}{K_{Cu}A_{Cu}} \tag{7.27}$$

This concept serves a similar role in the power transformer [see Eq. (5.4)].

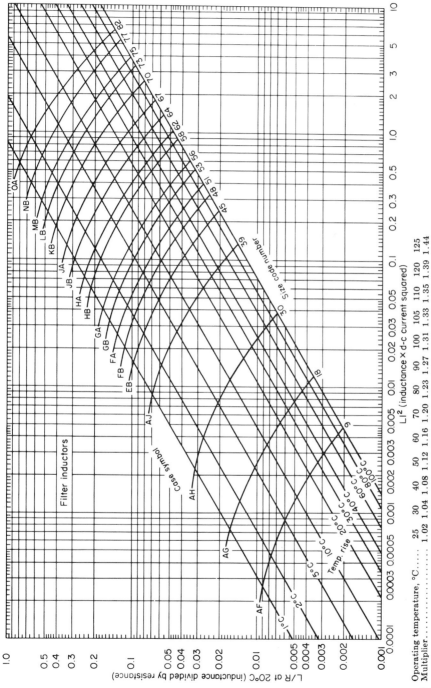

Operating temperature, °C...... 25 30 40 50 60 70 80 90 100 105 110 120 125
Multiplier................... 1.02 1.04 1.08 1.12 1.16 1.20 1.23 1.27 1.31 1.33 1.35 1.39 1.44

Fig. 7.4 LI^2 vs. L/R and temperature rise for each size of enclosure. Note: To determine resistance at the operating temperature, multiply resistance at 20°C by multiplier in the table. (*Courtesy of Raytheon Company.*)

In the scrapless E-I lamination with a square stack:

$$A_R = \frac{13.82(10)^{-6}}{K_{Cu}t} \qquad (7.28)$$

if we assume a mean-turn λ equal to $6t$. The copper space factor K_{Cu} will vary with the size of the core, the voltage gradients, and the winding practices deemed practical in a particular coil-manufacturing facility. The E-I 100 lamination, whose center limb t is 2.54 cm (1 in) will have a resistance factor of 14.7 microohms ($\mu\Omega$) when a layer type of coil winding results in a K_{Cu} of 0.37.

Coil magnetizing force can now be expressed in terms of dc copper power loss W_{Cu} and normalized resistance:

$$H_{dc} = \frac{0.4\pi\, NI}{l} = \frac{0.4\pi}{l}\sqrt{\frac{W_{Cu}}{A_R}} \qquad (7.29)$$

The power loss in the copper of the coil is related to the temperature rise of the coil by a formula similar to Eq. (4.49):

$$\theta = \frac{r W_{Cu}}{A_s} \qquad (7.30)$$

where r is thermal resistivity, defined in Eq. (4.31) of Chap. 4.*

We saw, in Fig. 7.4, that the time constant L/R of the linear inductor can be plotted as a function of energy LI^2 with volume (envelope size) as a parameter. A more succinct representation is possible if we combine L/R and LI^2 into one expression, thus:

$$[(LI^2)\, L/R]^{1/2} = \frac{LI}{\sqrt{R}}$$

We can also develop an expression for the ensemble L, I, and R, in the format $LI/\sqrt{R}$ as a function of variables of the coil, the core, and the geometry:†

$$\frac{LI}{\sqrt{R}} = 10^{-8}\frac{B_0}{l}\frac{K_{Fe}\,V_{Fe}}{\sqrt{A_R}} \qquad (7.31)$$

$$\frac{LI}{\sqrt{R}} = k_5 g_5 B_0 \qquad (7.32)$$

where the coefficients k_5 and g_5 are:

---

* An expression for LI^2 as a function of temperature rise is developed later (see Sec. 7.6b).

† These equations correspond to Eq. (7.72), derived in Sec. 7.6c.

$$k_5 = 10^{-8} K_{Fe}\left(\frac{K_{Cu}}{\rho}\right)^{1/2} \tag{7.33}$$

$$g_5 = A_{Fe}\left(\frac{A_{Cu}}{\lambda}\right)^{1/2} \tag{7.34}$$

The relation between core volume V_{Fe} and $LI/\sqrt{R}$ is made clearer when we substitute Eq. (7.27) into Eq. (7.31) and rearrange their terms. We obtain:

$$V_{Fe}{}^{5/6} = (\text{const}) \frac{LI_{dc}}{\sqrt{R}} \tag{7.35}$$

A more useful form of this equation is obtained if, for V_{Fe}, we substitute V_o, an "ideal" volume, defined as the volume of a rectangular envelope which snugly circumscribes the inductor.[15] The ideal volume of the important scrapless E-I lamination, whose proportions are fixed, is:

$$V_o = 7t^3(1+s) \tag{7.36}$$

where t is the tongue width and s is the stack-tongue ratio. The ratio between the size of the envelope and the volume of iron of the scrapless lamination is:

$$\frac{V_o}{V_{Fe}} = \frac{7}{6}\left(1 + \frac{1}{s}\right) \tag{7.37}$$

a. The Bias Operating Point

Let us suppose that the negative air gap line in Fig. 7.3 intersects the B_{dc} ordinate at 1.55 Tesla (T) (100 kilolines/sq in), a suitable choice for oriented silicon steel. An earlier assumption [Eq. (7.14)], that $\beta \gg 1/\mu_{dc}$, permits us to simplify Eq. (7.9) and obtain:

$$\beta = \frac{H_{dc}}{B_o} \tag{7.38}$$

In Table 7.1, line 15, the gap ratio for each of the scrapless E-I laminations listed is calculated from this formula. The gap is then simply β_l. An expression for β in terms of power loss in the copper and of geometry is obtained if we substitute the value of H_{dc} given in Eq. (7.29) into Eq. (7.38). We obtain:*

$$\beta = \frac{0.4\pi}{B_o l}\left(\frac{W_{Cu}}{A_R}\right)^{1/2} = \frac{k_7 g_7 W_{Cu}{}^{1/2}}{B_o} \tag{7.39}$$

---

* A similar form for the high-frequency inductor is stated later as Eq. (7.75).

TABLE 7.1   Polarized Inductor Ratings, E-I Laminations*

Line (1)	Parameter (2)	Unit (3)	625 (4)	75 (5)	87 (6)	100 (7)	112 (8)	125 (9)	138 (10)	150 (11)	175 (12)
1	t	in	0.625	0.75	0.875	1.00	1.125	1.25	1.375	1.50	1.75
		cm	1.59	1.91	2.22	2.54	2.86	3.18	3.49	3.81	4.45
2	l	in	3.75	4.50	5.25	6.00	6.75	7.50	8.25	9.00	10.50
		cm	9.53	11.43	13.34	15.24	17.15	19.05	20.96	22.9	26.7
3	λ	in	3.75	4.50	5.25	6.00	6.75	7.50	8.25	9.00	10.5
		cm	9.53	11.43	13.34	15.24	17.15	19.05	20.96	22.9	26.7
4	A_{Fe}	sq in	0.351	0.563	0.766	1.00	1.27	1.56	1.89	2.25	3.06
		sq cm	2.52	3.63	4.94	6.452	8.17	10.08	12.2	14.52	19.8
5	A_{Cu}	sq in	0.293	0.422	0.574	0.75	0.95	1.17	1.42	1.69	2.30
		sq cm	1.89	2.72	3.70	4.84	6.12	7.56	9.15	10.89	14.8
6	A_s	sq in	7.31	10.5	14.3	18.71	23.7	29.2	35.4	42.1	57.3
		sq cm	47.2	67.9	92.4	120.7	153	189	228	272	370
7	V_{Fe}	cu cm	24.0	41.5	65.9	98.3	140	192	256	332	527
8	g_1	cm$^{0.5}$	0.321	0.352	0.380	0.406	0.430	0.455	0.477	0.498	0.538
9	g_5	cm$^{2.5}$	1.12	1.77	2.61	3.64	4.83	6.38	8.07	10.0	14.7
10	g_6	cm$^{3.5}$	7.71	14.6	25.0	39.93	59.4	87.7	122	165	283
11	I^2R	W	2.77	3.99	5.44	7.10	8.99	11.09	13.42	15.98	21.74
12	A_R	$\mu\Omega$	23.54	19.61	16.81	14.71	13.08	11.77	10.70	9.81	8.41
13	A_L	μH	0.85	1.02	1.19	1.36	1.53	1.70	1.87	2.04	2.38
14	H_{dc}	Oe	45.3	49.6	53.6	57.3	60.8	64.1	67.2	70.2	75.8
15	β	10^{-3}	2.93	3.20	3.46	3.70	3.91	4.14	4.34	4.53	4.89
16	μ_e	—	342	312	289	270	256	242	230	220	204
16A	$\mu_e \times 0.85$	—	291	265	246	230	218	206	196	187	173
17	LI^2	J	0.128	0.241	0.414	0.661	1.0	1.44	2.02	2.73	4.69
18	LI^2/V_{Fe}	mJ/cu cm	5.33	5.81	6.29	6.72	7.14	7.50	7.89	8.22	8.90
19	$LI/\sqrt{R}$	×1.0	0.077	0.121	0.178	0.248	0.333	0.433	0.550	0.683	1.00
19A	$LI/\sqrt{R}$	×0.85	0.065	0.103	0.151	0.211	0.283	0.368	0.467	0.581	0.85

* $B_o = 1.55$ T, $\theta = 50°$C, $\rho = 1.724$ $\mu\Omega \cdot$ cm (at 20°C), $r = 850$. $I^2R/A_s = 58.8$ mW/sq cm (0.379 W/sq in), $K_{Cu} = 0.37$, $K_{Fe} = 0.95$. A_s is based on Eq. (4.42) and β is based on Eq. (7.65).

where $k_7 = 0.4\pi \, (K_{Cu}/\rho)^{1/2}$ (7.40)

and $g_7 = (A_{Cu}/l^2\lambda)^{1/2} = V_{Cu}^{1/2}/l\lambda$ (7.41)

We have, in effect, said that a practical bias operating point is at $H_{Fe} = 1$ and $B_o = 1.55$ T. The location of the bias point becomes more accurate when we take into account that the flux density B_o' in the core is somewhat lower than B_o, and is more precisely stated as:

$$B_o' = B_o \left[1 - \left(\frac{H_{Fe}}{H_{dc}} \right) \right] \left.\begin{array}{c} \\ \\ \\ \end{array}\right\}$$
$$B_o' = H_{dc} \left[\frac{1 - (H_{Fe}/H_{dc})}{\beta} \right]$$

(7.42)

Hence the assumption that $H_{dc} \geqq 10 \, H_{Fe}$ means that our choice of 1.55 T actually results in a bias operating point in the core about 10 percent lower, i.e., at 1.5 T.

The choice of an operating point (and air gap) should also satisfy any constraint set on the variation ΔL, that is, the linearity of L within a range of load current $\Delta I_{dc} = I_{max} - I_{min}$. A common requirement is that ΔL not vary more than 20 percent from no load to full load. But to satisfy this constraint, the gap ratio must exceed the minimum value β_{min}. This value can be calculated as follows:

$$0.80 = \frac{\mu_e}{\mu_\Delta} = \frac{1}{1 + \mu_\Delta \beta_{min}} \tag{7.43}$$

$$\beta_{min} = \frac{1/0.8 - 1}{\mu_\Delta} \tag{7.44}$$

These equations have important practical implications. Consider the situation when $\mu_\Delta = 1,000$, and therefore β_{min} must exceed $0.2 (10)^{-3}$. This requirement is easily satisfied in laminated transformers since a butt stack assembly results in a gap ratio of about 10^{-3} (see Table 6.4).

When ac flux density B_{ac} is appreciable, the operating point B_o' on the dc magnetization curve must be reduced in order to avoid saturation.[16] The new operating point is reduced by the factor:

$$\frac{B_o'}{B_o} = 1 - \frac{B_{ac}}{B_o} \tag{7.45}$$

and, therefore, the gap ratio β must be increased by the same factor. Note that the required inductor rating $LI\sqrt{R}$ must be increased by the reciprocal of this factor, B_o/B_o', when ac flux density is sufficiently large and is to be taken into account.

We now are able to express either LI^2, or $LI/\sqrt{R}$, as a function of the volume of the core, the operating dc flux density, the power loss in the core, the normalized resistance, and the geometry of the core. This has been done in Table 7.1, a list of inductor ratings of the scrapless E-I lamination, based on a uniform thermal flux density. The values chosen, $W_{Cu}/A_s = 58.8$ mW/sq cm (or 0.379 W/sq in) and $r = 850$ are suitable for open type construction.

Also entered in Table 7.1 (line 16A) are average values of μ_e for each lamination, which are 15 percent lower than the approximate value given by Eq. (7.18). This has been done to correct for values (8 to 13 percent lower than line 16) based on a more accurate computation of the operating bias point on the magnetization curve.[4] Correcting for this discrepancy leads to the simple, pragmatic formula in the linear biased inductor:

$$\mu_e \cong \frac{0.85}{\beta} \tag{7.17}$$

Since the gap ratio steadily but slowly increases with size, the behavior of the component becomes more linear as its size is increased.

We have indicated two values of $LI/\sqrt{R}$ in the table. One, (on line 19), is based on the simple assumption that $\mu_e = 1/\beta$; the other, lower, value (on line 19A) is based on an empirical, conservative reduction of 15 percent as indicated by Eq. (7.17). We selected the scrapless E-I lamination because of its widespread use and because each column of entries can be scaled by a simple ratio of dimensions.*

b. Example of Choke Design

The usefulness of the foregoing ideas is best demonstrated with an example. Suppose the specifications for a choke read: "3 H, 0.2A dc, 15 Ω, 290 volts ac, 120 Hz, 50°C rise max. Small size and low cost are desired. Only a tentative design is required in order to establish size and approximate cost."

We proceed as follows: To ascertain copper power loss, we first compute the ac flow due to inductance:

$$I_{ac} = \frac{290}{2\pi \times 120 \times 3} = 128 \text{ mA}$$

Current in the coil (excluding any loss in the core) is:

$$\sqrt{0.2^2 + 0.128^2} = 0.238$$

and $W_{Cu} = 0.238^2 \times 15 = 0.85$ watts (W). Consulting Table 7.1, we see that any core is thermally suitable.

We now compute $LI/\sqrt{R}$. $LI/\sqrt{R} = 3 \times 0.2/\sqrt{15} = 0.155$, a lower rating than that of a square stack of the E-I 87 lamination. Because of the possibility of large ac induction, we will initially choose the larger E-I 100. We use a worksheet such as Table 7.2, which refers to the equations used in our calculations.

Our initial step is to calculate the air-gap ratio from Eq. (7.38), $\beta = H_{dc}/B_o$. But in this example, H_{dc} is predicated on a resistance limitation, so we do not use the value of H_{dc} in Table 7.1, which is based on a dissipation limitation. Instead, we proceed to calculate β by finding the turns N, then H_{dc}, and finally β, as follows:

$$N = \sqrt{15/A_R} = 1{,}010 \qquad \text{(line 16)}$$

$$H_{dc} = 0.4\pi(1{,}010)0.2/15.24 = 16.66 \qquad \text{(line 15)}$$

$$\beta = 16.66/15.5(10)^3 = 1.07(10)^{-3} \qquad \text{(line 11)}$$

---

* The geometric coefficients of the scrapless lamination are summarized in Eq. (5.10), Sec. 5.4c.

TABLE 7.2 Worksheet for a Low-Frequency Choke Design

Line	Relevant equation	Parameter — Name, units	Symbol	Given value	Trial 1 E-I 100 29M6	Trial 2 E-I 125 29M6	Trial 3 125 × 1.08 29M6
1			L	3.0			
2		avg dc	I	0.2			
3		ripple	I_{ac}	0.128			
4		rms current	I_{rms}	0.238			
5		resistance, Ω	R	15	15	14.9	14.2
6	(7.53)	I^2R	W_{Cu}		0.85	0.85	0.81
7	(7.30)	temp rise	θ				
8		flux density	B_m, T	1.55	(2.41)	1.47	1.55
9	(7.68)	energy	LI^2	rating	0.695	1.52	
10	(7.32)		$LI/\sqrt{R}$	rating	0.248	0.433	
10A			$LI/\sqrt{R}$	0.155	0.155	0.348	
11	(7.38)	10^{-3}	β		1.07	1.55	1.55
12	(7.9), (7.22)	air gap, cm	l_a			0.030	0.030
13	(7.18)	eff. perm.	μ_e		930	645	
13A	(7.17)		$\mu_e \times 0.85$		791	548	473
14	(6.60)	μH/turn	A_L	rating	(1.36)	(1.70)	
14A			A_L	calc.	4.68	(3.21)	
15	(7.38)	Oe	H_{dc}	calc.	16.7	14.88	
16	(7.27)	turns	N		1,010	1,124	1,124
17			NI		202	224	
18	(7.8)	induct.	L		4.80	5.14	
18A			$L \times 0.85$		4.08	4.37	3.79
19		wire diam, cm	d				0.050
20		wire gauge	AWG				25
21	(7.27), (7.28)	$\mu\Omega$/turn	A_R		14.7	11.77	11.27
22	(7.27)	resistance, Ω	R	calc.	15.0	14.87	14.24
23	(7.30)	°C	θ	calc.	5.98	5.98	5.72
24	(7.53)	I^2R	W_{Cu}	calc.	0.85	0.85	0.81
25	(7.45)	Tesla	B_o, T		1.55	0.96	0.96
26	(1.8)		B_{ac}, T		0.86	0.51	0.59
27	(7.7)	max B	B_m, T		2.41	1.47	1.55

Our calculation of the gap ratio β (line 11) shows that $\beta = 1.07$. This initiates a sequence of steps culminating in values of A_L, L, and B_{ac}. Note that the values H_{dc} (line 5) and A_L (line 14A) differ from those in Table 7.1. This is because our design is predicated on a resistance limitation, whereas H_{dc} and A_L in Table 7.1 are based on a dissipation limitation. We see that we have a satisfactory value, on paper, of $L = 4.08$ H (line 18A). But we also note that the value of B_{ac}, 0.86 T, is excessive. Actually, therefore, a design predicated on a small value of β would produce core saturation and L *would not* meet the specification, 3 H.

A second trial can be initiated using a smaller value of B_o':

$$B_o' = B_o - B_{ac} = 1.55 - 0.86 = 0.69 \text{ T}$$

But a fresh calculation, using Eq. (7.45), shows the need for a larger rating of $LI/\sqrt{R}$:

$$\frac{B_o}{B_o'} \frac{LI}{\sqrt{R}} = \frac{1.55}{0.69} (0.155) = 0.348$$

As an exercise we could proceed to use a larger air gap, but this would result in too small an inductance. We could begin a second trial with the next larger core, E-I 112 ($LI/\sqrt{R} = 0.283$), but instead we choose the still larger E-I 125 with its ample $LI/\sqrt{R}$ rating of 0.433. The larger core area permits a reduction in B_{ac} to 0.59 T (line 26), and therefore a change of B_o' to:

$$B_o' = 1.55 - 0.59 = 0.96 \text{ T} \qquad \text{(line 25)}$$

The new air-gap ratio can be calculated from either Eq. (7.38) or (7.39). Alternatively, we can arrive at the new, larger gap ratio by scaling. We note that, in Eq. (7.39), β is a function of the geometric coefficient g_7, which changes in the ratio $(1/1.25)^{1/2}$ when the lamination changes from E-I 100 to E-I 125. Accordingly, we compute:

$$\frac{B_o}{B_o'} \left(\frac{t}{t'}\right)^{1/2} = \frac{1.55}{0.96} \left(\frac{1}{1.25}\right)^{1/2} = 1.44$$

The new air-gap ratio is now estimated as:

$$B' = 1.44 \times 1.07(10)^{-3} = 1.55(10)^{-3}$$

and is entered as item 11 in the column of Table 7.2 marked Trial No. 2.

A new train of calculations (items 11–18A) concludes with the value of 4.38 H, much higher than the desired 3.0 H. But the calculated ac flux density is 0.51 T, lower than the estimate of 0.59 T.

We will therefore, in Trial No. 3, reduce both the stack and the inductance by the ratio 0.51/0.59. Thus:

$$\text{adjusted stack} = (0.51/0.59)\,1.25 = 1.08 \text{ in}$$

$$\text{adjusted } L = (0.51/0.59)\,4.38 = 3.79 \text{ H}$$

We are now confident of having avoided saturation. (An alternative design is possible, using the smaller E-I 112 lamination with a large stack). Three or more trials (iterations) were necessary because ac flux density could not be ignored in this example. Nevertheless, further refinement of the design is clearly indicated.

The next phase of the design process can be to sketch out a manufacturing specification which contains the practical details of a workable design. But the final phase is optimization, in which the designer reexamines the preliminary design from the point of view of minimum size or minimum cost. In this phase of design, we examine alternatives to our initial choice of core material and core shape.

We have chosen this example in order to show that the synthesis of a design is not quite so direct when saturation and excessive heating are possible. We also hope to convey the idea that the simplicity of the single-winding inductor is deceptive. Note also that a complete set of specifications was used. However, we neglected considerations such as adequate insulation and mechanical design, not because they are unimportant but because we wished to focus on the electrical variables. A thoroughgoing design entails consideration of *all* factors— mechanical and dielectric as well as electrical, thermal, and financial.

7.6. HIGH-FREQUENCY INDUCTORS

The smoothing choke in the low-frequency dc power supply will typically be specified in henries (H) and has a sizeable ac flux density. It is large and may run hot. In the high-frequency power supply, the series storage inductor is typically specified in millihenries (mH) or microhenries (μH). Ac flux density is small and core loss is negligible. Losses of power are predominantly in the copper. Let us see why.

a. Switch-Mode Inductors

In the high-frequency dc power supply, a high degree of efficiency is achieved in switch-mode conversion circuits. As we noted in Sec. 2.10. the inductor can be placed in one of three branches of the circuit, depending on the topology. Hence, as shown in Fig. 7.5, the inductor can be located in the series branch (a or b) or in the shunt branch

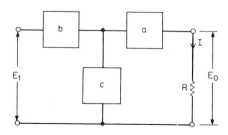

Fig. 7.5 Storage inductor topology. Inductance L is in branch a forward (buck) circuit, b boost circuit, or c shunt (flyback) circuit.

(c). We choose for our discussion the series inductor.* When the inductance L is in branch a, we have the forward converter circuit (Fig. 2.12a), and when L is in branch b of Fig. 7.5, we have the boost circuit (Fig. 2.12b).

b. Analysis

There are several approaches to the analysis of the energy storage inductor. It is logical to begin with the general topology of the converter circuit (Fig. 7.5) and stress the objective of *minimum power loss* in the inductor.[17] It is also certainly desirable to establish the relationship of the weight of the inductor to the switching frequency and output power.[18]

One approach to the analysis of the storage inductor, the one used in this section, begins with the supposition that the power loss in the series inductor is principally in the copper. That is, $W_{Cu} \gg W_{Fe}$.

To make this assumption plausible, we begin with Fig. 7.6, which depicts the wave shape of the ripple voltage and ripple current in the series-connected inductor. Current flow consists of ΔL, a triangular ripple current (measured from peak to peak) superimposed on the average value of direct current I_o. Ac voltage e_L across the inductor is rectangular, and ripple voltage across a resistive load R is triangular. The relations between inductance L, the current I_{pk} (which equals $I_o + \Delta I_o/2$), and flux density ΔB were described in Sec. 2.10, but we will now extend that discussion.[19] The product of inductance and ripple current in Eq. (2.24) can be expressed in terms of the time period T and the duty ratio $d = t_{on}/T$:

$$L\,(\Delta I) = E_{on}t_{on} = E_{off}t_{off} = E_oT(1-d) \qquad (7.46)$$

It is interesting to note that Eq. (7.46) can be equated directly with the volt-second product and total flux linkages ϕ of the core:†

$$L\,\Delta I - E_{on}t_{on} = \Delta B\,A_eN10^{-8} = \phi \qquad (7.47)$$

This equation permits the designer to estimate flux density ΔB in an inductor whose core area A_e and number of turns N are known.

The fluctuation of ripple current $r = \Delta I/I_{pk}$ affects the size of the root mean square (rms) current I, which can be calculated from:

$$I + I_{pk}\sqrt{1 - r + \frac{1}{3}r^2} \qquad (7.48)$$

---

* The most common application is the forward converter, Fig. 2.12a. When L is in branch c, we have the flyback circuit, discussed briefly in Sec. 5.7a.

† Compare with Chap. 10, Eq. (10.46).

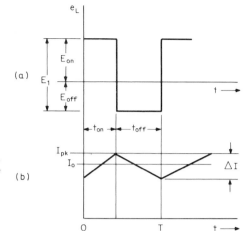

Fig. 7.6 Inductor waveforms.
(*a*) Ripple voltage. (*b*) Ripple current.

where r is the fraction of ripple $\Delta I/I_{pk}$. When ripple is maximum, $r = 1$, $I = 0.577 I_{pk}$, and the maximum value of the crest factor F_c will be:

$$F_c = \frac{I_{pk}}{I} = \sqrt{3} \tag{7.49}$$

In the inductor, the stored energy LI^2 is proportional to the output power P_o:

$$LI^2 = P_oT(1-d)F_c \tag{7.50}$$

$$P_o = E_oI \tag{7.51}$$

We can associate each of these equations with the core structure of an inductor [see (Eq. 2.6)] and a transformer, respectively. Thus:

$$LI^2 = JB_{ac}A_eA_c \, 10^{-8} \tag{7.52}$$

$$EIT = kJB_mA_eA_c \, 10^{-8} \tag{5.23}$$

where J is the current density, B_{ac} (which equals ΔB) and B_m are ac flux density, A_eA_c is the net area product, and k equals 1 or 2, depending on the type of converter circuit.

From an examination of these equations, we infer that the area product, and therefore the size, of both the inductor and transformer are proportional to the time period T. And since $T = 1/f$, we conclude that the size of the inductor, as well as its transformer counterpart, can be drastically reduced when f is large. Since the inductor can be made very small, the volume of its core and the surface of its area also shrink. Note, however, that while the magnitude of L can be reduced in proportion to $1/f$, a corresponding reduction in the core's area A_e is not possible. This is so because the surface area of the coil (which is proportional to A_e) must satisfy a restriction set on the temperature rise in the coil.

To illustrate the restriction, let us compare LI^2 in two cases: 120 hertz (Hz) ripple and 24 kHz ripple from a high-frequency switch converter. The ratio of LI^2 is in the order of:

$$\frac{\text{Low-frequency } LI^2}{\text{High-frequency } LI^2} = \frac{120}{24,000} = \frac{1}{200}$$

But if the reduction in the size of the core is restricted by a factor of 10, then the ac flux density B_{ac} can be reduced only by a factor of $200/10 = 20$. If a typical low-frequency inductor has an ac induction of 2,000 G, its high-frequency counterpart will operate at an ac induction of $2,000/20 = 100$ G. We also therefore conclude that, since B_{ac} is small, the loss of power in the core is negligible compared with that in the copper.

Another important difference between the low-frequency and high-frequency inductor is that in the latter, the gap ratio is large. There are two reasons for this. One is that core materials are operated at lower dc flux density than 1.5 T when the frequency is high—in the range of 0.2 to 0.6 T. Hence, in accordance with Eq. (7.38) β will be about two to seven times larger than in the low-frequency inductor. The second reason is the need for stable inductance over a wide range of temperature. This is particularly evident with ferrite materials, where, as we saw in Eq. (6.39), stabilization of permeability is a major requirement of the design. The net effect of large β is to justify the approximation made earlier in the chapter:

$$\mu_e \cong \frac{1}{\beta} \qquad (7.18)$$

We continue our analysis with the five expressions which connect energy LI^2, temperature rise, and power loss in the copper. It will be useful to express the relations in terms of the variables A_L, A_R, and μ_e, which are often listed in the catalogs of the core manufacturer:

$$A_L = \frac{L}{N^2} = \frac{\mu_0 \mu_e A_e}{l} \qquad (6.60)$$

$$A_R = \frac{R}{N^2} = \frac{\rho \lambda}{K_{Cu} A_{Cu}} \qquad (7.27)$$

Also:

$$W_{Cu} = I^2 R = \frac{I^2 N^2}{A_R} \tag{7.53}$$

$$LI^2 = \frac{W_{Cu} A_L}{A_R} \tag{7.54}$$

and

$$\theta = \frac{r W_{Cu}}{A_s} \tag{7.30}$$

where $\mu_o = 4\pi \, 10^{-9}$, $A_e = K_{Fe} A_{Fe}$, $A_C = K_{Cu} A_{Cu}$, A_s is the effective surface area of the inductor, ρ is the resistivity of the copper, and r is the thermal resistivity.

The last two equations enable us to express LI^2 as a function of permeability, temperature rise, and geometry:

$$\frac{LI^2}{A_{Fe} A_{Cu}} = \frac{\mu_o \mu_e \, \theta K_{Cu}}{\rho r} g_4 \tag{7.55}$$

$$\frac{LI^2}{V_{Fe}} = \frac{\mu_o \mu_e \, \theta K_{Cu} g_1^2}{\rho r} \tag{7.56}$$

where geometry is formulated in terms of the area product $A_P = A_{Fe} A_{Cu}$, iron volume $V_{Fe} = A_{Fe} l$, and the two shape coefficients are:*

$$g_4 = A_s l \lambda \tag{7.57}$$

$$g_1 = \left(\frac{A_s A_{Cu}}{l^2 \lambda} \right)^{1/2} \tag{7.58}$$

c. The Bias Operating Point

In Sec. 7.5a, we noted that a major consideration is that the dc current not saturate the core. This objective is achieved by adding an air gap l_a (in series with the magnetic path l) and, because ac flux density is very small, we can select a value for B_o quite close to the saturation density B_s. We now assume that the gap's reluctance is much larger than the iron's reluctance, which means that:

$$\frac{H_{dc}}{B_s} \gg \frac{H_{Fe}}{B_s} \qquad B_o \cong B_s$$

---

* Coefficients g_1 and g_4 were introduced in Chap. 4 as Eqs. (4.10) and (4.21), respectively.

This statement implies that the slope of the air gap line is much smaller than that of a line from origin to the intersection (H_{Fe}, B_o) with the B-H curve, as can be seen in Fig. 7.3b. Under these conditions, the gap ratio of Eq. (7.9) becomes simply:

$$\beta = \frac{l_a}{l} \cong \frac{H_{dc}}{B_s} \qquad (7.59)$$

which is consistent with the approximation:

$$\mu_e \cong \frac{l}{\beta} \qquad (7.18)$$

Next, we wish to equate maximum energy, LI^2_{max} with the important parameters of saturation density and the air gap ratio. In order to emphasize the feasibility of using a maximal value of dc flux density, we set B_o equal to B_s and proceed as follows:
Since

$$LI^2_{max} = (NI)^2 A_L$$

and

$$NI = \frac{B_s l}{0.4\pi \mu_e} = \frac{\mu_o B_s}{0.4\pi} \frac{A_e}{A_L}$$

then

$$LI^2_{max} = \frac{10^{-16} B_s^2 A_e^2}{A_L} \qquad (7.60)$$

When the inductance factor A_L from Eq. (6.60) is substituted into Eq. (7.60), we can obtain an expression for energy density:

$$\frac{LI^2}{V_{Fe}} = \frac{10^{-8}}{0.4\pi} \frac{B_s^2}{\mu_e} \qquad (7.61)$$

Or, in terms of the air-gap ratio β:

$$\frac{LI^2}{V_{Fe}} = \frac{10^{-8}}{0.4\pi} \beta B_s^2 \qquad (7.62)$$

Energy can also be related to the inductance factor A_L and temperature rise θ by means of:

$$LI^2 = A_L \frac{\theta}{\rho r} \frac{A_s A_c}{\lambda} \qquad (7.63)$$

where A_s is the effective surface area of the inductor.
We pause here to note that a wide variety of core shapes and discrete

air gaps is available. This results in corresponding discrete values of effective permeability μ_e and the inductance factor A_L. In consequence, catalogs that list such cores may assign two distinct values for energy: a dissipation limit based on Eq. (7.56) and a saturation limit based on Eq. (7.61) or (7.62). Notice that μ_e appears in the numerator of Eq. (7.56) and in the denominator of Eq. (7.61). Hence, a change in μ_e will increase LI^2 in one case and decrease it in the other.

With two distinct limits and two formulas, it is possible for ambiguities to arise. It will be useful to reconcile the two equations so that a single value of the permeability μ_e, and therefore of β, will satisfy both equations and result in one expression for LI^2. This can be done if we first set Eq. (7.56) equal to Eq. (7.62) and make β an explicit function of θ. When this is done, we have:

$$\frac{1}{\mu_e} = \beta = \frac{0.4\pi g_1}{B_s}\left(\frac{\theta K_{Cu}}{\rho r}\right)^{1/2} \tag{7.64}$$

In more succinct form:

$$\beta = \frac{k_8 \theta^{1/2} g_1}{B_s} \tag{7.65}$$

where

$$k_8 = 0.4\pi\left(\frac{K_{Cu}}{\rho r}\right)^{1/2} \tag{7.66}$$

and

$$g_1 = \left(\frac{A_s A_{Cu}}{l^2 \lambda}\right)^{1/2} \tag{7.58}$$

Now, when Eq. (7.64) is substituted into Eq. (7.62), we obtain a single expression for LI^2 as a function of *both* a saturation limit B_s and a dissipation limit ($\theta\ °C$):

$$\frac{LI^2}{V_{Fe}} = 10^{-8} B_s g_1\left(\frac{\theta K_{Cu}}{\rho r}\right)^{1/2} \tag{7.67}$$

Equation (7.67) can be stated even more succinctly:

$$LI^2 = k_6 B_s\ \theta^{1/2}\ g_6 \tag{7.68}$$

where

$$k_6 = 10^{-8} K_{Fe}\left(\frac{K_{Cu}}{\rho r}\right)^{1/2} \tag{7.69}$$

and

$$g_6 = V_{\text{Fe}}g_1 = A_p \left(\frac{A_s}{V_{\text{Cu}}}\right)^{1/2} \tag{7.70}$$

The preceding equations provide us with the basis for a fairly simple design procedure when allowable temperature rise is a limiting factor.

Please note that a temperature rise limit is implicit even when there are explicit requirements for maximum copper power loss W_{Cu} or dc resistance R. This fact permits us to substitute $I_2R = W_{\text{Cu}}$ for θ in Eq. (7.67) and obtain LI^2 as a function of A_R and copper power loss. When this is done, we obtain an expression similar to Eq. (7.26):

$$\frac{LI^2}{V_{\text{Fe}}} = \frac{10^{-8}B_s}{l}\left(\frac{W_{\text{Cu}}}{A_R}\right)^{1/2} \tag{7.71}$$

We also obtain, for $LI/\sqrt{R}$, an expression similar to that for the low-frequency inductor, Eq. (7.32):

$$\frac{LI}{\sqrt{R}} = \frac{10^{-8}B_sA_el}{l\sqrt{A_R}} = k_5B_sg_5 \tag{7.72}$$

where

$$k_5 = 10^{-8}K_{\text{Fe}}\left(\frac{K_{\text{Cu}}}{\rho}\right)^{1/2} \tag{7.73}$$

and

$$g_5 = A_{\text{Fe}}\left(\frac{A_{\text{Cu}}}{\lambda}\right)^{1/2} = \frac{A_{\text{Fe}}V_{\text{Cu}}^{1/2}}{\lambda} \tag{7.74}$$

The air-gap ratio is essentially the same as that for the low-frequency inductor, Eq. (7.39):*

$$\beta = \frac{k_7 W_{\text{Cu}}^{1/2}g_7}{B_s} \tag{7.75}$$

where

$$k_7 = 0.4\pi \left(\frac{K_{\text{Cu}}}{\rho}\right)^{1/2} \tag{7.40}$$

and

$$g_7 = \left[\frac{A_{\text{Cu}}}{(l^2\lambda)}\right]^{1/2} = \frac{V_{\text{Cu}}^{1/2}}{l\lambda} \tag{7.41}$$

---

* B_s has been substituted for B_o, as stated earlier.

The foregoing equations establish the basic relations among the magnetic, thermal, and geometric variables. Their usefulness will be demonstrated in the discussions of geometry and design that follow.

d. Geometry

It is clear that the geometry of the inductor depends on the specifications for L, I, R, and θ. This can be seen in the equations derived in the preceding section:

Resistance limit:
$$\frac{LI}{\sqrt{R}} = k_5 B_s g_5 \tag{7.72}$$

Resistance limit:
$$\beta = \frac{k_7 W_{\text{Cu}}^{1/2} g_7}{B_s} \tag{7.75}$$

Temperature limit:
$$LI^2 = k_6 B_s \theta^{1/2} g_6 \tag{7.68}$$

Temperature limit:
$$\beta = \frac{k_8 \theta^{1/2} g_1}{B_s} \tag{7.65}$$

Several conclusions can be drawn from a study of these equations:

1. The size and weight of the inductor are inversely proportional to the saturation density B_s of the core. The choice of core material has a direct bearing on size, since it establishes an upper limit for saturation density B_s. It is good practice to choose an operating value of $B_a \cong B_o$ somewhat below saturation density B_s but above the knee of the B-H normal dc magnetization curve. Typical values of B_a are in the following ranges:

Material	B_a, T
ferrite (Mn-Zn)	0.20 to 0.35
powder permalloy*	0.25 to 0.5
powder iron (MSS)	0.6 to 0.8
silicon-iron	1.2 to 1.55

2. The *optimum shape* can depend on whether it is the resistance or the temperature rise that sets limits on the size of the inductor. This can be inferred from the fact that the shape coefficients g_5 and g_6 differ in Eqs. (7.74) and (7.70).

3. When the temperature rise θ is the limiting specification, then a large surface area is desirable. And when virtually all power loss

---

* An alternative powder nickel-iron alloy, HF (50 percent nickel) is available from Magnetics, Inc. Its flux density is rated in the order of 1.2–1.4 T.

is in the copper, it is natural for the designer to favor a geometry whose configuration results in a small thermal resistivity r. This is likely to occur if we choose a structure whose coil is exposed to the air. Hence, a natural choice for the core is the toroid, which is completely enveloped in turns of copper. But when cost of manufacture must be minimized, the E geometry or the open pot core are likely alternatives. In many instances, the pot core is the preferred core geometry, simply because it is available with a square perimeter and results in a compact or felicitous arrangement of components on a chassis (see Figs. 2.3e and f).

4. There is an *optimum air gap* and therefore an *optimum permeability* for a design.[18] And these optimum values can change, depending on whether it is dc resistance or temperature rise that has been specified. We infer this conclusion from the fact that the geometric coefficients g_1 and g_7 of Eqs. (7.65) and (7.75) are not the same.

e. A Design Procedure

There are several ways of designing a storage inductor. We can make use of nomograms, charts, and graphs provided by the manufacturer of cores. Or we can utilize a computer program specifically designed for the polarized inductor. Chen, Owen, and Wilson show that, as an alternative to a computer program, one can use a table-aided design procedure. The employment of a screening rule and a special table of parameters results in an orderly search for cores which will satisfy the requirement for transfer of energy.[20]

In the examples used in this chapter, we have made use of charts of inductor ratings based on the equations developed earlier. We can construct such a table for any specific core shape and material. For Table 7.3 we have arbitrarily chosen, for illustration, the ferrite E-C core configuration. (Data for this core have also been used in Table 5.7, in connection with the high-frequency converter transformer.) Table 7.3 was derived after consideration of five sets of parameters:

1. Size (V_e, A_p, A_e)
2. Geometric (dimensions, g factors)
3. Magnetic (μ_e, A_L, β, B_a, H_{dc}, NI)
4. Thermal (θ, r, A_s)*
5. Dissipation (A_R, ρ, K_{Cu}, W_{Cu}, R)

---

* See Sec. 4.4a and Eq. (4.49) for comments on the magnitude of thermal resistivity r.

TABLE 7.3 Storage Inductor Ratings for E-C Ferrite Cores*

Line	Relevant equation	Parameter	Unit	E-C 35	E-C 41	E-C 52	E-C 70
1		A_P	cm⁴	1.299	2.474	5.616	17.24
1A		d	cm	0.98	1.19	1.375	1.68
2		l	cm	7.74	8.93	10.5	14.4
3		λ	cm	5.3	6.2	7.4	9.7
4		A_{Fe}	sq cm	0.843	1.21	1.80	2.79
4A		A_d†	sq cm	0.754	1.15	1.49	2.22
5		A_{Cu}	sq cm	1.541	2.045	2.98	6.18
5A		$0.5A_{Cu}$	sq cm	0.771	1.02	1.49	3.09
6	(4.49)	A_s	sq cm	43.5	59	91	150
7		V_e	cu cm	6.53	10.81	18.8	40.1
8	(7.58)	g_1	cm⁰·⁵	0.459	0.494	0.577	0.679
9	(7.74)	g_5	cm²·⁵	0.445	0.695	1.14	2.22
10	(7.70)	g_6	cm³·⁵	3.00	5.34	10.9	27.2
11	(7.53)	I^2R	W	1.28	1.73	2.68	4.41
12	(7.27)	A_R	$\mu\Omega$	13.7	11.5	9.44	5.96
13	(6.60)	A_L	μH	53.9	62.0	67.7	65.0
14	(7.13)	H_{dc}	Oe	50.8	54.6	63.8	75.0
15	(7.65)	β	10⁻³	25.4	27.3	31.9	37.5
16	(7.18)	μ_e		39.4	36.6	31.3	26.6
17	(7.68)	LI^2	mJ	5.28	9.40	19.2	47.9
18	(7.61)	LI^2/V_e	mJ	0.809	0.870	1.02	1.19
19	(7.72)	$LI/\sqrt{R}$	10⁻³	4.67	7.13	11.7	22.8

* B_o = 2,000 G, θ = 25°C, ρ = 1.9 $\mu\Omega$ cm (at 45°C), r = 850, W/A_s = 29.4 mW/sq cm, K_{Cu} = 0.5, K_{Fe} = 1.0.
† Center post (d) has smaller sectional area than A_{Fe}.

Two basic assumptions are made in deriving a table such as 7.3: the temperature rise (e.g., 25°C) and the operating flux density [e.g., 0.2 Tesla (T)]. Both values are arbitrary: higher, less conservative, values change the results by simple scale factors.

f. Example of an Inductor Design

Assume that an inductor is required and that the specifications are as follows:

$L = 25\ \mu H$
$I_0 = 10$ A average, nominal; $\Delta I/2 = 3$ A; $I_{pk} = 13$ A
$I_0 = 14$ A maximum
Power loss = 1 W nominal, 1.4 W maximum
Temperature rise = 25°C maximum

We first calculate the ripple factor and then its rms current I from Eq. (7.48):

$$I/I_{\text{pk}} = \left(1 - \frac{6}{13} + \frac{(6/13)^2}{3}\right)^{1/2} = 0.78$$

$I = .078\ (13) = 10.15$, which is only slightly higher than the nominal average, 10 A.

We calculate the two values of each of the following:

Parameter	Nominal	Maximum
LI^2	2.5 Millijoule (mJ)	4.9 mJ
$R = I^2/14$	5.10 milliohms (mΩ)	7.14 mΩ
$LI/\sqrt{R}$	4.14(10)$^{-3}$	4.90 (10)$^{-3}$
	(large R)	(small R)

Any of several cores can be used, but we will select a standard ferrite E-C core to illustrate the design sequence. From Table 7.3 (line 17) we see that E-C 35 with $LI^2 = 5.28$ mJ satisfies the higher of the two limits. We also note that $LI/\sqrt{R} = 4.90(10)^{-3}$ exceeds the $4.66(10)^{-3}$ value in the table. However it is our intention, in our first trial, to comply with the temperature limitation.

A worksheet such as Table 7.4 is useful, since it establishes an orderly sequence of steps and refers to the equations which define each parameter.

1. We assume a dc flux density of 2,000 G, knowing that, if necessary, we have an ample ceiling (3,000 G max) if alterations in the design become necessary.

2. The air-gap ratio can be based on a prescribed value of copper power loss, Eq. (7.75) or on *maximum temperature rise*, Eq. (7.65). We choose the latter course, and therefore enter the value given in Table 7.3 (line 15), which is $\beta = 25.4(10)^{-3}$, onto Table 7.4 (line 11).

3. The choice of β sets off a chain of calculations in the worksheet concluding with the wire diameter (line 19) and its corresponding gauge number (line 20) (chosen from a wire table or formula).

4. We can now calculate $\Delta B/2$, in order to confirm that ac flux density is small and, therefore, that power loss in the core can be neglected. To do this, we equate $L\Delta I$ with the flux linkages ϕ in the core:

$$L\ \Delta I = E_{\text{on}}t_{\text{on}} = N\ \Delta B\ A_e\ 10^{-8} = \phi \qquad (7.48)$$

Since $L\Delta I = 150\ (10)^{-6}$:

$$\frac{\Delta B}{2} = \frac{150\ (10)^{-6}\ 10^{-8}}{2\ (22)\ 0.843} = 404\ \text{G}$$

Although 404 G (line 26) is much less than $B_o = 2,000$ G, it will result in a slightly higher intercept on the dc magnetization curve of Fig. 7.3.

TABLE 7.4 Worksheet for a High-Frequency Inductor Design

Line	Relevant equation	Parameter Name, units	Symbol	Given value	Trial 1 E-C 35 ferrite	Trial 2 E-C 35 ferrite
1		inductance	L	25 μH		
2		avg dc	I	10, 14		
3		ripple	ΔI	6		
4		A	I_{rms}	10.15		
5		resistance, $\mu\Omega$	R	5.10, 7.14		
6	(7.53)	I^2R	W_{Cu}	1, 1.4		
7	(7.30)	temp rise	θ	25		
8		B_m, G			2,404	2,683
9	(7.54), (7.68)	energy	LI^2	4.9	4.9	4.98
10	(7.72)	X10^{-3}	$LI/\sqrt{R}$	rating	4.67	4.90
10A		X10^{-3}	$LI/\sqrt{R}$	4.14, 4.90	4.67	4.98
11	(7.65)	10^{-3}	β		25.4	20.5
12	(7.9)	air gap	l_a, cm		0.197	0.159
13	(7.18)	eff. perm.	μ_e		39.4	48.9
14	(6.60)	nH/turn	A_L	rating	53.9	
14A			A_L	calc.		66.8
15	(7.38)	Oe	H_{dc}		50.8	45.7
16	(6.60), (7.27)	turns	N		21.5	19.5
17			NI		301	273
18	(6.60)	L, μH			25.0	25.4
19		wire diam, cm	d		0.187	0.199
20		wire gauge	AWG		14HF	13HF
21	(7.27)	$\mu\Omega$/turn	A_R		13.7	13.7
22	(7.27)	resistance, $\mu\Omega$	R		6.33	5.10
23	(7.30)	°C	θ		24.2	19.4
24	(7.53)	I^2R	W_{Cu}		1.24	1.0
25	(7.75)	G	B_o, G	2,000	2,228	
26	(7.48)	B_{ac}	$\Delta B/2$		404	455
27	(7.7)	max B	B_m		2,404	2,683

5. Calculated values of resistance (line 22), copper loss (line 24), and temperature rise (line 23) are compared with the desired values. We see that L, R, W_{Cu}, and θ comply with the maximum values of the specification.

6. We also see that R and W_{Cu} *exceed* the nominal values of 5.10 $\mu\Omega$ and 1.0 W. If it is necessary to comply with these values, that is, if there is a resistance limitation, we initiate a second trial.

7. The second iteration (in the trial 2 column) can begin by decreasing the copper power loss by a factor of:

$$1.24 = \frac{1.24 \text{ calculated value}}{1.0 \text{ desired value}}$$

8. This is readily accomplished. First we increase flux density B_0 by the factor $\sqrt{1.24} = 1.114$ as indicated by Eq. (7.75).

9. The decrease in W_{Cu}, and a concurrent increase in B_a to $1.114 \times 2000 = 2{,}228$ G (line 25), require a corresponding decrease in the gap ratio of $1.114^2 = 1.24$. This can be inferred from Eq. (7.62).

10. Thus, β is reduced in value to $25.4/1.24 = 20.5\ (10)^{-3}$.

11. The new value of β (entered in line 11) initiates the same train of steps described above. We obtain new parameters altered by a scale factor, and conclude with the desired values of $R = 5.10$ and $W_{Cu} = 1.03$.

12. Actually, the iteration would not have been necessary had we noted, at the outset, that the low and high watt conditions straddle the rating $LI/\sqrt{R} = 4.66\ (10)^{-3}$ of the core; that is, the specification of small resistance requires that B_0 be increased by the factor calculated in Step 8.

13. We also note that had we chosen a different type of core, one with higher B_s, a smaller inductor would be possible.

7.7. OPTIMIZATION

A later phase of the design process is optimization, in which the objective is minimum total power loss, minimum weight, or minimum cost.[17,18] Close scrutiny is given to geometry, which entails a comparison of various cores and their shapes. In a careful study, the designer also considers the relative cost of iron and copper and evaluates the cost of alternative techniques of winding and assembling the inductor. The point of view taken in this chapter has been to comply with an explicit specification of maximum temperature rise or dc resistance.

If cost and packaging are set aside, we can say that the optimum geometry is one which results in a maximum value of a shape factor such as g_6. The geometric coefficients are useful in the construction of charts and also in computations with a programmable calculator. We therefore collect them here for convenience.* The first two coefficients relate to the inductor with a resistance limitation; the second pair of coefficients relates to the inductor with a thermal limitation:

$$g_5 = A_{Fe}\left(\frac{A_{Cu}}{\lambda}\right)^{1/2} = \frac{A_{Fe}\,V_{Cu}^{1/2}}{\lambda} \qquad (7.74)$$

$$g_7 = \left(\frac{A_{Cu}}{l^2\,\lambda}\right)^{1/2} = \frac{V_{Cu}^{1/2}}{l\,\lambda} \qquad (7.41)$$

---

* See also Sec. 5.4 and Eq. (5.10).

$$g_6 = V_{Fe}g_1 = A_p \left(\frac{A_s}{V_{Cu}}\right)^{1/2} \tag{7.70}$$

$$g_1 = \left(\frac{A_s A_{Cu}}{l^2 \lambda}\right)^{1/2} \tag{7.58}$$

When we lay components out on a chassis, our objective is to obtain the highest LI^2 or $LI/\sqrt{R}$ rating per unit of *gross* volume. Our choice of core and coil geometry is as relevant in the design of an inductor as it is when we design a power transformer.

When minimum resistance is mandatory, the core type of geometry with two coils (one on each leg) may be used to advantage. The resulting large surface area is also advantageous if a minimum temperature rise in the coil is desired. A nonsquare stack is often used because the magnetic path remains constant while the core area increases and because the increase in resistance is matched by the increase in the surface area of the coil. A long, narrow window also favors a small thermal drop within the coil as well as a small mean turn λ.

To assess the merit of a core from a thermal point of view, we can evaluate a shape factor such as $g_4 = A_s/l\lambda$, Eq. (7.57). The g_4 factor, because it is numeric and therefore not a function of linear dimensions, can serve as a useful index of merit when comparing one core shape with another. Or we can compute the quotient of surface area A_s and $\sqrt{A_{Fe}A_{Cu}}$. For the laminated scrapless E-I lamination, the merit index m would be:

$$m = \frac{A_s}{A_p^{1/2}} = \frac{(11s + 7.71)t^2}{(0.75\ st^4)^{1/2}} = \frac{11s + 7.71}{(0.75s)^{1/2}} \tag{7.76}$$

A table relating multiples of the stack ratio s (= stack/center limb) to the index m would be as follows:

Stack/tongue	0.5	1	2	3
m	21.57	21.60	24.26	27.14
m'/m	0.999	1.00	1.123	1.256

A 12 percent improvement results from changing from a square to a double stack, but further study may indicate that the increment in the cost of manufacture does not justify a rectangular stack.

The criterion of minimum total power loss is likely to be used in the flyback (buck-boost) circuit, where the inductor is in the shunt branch c of Fig. 7.5.* In that case, the coefficient:

---

* Branch c corresponds to the primary inductance of the flyback transformer discussed in Sec. 5.7a.

$$g_2 = \left(\frac{A_{Cu}A_{Fe}}{\lambda l}\right)^{1/2} \tag{6.67}$$

which is used in the design of high-Q inductors can serve as a useful index of merit in the selection of a core shape.

7.8. THE CONTROL OF BIAS

It is clear from our previous discussion that the biased inductor or transformer is likely to be substantially larger than it would be if no direct current were present. Because of this, practical polarized transformers are difficult to design. Center-tapped (push-pull) transformers and chokes are therefore very desirable.

In this section we will summarize various attempts to reduce or control the net amount of polarizing bias in the core.

a. Unbalance DC in Transformers

A symmetrical push-pull circuit, such as Fig. 2.11, driven by transistors with identical characteristics, will produce zero direct current in the center-tapped winding. In practice, this does not happen. Some direct current does flow in the winding, resulting in a bias depicted as H_{dc} in Fig. 7.7. In this figure we have assumed that the amount of unbalance direct current is 10 percent of the peak value of the ac magnetizing force H_m. The bias (vertical) ordinate intersects the dc B-H curve at B_o and, as a result, the positive excursion of flux density is restricted to:

$$\Delta B/2 = B_m - B_o \tag{7.77}$$

which is much smaller than the normal B_m swing in a reasonably symmetrical circuit [see Eq. (7.7)]. Hence a transformer designed to operate at a flux density $\Delta B/2 = B_m$ will become saturated during a fraction of its cycle, resulting in a sharp increase in magnetizing current at the end of the interval t_{on}.

One solution to this problem is to design the transformer to operate at reduced flux density, $\Delta B/2 \ll B_m$.

Another solution is to introduce an air gap into the magnetic circuit. If the gap is large, the total magnetizing current will increase markedly (see Fig. 6.7) and may require an increase in the drive power available from the transistors. A practical solution is to provide a small gap which would produce only a 10 percent reduction in the net B_m swing. We depict this condition in Fig. 7.7 by drawing the gap load line

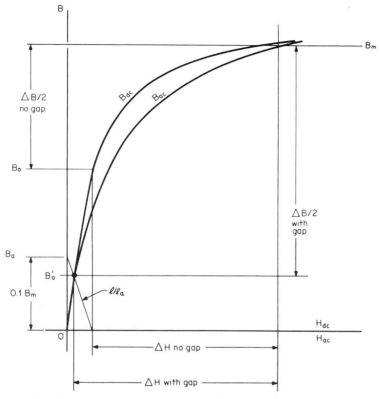

Fig. 7.7 Unbalance dc in transformer winding.

from $H_{dc} = 0.1 H_m$ on the abscissa to the intersection of the B ordinate at $B_a = 0.1 B_m$. The air gap l_a is:

$$l_a = \frac{0.1 \, H_m l}{0.1 \, B_m} = \frac{0.1 \, (0.4 \, \pi \, N I_m)}{0.1 \, B_m} \tag{7.78}$$

The new vertical intercept of the B_{dc} curve is B_o', and the permissible swing is now increased to:

$$\Delta B'/2 = B_m - B_o' \cong 0.9 B_m$$

Since dc and ac permeability is approximately equal near the origin of the B-H curve, we can estimate the size of the required air gap when the unbalance is moderate:

$$l_a = \frac{l}{\mu_{dc}} \tag{7.79}$$

Thus, for example, in a magnetic circuit with a core path of 10 cm (3.94 in) and an initial μ_{ac} of 1,000, the gap should be 0.1 mm (4 mil).

In interleaved laminations there is a residual gap, and the addition of a physical gap spacer may not be required (see Table 6.4). However, in cores with lapped mating surfaces, adding a small discrete gap may be the simplest way to obtain maximum flux excursion without causing excessive dissipation in the transistor switch.

b. Integrated Magnetics

When several chokes are used in the same dc rectifier circuit, their size can be reduced by magnetically coupling each inductor onto the same core frame. The resulting multiple windings are then proportioned in such a way that the dc currents and ac voltages of the original chokes are in the proper phase and turns ratio.[21]

A similar integration of inductors onto one core can be effected in the dc-to-dc converter circuit. Consider the forward converter, Fig. 2.12a, in which a discrete inductor is either at the input or output of the circuit.

As S. Ćuk has shown, by topological analysis, the single inductor can be split into two discrete inductors, one in the input circuit and the other in the output circuit (see Fig. 7.8a). Integration is obtained when the two inductors, each with proper phasing, are placed on the same core frame. A further level of integration can be achieved in

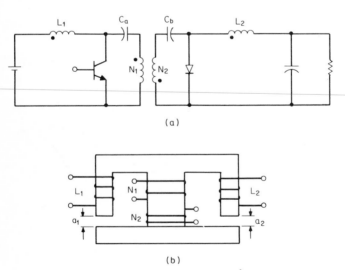

(a)

(b)

Fig. 7.8 Integrated coupled inductors. (a) Ćuk circuit. (b) E-I core and windings.

the isolation circuit of Fig. 2.13a when both inductors are integrated into the core structure of the coupling transformer.   In Fig. 7.8b, each inductor coil is placed on the outer limb of an E-I structure, the center limb of which is circled by the transformer windings. The gap in series with each outer leg is adjusted for minimum ripple current at both the input and output.[22]

c. Bias Reset

In Chap. 11, we shall see that the problem of small permeability is especially acute in high-voltage unidirectional pulse transformers with large excitation voltages.   A large swing in ΔB necessitates a gap and, consequently, small permeability.   Large size results, but this is undesirable because it entails large capacitance and a longer rise time.

When a balanced winding is not feasible, it may be possible to introduce a compensating bias $-H_p$ to offset $+H_{dc}$.   One solution is to introduce a reset bias, shown as $-H_p$ in Fig. 7.1a.   (Also see Fig. 11.3.) In the forward converter circuit (Fig. 7.9), a negative bias can be obtained by adding a demagnetizing winding N_3 and a reset diode D_2. Current flows in this winding only during the off period and resets the core close to $-H_{dc}$.   A total swing of about twice ΔB becomes feasible during the on period.   The dc ampere turns of the output and reset windings are made equal, but the copper area of the reset winding is kept at a minimum by using fine magnet wire.   Here the designer's approach is that the increase in complexity and the copper power loss are compensated for by the reduction in the core's size.

Other techniques include adding either a permanent magnet in the air gap or one or more magnets external to the core.[23,24]   Some techniques are borrowed from the design of saturable reactors, in which either a bias winding or a bias magnet is used.   The addition of a control winding then permits the inductance to be controlled by re-

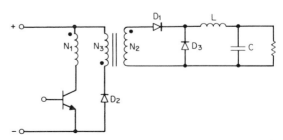

Fig. 7.9   Forward converter with demagnetizing winding and reset diode D_2.

mote dc.[25] Although all these techniques complicate the circuit or add to its cost, they deserve serious consideration from the innovative designer.

REFERENCES

1. MIT Electrical Engineering Staff, *Magnetic Circuits and Transformers* (New York: John Wiley & Sons, 1943), p. 72.
2. C. Hanna, "Design of Reactances and Transformers Which Carry Direct Current," *Transactions AIEE,* Vol. 46 (February 1927): 155–60.
3. P. McElroy, "The Design of Iron-Cored Chokes Carrying D.C.," *General Radio Experimenter,* Vol. 31, No. 10 (March 1957).
4. I. Richardson, "New Procedure for Designing Linear and Swinging Chokes," *Electrical Manufacturing,* Vol. 60, No. 6 (December 1957).
5. H. Kervoes, "Divided-Gap Choke is a Real Swinger," *Electronic Engineer* (September 1969): 61–62.
6. J. Dishman, D. Kressler, R. Rodriguez, "Characterization, Modeling and Design of Swinging Inductors for Power Conversion Applications," *Proceedings Powercon 8, 1981.*
7. F. Schwartz et al., "A Method for the Calculation of an Apparent Time Invarying Inductance for an Iron Core Inductor," *IEEE Transactions on Industrial Electronics,* Vol. IEC I-25, No. 2 (May 1978): 117–25.
8. R. Carter, D. Richards, "The Incremental Magnetic Properties of Silicon-Iron Alloys," *Proceedings IEE,* Vol. 97, Part 2 (April 1950): 199–214.
9. V. Legg, "Optimum Air Gap for Various Magnetic Materials in Cores of Coils Subject to Superposed Direct Current," *Transactions AIEE,* Vol. 64 (October 1945): 709–12.
10. N. Crowhurst, "Design of Iron Cored Inductances Carrying D.C.," *Electronic Engineering,* Vol. 22 (December 1950): 516–23.
11. A. Yair, M. Steinkoller, "New Approach to Design of Iron-Cored Inductances Carrying Direct Current," *Proceedings IEE,* Vol. 125, No. 10 (October 1978): 1,006–08.
12. R. Ray, E. Sartori, "Computer Design of an Inductor Carrying D-C," *IEEE Transactions on Magnetics,* Vol. MAG-7, No. 3 (September 1971): 453–55.
13. A. Ohri, T. Wilson, H. Owen, "Design of Air Gapped Magnetic Core Inductors for Superimposed D-C and A-C," *IEEE Transactions on Magnetics,* Vol. MAG-12, No. 5 (September 1976).
14. S. Szuba, "Computer-Aided Design of Air-Gapped Magnetic Core Inductors with Minimum DC Winding Resistance," *IEEE Transactions on Magnetics,* Vol. MAG-15, No. 3 (May 1979): 1,085–96.
15. R. Luebben, "How to Design Inductors for Electronic Equipment," *Electronic Equipment Engineer* (September 1959): 40–44.
16. Reuben Lee, *Electronic Transformers and Circuits,* 2d ed. (New York: John Wiley & Sons, 1955), p. 95.
17. J. Dishman, F. Dickens, R. Rodriguez, "Optimization of Energy Storage

Inductors for High Frequency dc to dc Converters," *Proceedings Powercon 7, 1980.*

18. R. Wong, H. Owen, and T. Wilson, "Parametric Study of Minimum Reactor Mass in Energy Storage DC to DC Converters," *IEEE Power Electronics Specialists Conference, 1981,* Vol. 81CH 1,652–7.

19. J. Watson, *Applications of Magnetism* (New York: John Wiley & Sons, 1980), pp. 262–66.

20. D. Chen, H. Owen, Jr., T. Wilson, "Table Aided Design of the Energy Storage Reactor in DC-to-DC Converters," *IEEE Transactions on Aerospace and Electronic Systems,* Vol. AES-12, No. 3 (May 1976): 374–86.

21. A. Lloyd, "Choking up on LC Filters," *Electronics* (August 21, 1967): 93–97.

22. S. Ćuk, "A New Zero-Ripple Switching DC-to-DC Converter and Integrated Mechanics," *IEEE Power Electronics Specialists Conference, 1980,* Vol. 80CH 1,537–0: 12–32.

23. J. Ludwig, "Inductors Biased with Permanent Magnets: I, II," *Transactions AIEE,* Paper 60–198 (July 1960).

24. S. Shiraki, "Reverse-Biased Ferrite Cores," *Electronic Design* (July 19, 1978): 86–89.

25. R. Segworth, S. Dewan, "A Hybrid Magnetic Solid-State Control," *IEEE Transactions on Magnetics,* Vol. MAG-6, No. 3 (September 1970): 668.

PART 3

The Transmission of Information

CHAPTER 8

The Wide-Band Transformer: Analysis

So far in our discussion we have dealt with transformers whose principal function is the *transfer of power*. Specific requirements such as efficiency, miniaturization, and cost dominate the design. We pause at this juncture to introduce the final part of our book, which is devoted to wide-band and pulse transformers. The principal function of such transformers is the *transfer of information*. Information in the form of energy should be transferred efficiently, but this requirement is viewed in a special sense in electronic circuits. Efficiency in the conventional sense (the ratio of power output to power input) is supplanted by the concept of the ratio between signal and noise. The focus is on the undistorted or accurate transmission of information in the form of a signal. The signal is typically a complex waveform or a coded train of pulses. We consider the twin aspects of the transmission of signals in four chapters: the frequency domain (Chaps. 8 and 9) and the time domain (Chaps. 10 and 11).

It will become clear, as we proceed, that the distinction between a wide-band and a pulse transformer is an artificial one. Because the transfer of energy, like every event, occurs only in time, a more apt name for our subject is the *transmission transformer*. However, a consideration of frequency yields mathematical tools useful in discussing events that are occurring in time. We begin our analysis, therefore, by considering frequency.

A transformer specifically designed to pass a band of frequencies exceeding a few decades is called a *wide-band transformer*. The signal passed by a wide-band transformer is usually complex, not sinusoidal like the signal passed by the conventional power transformer. Voice, music, a train of pulses—these are typical of such a signal.

Wide-band transformers may be classified in several ways. In terms of the specific portion of the *frequency spectrum* they are designed to pass, they are called subsonic, audio, ultrasonic, high, very high, and ultra-high frequency transformers. In terms of the *system* in which they are used, they are known as control-system (or servo), carrier, video, and data-transmission transformers. In terms of their specific *function* in a circuit, they are called line, coupling, driver, output, modulation, bridge, and hybrid transformers.

8.1. SCOPE OF THE DISCUSSION

Although the wide-band transformer is required to pass a broad band of frequencies, the distortion must be constrained. The various factors to be taken into account can be inferred from a complete set of specifications. The requirements parallel those for an ordinary power transformer, with additional constraints on the variation of the amplitude and the phase shift with the frequency, the amount of harmonic distortion, the degree of unbalance, and the vulnerability to stray fields. The effect of the external circuit (e.g., the nature of the generator and the amount of feedback) must also be taken into account.

Our survey comprises two chapters, analysis (Chap. 8) and synthesis (Chap. 9). In this chapter, we know what is in the black box (i.e., we know the constraints of the transformer), and our approach is analytic. In Chap. 9 we approach the subject from a more general point of view, that of synthesis. We know the specifications, and our objective is to design the black box, i.e., the physical transformer.

8.2. EQUIVALENT CIRCUITS

Our analysis of the wide-band transformer begins with the lumped-sum equivalent circuit. For convenience, assume that the shunt inductance L_P and the external circuit elements are constant.

The performance of a wide-band transformer cannot be dissociated from the circuit in which it is used. In Fig. 8.1a, the complete circuit, the generator and load resistances and the capacitances external to the transformer are included. (In the low-frequency power transformer only the inductive elements and the winding resistances must be accounted for.) Note, in Fig. 8.1a and b, the pi network of the capacitances. (The subscripts g, p, s, and l refer, respectively, to the capacitances of the generator, primary, secondary, and load; C_B is the external capacitance which bridges the input terminals and the output terminals. The turns ratio n shown in Fig. 8.1 equals the number of turns in the primary winding divided by the number of turns in the secondary winding.

A rigorous analysis and a general solution of this network are unnecessarily complex for most purposes. It is usually sufficient to evaluate the complete circuit and search out the essential relations, thus simplifying the circuit and the analysis.

The reactive elements, inductance and capacitance, have a dominating effect on the circuit performance and must be retained. Input capacitances can be combined into $C_1 = C_g + C_p$, and output capacitances into $C_2 = (C_s + C_L)/n^2$. The bridging capacitance $C_3 = C_B/n$ may or may not be essential; its elimination is sometimes justified on the ground that it is mathematically possible to assign a portion of C_B to the input and output capacitances. The series resistive elements, invariably small, can be absorbed into the source and load resistances or omitted. The power loss in the core (represented by the

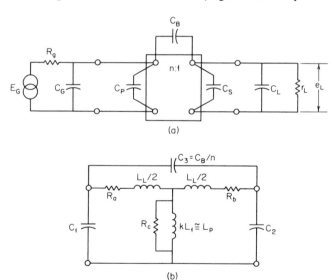

(a)

(b)

Fig. 8.1 Complete equivalent circuits. (a) The complete circuit. Inductive elements are inside the box. (b) The unity-ratio transformer comprising lumped capacitances together with inductive and resistive elements.

shunt resistance R_c) is not omitted when a precise calculation of power loss or impedance is necessary. However, if the power level is high, it can be lumped with the load resistance R_L; if the power level is low, the resistance of the shunt path can be ignored.

If now we divide the resulting unity-ratio circuit into three frequency ranges—low, middle, and high—we can devise a set of simpler circuits for each band of frequencies and analyze these.

It is advantageous, in analyzing these circuits, to express the transfer ratio or gain in a form which permits a simple transition between the concepts of a constant-current generator I_G and a constant-voltage generator E_G. From Fig. 8.2 we can infer that the voltage gain G may be expressed in normalized form by the ratio:

$$G = \frac{ne_L}{RI_G} = \frac{ne_L}{aR_GI_G} = \frac{E_L}{aE_G} \tag{8.1}$$

where n is the step-down turns ratio, e_L is the output voltage, E_L is the output voltage reflected to the primary, $R = R_LR_G/(R_L + R_G)$ is the parallel combination of source and reflected load resistance R_L, and $a = R_L/(R_G + R_L)$ represents the component of attenuation due to the presence of both source and winding resistances.

The loss of power caused by inserting the transformer into the circuit is also easily visualized. Since $aE_G/n = e_L$ is the voltage at the output of a purely resistive transformer circuit, Eq. (8.1) is a ratio which enables us to compare the actual output with the attenuated output of a real transformer having no reactive elements (Fig. 8.2c).

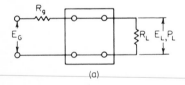

(a)

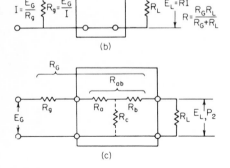

(b)

(c)

Fig. 8.2 Mid-band circuits. (a) Constant-voltage circuit; transformer winding resistances are ignored. (b) Constant-current equivalent of (a). (c) The transformer circuit with only winding resistances accounted for.

8.3. MID-BAND RESPONSE

An exceedingly simple response is obtained if we define a middle, or mid-band, frequency ω_m:

$$\omega_m = \sqrt{\omega_1 \omega_2} \tag{8.2}$$

which is the geometric mean of the low-frequency cutoff ω_1 and the high-frequency cutoff ω_2. Because $\omega_m \gg \omega_1$, the flux density is low in the mid-band circuit (Fig. 8.2c), and R_c, the shunt core-loss resistance, can be neglected.

The normalized gain G at ω_m provides a convenient basis for comparison with the low- and high-frequency responses:

$$G = 1 \tag{8.3}$$

When Eq. (8.1) is stated in the form:

$$\frac{ne_L}{E_G} = \frac{n^2 r_L}{R_G + n^2 r_L} = a \tag{8.4}$$

we see that the transformer behaves like a simple resistive attenuator, with zero phase shift. When the source resistance is fixed, the maximum overall gain e_L/E_G occurs when $n^2 = R_G/r_L$—that is, when the load resistance is matched to the source resistance.

Because the circuit behavior is least complicated at ω_m, it is customary to specify several important transformer characteristics at the mid-band frequency: (1) the rated load power; (2) either the turns ratio or the desired impedance at the input when the secondary winding is loaded; (3) the desired minimum efficiency, or the maximum loss of power resulting from the insertion of the transformer into the circuit (called the *flat loss* or the *insertion loss*); and (4) a measure of the mismatch between load and source impedances, expressed as *reflection loss* and, in the high-frequency spectrum, as the *standing-wave ratio*.*

For class A and class B amplifier circuits, it is vital to calculate the load correctly, bearing in mind that the theoretical efficiency of the class A circuit is only 50 percent and that of the class B circuit is 78.5 percent.† In addition to allowing for the estimated mid-band efficiency of the transformer, we must realize that volt-amperes (VA) must be available to the load at both low and at high frequencies, because the input impedance decreases as it becomes reactive.

In audio, carrier, and other wide-band circuits, it is common practice to specify the power level in decibels (dB). Such ratings are precise

---

* The standing-wave ratio is discussed in Sec. 11.3a.
† See also Sec. 8.7 and Table 8.2.

only if both the reference, or zero, level and the impedance are explicitly stated. A preferred nomenclature is the *dBm* rating, defined so that 0 dBm equals 1 milliwatt (mW) developed in 600 ohms (Ω), the standard impedance of telephone circuits.

a. Matching and Regulation

When an accurate estimate is required of the full-load input impedance (i.e., the impedance seen at the input with the output loaded), winding resistances cannot be ignored. In Fig. 8.2c, for example, the input impedance at the transformer terminals will exceed $n^2 r_L$ by $R_a + n^2 r_b = R_{ab}$. (Here the winding resistances of primary and secondary are, respectively, R_a and r_b).

When the specifications require that the input impedance at ω_m be exactly equal to R_L, it may be necessary to alter (i.e., decrease) the theoretical turns ratio $n = \sqrt{R_L/r_L}$ to n' in accordance with:

$$\frac{n}{n'} = \left(1 + \frac{R_{ab}}{R_L}\right)^{1/2} \cong 1 + \frac{R_{ab}}{2R_L} \tag{8.5}$$

At ω_m, the transformer *copper efficiency* η may be determined from the regulation; that is:

$$\frac{P_1}{P_2} = \frac{1}{\eta} = 1 + \text{reg} \tag{8.6}$$

where P_1 is input power, P_2 represents the power expended in the load with the winding resistances present (Fig. 8.2c), and the definition of regulation (reg), R_{ab}/R_L, is consistent with Eq. (5.5).

To determine the *flat insertion loss* (IL) due to the transformer, we form the power ratio $P_L P_2$, where P_L represents the power expended in the load in the absence of winding resistances (Fig. 8.2a) and P_2 represents the power expended in the load with the resistances present. The resulting computation yields:

$$\text{IL} = \frac{P_L}{P_2} = \left(1 + \frac{R_{ab}}{R_G + R_L}\right)^2$$

The insertion loss IL is expressed in decibels (dB) as A_i:

$$A_i(\text{dB}) = 10 \log \frac{P_L}{P_2} = 20 \log \left(1 + \frac{\text{reg}}{1 + 1/m}\right) \tag{8.7}$$

when $m = R_L/R_G$ is the *matching factor*. Note that when the source resistance is very small ($R_G \ll R_L$), the constraint on the transformer regulation becomes great. Equation (8.7) serves to establish the maximum copper regulation of the transformer when the maximum IL is specified.

b. Mismatch and Reflection

An important branch of the classical theory of transmission circuits and filter networks is based on the condition of maximum transfer of power, which occurs when source and load impedances are matched,[1] and it is a common requirement that a wide-band transformer serve to match the load impedance to that of the generator (i.e., $R_L = R_G$).

How much power is lost if $R_L \neq R_G$? The practical consequences of mismatching can be expressed in terms of the *mismatch loss* (ML):

$$\text{ML (dB)} = 20 \log \frac{1 + m}{2 \sqrt{m}} \tag{8.8}$$

If we substitute several typical values of $m > 1$ into this formula, we see that a large mismatch results in only a small loss of power. The loss is expressed in decibels: for example, a 100 percent mismatch ($m = 2$) results in a loss of 0.5 dB. At first sight, therefore, exact matching does not seem important. However, in a system with many transformers or a tandem iteration of transformers (e.g., long-distance telephone circuits), mismatch can become important. A string of five transformers, each with the same mismatch of 100 percent, results in a large power loss of 2.5 dB. To achieve the same *overall* power loss of only 0.5 dB, the mismatch must be reduced to 35 percent ($m = 1.35$, IL $= 0.1$).

We have stated that the flat insertion loss A_i (Eq. (8.7)) is a measure of the dissipation of power introduced by connecting a transformer between the source and load; that is:

$$A_i = 10 \log \frac{\text{load power without transformer (Fig. 8.2}a)}{\text{load power with transformer (Fig. 8.2}c)} \tag{8.9}$$

The concepts of effective loss and return (or reflected) loss yield a more sensitive measure of the extent to which practical circuits depart from the "ideal" or matched condition. First, we define an ideal transformer as one whose transformation ratio* $r:1$ is such that the source R_g and load impedance r_L of Fig. 8.1 are matched when:

$$r = \sqrt{\frac{R_g}{r_L}} \qquad R_g = R_L$$

---

* In practice, the transformation ratio r is not synonymous with the turns ratio n.

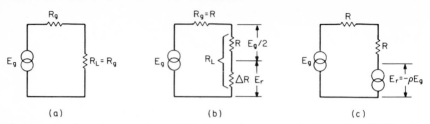

Fig. 8.3 Match and mismatch. (*a*) Ideal transformer match, $R_L = R_g$. (*b*) Nonideal transformer results in mismatch, $R_L = R + \Delta R$. (*c*) Equivalent viewed in terms of returned power.

Effective loss A_e is the result of substituting a nonideal transformer for an ideal transformer:

$$A_e = 10 \log \frac{\text{load power with ideal matching transformer (see Fig. 8.3a)}}{\text{load power with nonideal transformer (see Fig. 8.3b)}} \tag{8.10}$$

In the nonideal transformer (Fig. 8.3*b*) the reflected load resistance R_L is the sum of two components: a matched resistance $R = R_g$ (the generator resistance) and a mismatched component $\Delta R = R_L - R_g$. Consider now the circuit of Fig. 8.3*c*, which contains two generators. In addition to a "send" or transmitting generator E_g in series with a matched load R, there is also a return or reflection generator E_r. E_g furnishes a matched amount of power to R, while the (hypothetical) E_r returns a fraction ρ^2 of the matched power back to the source; that is:

$$E_r = -\rho E_g \qquad P_r = \rho^2 \frac{(E_g/2)^2}{R_g} \tag{8.11}$$

where P_r is the power reflected back to the source.

The *reflection coefficient* ρ is that fraction of voltage E_r which results in the return (reflection) of current and power P_r to the generator:

$$\rho = \left| \frac{R_L - R_g}{R_L + R_g} \right| = \frac{m - 1}{m + 1} \tag{8.12}$$

where m is the matching factor.

Return loss A_r is expressed in decibels as:

$$A_r = 10 \log \frac{\text{applied power}}{\text{reflected power}} \tag{8.13}$$

Stated in terms of the reflection coefficient ρ:

$$A_r = 20 \log \left| \frac{1}{\rho} \right| dB \qquad (8.14)$$

Note that a small percentage of reflection results in large values of A_r:

%ρ	10	1	0.1	0.01
A_r	26	46	66	86

Hence, the reflection coefficient is a sensitive measure of the degree to which a transformer accurately matches a load to its source.*

8.4. LOW-FREQUENCY RESPONSE

At frequencies much lower than ω_m, shunt inductance must be reckoned with and the circuit in Fig. 8.4 becomes applicable. However, let us first consider the simpler circuit in Fig. 8.5, which is based on the assumptions that power losses in the winding and core are negligible and that L_p is constant. The response at $\omega < \omega_m$, relative to the mid-band gain, can then be shown to be:

$$G = \frac{1}{H_L} \qquad H_L = 1 - j\frac{R}{X_p} \qquad (8.15)$$

where H_L is the transmission factor and primary reactance $X_p = \omega L_p$.
In terms of magnitude, or absolute value:

$$|G| = \frac{1}{\sqrt{1 + (R/X_p)^2}} \qquad (8.16)$$

The frequency ω_1, at which $G = 1/\sqrt{2}$ and $R/X_p = 1$, is called the cutoff frequency, the half-power frequency, or the 3-dB down frequency.

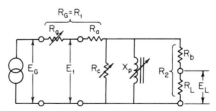

Fig. 8.4 Complete low-frequency equivalent circuit.

---

* The relation between reflection and the standing-wave ratio is discussed in Sec. 11.3*a*.

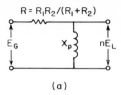

(a)

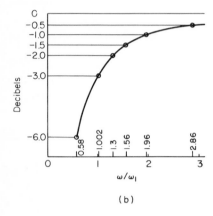

(b)

Fig. 8.5 Low-frequency response. (a) Equivalent high-pass or lead circuit. (b) Amplitude-frequency curve (first-order linear case).

Attenuation at *low frequencies* A_{LF}, expressed in decibels, is:

$$A_{\text{LF}} = 10 \log \left[1 + \left(\frac{\omega_1}{\omega} \right)^2 \right] \tag{8.17}$$

The cutoff frequency $\omega_1 = 2\pi f_1$ [in radians per second (rad/sec)] and the phase shift θ are:

$$\omega_1 = \frac{R}{L_P} \qquad \tan \theta = \frac{R}{X_P} \tag{8.18}$$

where f_1 is the cutoff frequency in hertz (Hz) and L_P is the shunt inductance.

As is shown in Fig. 8.5b, attenuation increases as either ω or L_P decreases. From Eq. (8.15) we can infer that the output voltage leads the input voltage, since θ is positive. Table 8.1 lists the attenuation, $20 \log |G|$, in decibels, as well as the phase shift accompanying significant values of X_P/R.

The exact behavior of the transformer at low frequencies is of great practical importance in electronic circuits. We are particularly interested in the connection between the key circuit parameters and the shunt inductance L_P, for an appropriate choice of L_P will assure us of compliance with the low-frequency specifications for attenuation A_{LF}, phase shift θ, and harmonic distortion.

TABLE 8.1 Low-Frequency Attenuation and Phase Shift

X_pR or ω/ω_1	Attenuation, dB	Phase shift, θ
206,000	—	$1''$
3,438	—	$1'$
573	—	$0.1°$
57.3	—	$1°$
28.6	—	$2°$
11.43	—	$5°$
6.6	0.1	$8.62°$
4.62	0.2	$12.2°$
4.11	0.25	$13.7°$
2.86	0.5	$19.3°$
1.96	1.0	$27.0°$
1.31	2.0	$37.4°$
1.002	3.0	$45°$
0.579	6.0	$60°$
0.260	12	$75.4°$

a. Low-Frequency Distortion

Distortion is particularly important in communications circuits. One important factor, ferromagnetic distortion, which is attributable to hysteresis in the core, was described in Sec. 6.5, but all of the following factors have a direct bearing on our choice of the amplitude attenuation A_{LF}, the cutoff frequency ω_1, and L_P:

1. The presence of a generator of limited current and VA handling capacity or of one which can exhibit nonlinear source resistance.
2. The phase angle ϕ of the inductive load which results from the fact that L_P is in parallel with the reflected load R_L.
3. The reduction in load impedance (manifested by a drop in the power factor PF $= \cos \phi$) as frequency decreases.
4. Ferromagnetic distortion D.
5. The effect of feedback on source resistance, phase shift, and rate of change of phase shift.

The factors which become prominent, other than hysteresis, result from the fact that the transformer is used in circuits with an active element such as a transistor. In such circuits, distortion depends on the extent to which the source can exhibit nonlinear resistance. Nonlinearity of resistance can be inferred from the characteristic curves; it is evident in the curvature and unequal spacing of the family of curves for collector current vs. collector voltage. At low frequencies, the transformer becomes an inductive load and produces an elliptical load line. As the cutoff frequency is approached, the slope of this load line increases and the load impedance decreases. The reduction in impedance, which may be equated with the power factor, produces a nonlinear change in the source resistance. Simultaneously, flux density in the core rises even though output voltage decreases. The power output drops as the distortion produced by nonlinear source resistance *and* hysteresis increases. Harmonic distortion will, therefore, depend on the type of generator used and on how the power factor of the load and the hysteresis of the core vary with the power level and frequency. Further complications ensue if the circuit includes a feedback loop.* We shall see that the cutoff frequency must often be substantially lower than that requested in the user's specifications.

In our analysis, we employ a definition of the impedance quotient which takes into account the central parameters of frequency, inductance, and load resistance:

$$K_L = \frac{\omega L_P}{R_L} \tag{8.19}$$

It is our intention to show that this factor is a function of many parameters, that is: $K_L = F(A_{LF}, \theta, VA, PF, D)$.

The choice of a suitable value of K_L is a simple matter if the generator has ample VA capacity, is linear, and operates at a low voltage level, for then K_L is a function only of the matching factor and the cutoff frequency. In such a low-level, linear circuit:

$$K_L = \frac{1}{(1 + m)\tan\theta} \tag{8.20}$$

If $m = 1$, and the LF response is to be 3 dB down at f_1, then $\tan\theta = 1$, and K_L should be $\frac{1}{2}$. We will see that higher values of K_L are necessary when low harmonic distortion is also specified.

(1) Generator Current for an Inductive Load. An important conclusion to be drawn from Eq. (8.16), which was based on a linear circuit, is that low-frequency attenuation may be made as small as we wish simply by making R/X_p approach zero. Thus, a small value of L_p is permissible, provided R is also made small.†

In practice, however, there is a lower bound on the magnitude of the shunt reactance X_p, and therefore on L_p. The reduction of X_p is intimately related to the maximum current available from the gener-

---

* Feedback, the slope of attenuation, and the phase shift are discussed in Sec. 9.1.

† When both positive and negative feedback are used in amplifier circuitry, R_G can approach zero. This, in turn, reduces R to approximately R_a, the resistance of the primary winding, and results in an exceptionally low value of R/X_p, with the consequent benefits of extended low-frequency response (see Sec. 9.1*b*).

ator and the maximum permissible variation of the reflected load impedance.

The important relationships are clarified in the low-frequency circuit, Fig. 8.4. The arrows through the symbols X_p and R_c (the core-loss resistance) emphasize the nonlinearity of these parameters; and (as shown in Sec. 6.1) they vary with the flux density B_m and frequency ω. The product $B_m\omega$ is a function of the input voltage E_1 furnished to the transformer's primary winding.

It is convenient to combine the nonlinear variables R_c and $X_p = \omega L_p$ into one variable Z_c, the *core impedance*, defined as the parallel combination of R_c and X_p. In an electronic circuit, the maximum current I_M that can be made available to the load R_L by the semiconductor generator is substantially less than the short-circuit current E_G/R_G; that is:

$$\frac{E_G}{R_L} < I_M \ll \frac{E_G}{R_G}$$

where the value I_M depends, in turn, on the power capability and regulation of the power supply.

Good design practice dictates that Z_c amply exceed R_L. Since, as a rule, $R_L \ll R_c$, then:

$$\frac{Z_c}{R_L} \cong \frac{X_p}{R_L} = K_L \tag{8.21}$$

An initial estimate of minimum permissible K_L may be arrived at by first noting that the power supply must be capable of furnishing an amount of VA in excess of the power P_L to be furnished to the load. Thus:

$$\left|\frac{VA}{P_L}\right| = \frac{1}{\cos\phi} = \sqrt{1 + \frac{1}{K_L{}^2}} \tag{8.22}$$

It is uneconomical, however, to allow the VA to exceed P_L by too large an amount. Typically, $VA/P_L = 1.41$, 1.12, and 1.05 when $K_L = 1$, 2, and 3, respectively.

As an example, assume that we wish to furnish 50 W to a transformer-coupled load $R_L = 5{,}000$, when m is 0.2. The insertion flat loss is to be 0.25 dB when f_M (the mid-band frequency) is 1 kilohertz (kHz), and we also wish to furnish 50 W when f is 50 Hz. From the copper efficiency, obtained by solving Eq. (8.7) for regulation and then Eq. (8.6) for efficiency, we determine that the necessary mid-band power will be 50/0.847, or 58 W.

At 50 kHz, however, the power supply must be designed to furnish either 82, 65, or 61.5 VA depending on whether K_L has been made equal to 1, 2, or 3. If, because of its high efficiency, a class B circuit is chosen,* then the large swing of current demanded from the power supply places strict constraints on its regulation, especially when K_L is small. If we elect to make K_L small, thus keeping the transformer's input inductance and size small, the VA capacity and the size of the power supply must be increased.

(2) **The Load Power Factor.** Even were L_p constant or linear, some degree of distortion is likely to occur if the circuit has an electronic generator. One reason is that an elliptical load line (rather than a linear, resistive one) results in a nonlinear variation of R_G over a sinusoidal cycle of collector or plate current. The distortion is usually estimated graphically because the width of the ellipse changes with the value of K_L. Reuben Lee finds, however, that distortion due to an elliptical load line ($\phi = 30°$) is less than that resulting from variation of the impedance of the load.[2]

The variation of the load impedance, as we observed in the previous section, affects the power supply. Furthermore, any *reduction* in this impedance from its mid-band value will increase the slope of the load line, increase collector or plate dissipation, alter the optimum loading, and hence increase distortion. As a result, the amount of available *undistorted* power at frequencies in the vicinity of cutoff is substantially less than in the mid-band range. Lee has shown that, in a typical class A amplifier where $R_2 = 2R_1$ (see Fig. 8.4), a 30 percent drop in load impedance, to $0.7R_2$, is sufficient to increase the second harmonic from 4 to 10 percent and the third-harmonic distortion from 1 to 4 percent. It is desirable, therefore, to ensure that the load power factor be held reasonably close to unity.

The shunt input-impedance ratio of the loaded transformer is:†

$$\frac{Z_L}{R_L} = \cos \phi = \frac{1}{\sqrt{1 + 1/K_L^2}} \tag{8.23}$$

from which we obtain:

$$K_L = \frac{\cos^2 \phi}{1 - \cos^2 \phi} = \cot \phi \tag{8.24}$$

When ϕ is 30° and *PF* is 0.87, the impedance is 13 percent lower than its mid-band resistance. Experience has shown this reduction to be acceptable.[3] Hence, $K_L = \sqrt{3}$, or 1.73. Our computations are

---

* See Table 8.2, Sec. 8.7.
† Shunt impedance Z_L is the parallel combination of core impedance and load impedance.

consistent with the standard, rule-of-thumb procedure when detailed circuit data are not available; that is, K_L is given values from 2 to 3 at the lowest useful operating frequency.

b. Ferromagnetic Distortion

In general, the level of the power output is correlated with the level of the flux density. High voltage and high power result in large B_m. And a large dynamic range of power reduces the effective power factor of the load and increases hysteresis in the core. Each of these effects produces distortion at the output. Here we examine the relationships among the circuit and transformer variables; these include the matching factor m, the impedance ratio K_L, and the operating flux density.

We noted, in Sec. 6.5, that the nature of the magnetic circuit is such that if R_G is small, E_1 and B_m are sinusoidal. Then the magnetizing current i_m is nonsinusoidal and will contain harmonic components. If, on the other hand, R_G is large, i_m is sinusoidal and then B_m and E_1 will be nonsinusoidal. In the first instance, the flow of harmonic components of current through R_G will result in a distorted voltage across R_G and, consequently, in a distortion of the output voltage E_L across the load. In the second, even though the magnetizing current is sinusoidal, the output voltage will be distorted.

The situation when E_1 and B_m are sinusoidal is of particular interest. If it were possible for the total source resistance $R_g + R_a$ to equal zero (and m to equal infinity), presumably there would be no distortion at the output even if the magnetizing current were greatly distorted. Consequently, when distortion must be small, low source impedance is very advantageous. This can be accomplished by using negative feedback to make R_g small and m large or by combining positive and negative feedback to achieve zero R_g. According to Eq. (6.87), ferromagnetic distortion will not exceed the maximum distortion factor $D_m = R/X_p$. This factor can be directly related to the phase shift and may be regarded as a basic parameter of distortion. We can then show that:

$$K_L = \frac{1}{D_m(1+m)} \tag{8.25}$$

The dependence of D_m or of the actual distortion D on flux density can be ascertained from a family of curves:

$$D_m > D = F\left(\frac{R}{Z_c}, B_m\right)$$

However, it is simpler to relate maximum distortion directly to the phase shift:

$$\frac{R}{Z_c} \cong \frac{R}{X_p} = D_m = \tan \theta \qquad (8.26)$$

Now $\tan \theta$ and B_m, which are readily computed, provide us with a measure of the distortion due to hysteresis. Suppose D_m is to be limited to 5 percent, a typical requirement. If we also want K_L to have the low value of 2, then we must mismatch and make m equal to 9. Under matched conditions, K_L would be 10 and the inductance would have to be five times greater than in the mismatched case.*

8.5. HIGH-FREQUENCY RESPONSE

The prediction and control of high-frequency response are of great practical importance. These pose problems which merit examination from several points of view.

We initiate the subject in this section by considering the dependence of the high-frequency response—its shape, slope, and phase shift—on the complexity of the model we choose for the equivalent circuit.

In the next chapter (Sec. 9.2), we consider high-frequency response from the point of view of network synthesis, in order to ascertain methods of achieving three objectives of circuit design—maximal flatness, fidelity of individual waveforms, and maximum bandwidth. In Chaps. 10 and 11, we discuss the connection between high-frequency response in the frequency domain and performance in the time domain. Rise time, maximally flat delay time (linear phase), and fidelity of group waveforms are important topics in these last two chapters.

a. Equivalent Circuits[4,5]

One of the major problems of transformer design is well described by T. O'Meara:[4] "It is not an easy matter to choose an equivalent circuit which is sufficiently complex to represent the physical transformer with reasonable accuracy, and yet is sufficiently simple to permit ready analysis or synthesis."

A lumped-parameter model of the transformer, if it is to be accurate at high frequencies, must include the bridging capacitance C_B between the primary and secondary windings (see Fig. 8.6a). We can simplify it somewhat by combining the input and output capacitances and by

---

* The relation between distortion and the size of the core is discussed in Sec. 9.3, where we consider core design.

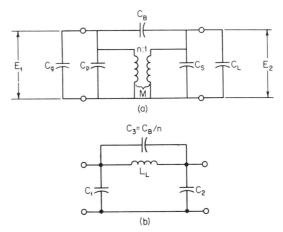

Fig. 8.6 Lumped-parameter, high-frequency circuit. (*a*) Complete circuit. (*b*) Unity-ratio circuit.

referring all elements to the primary winding. The result is the unity-ratio circuit (Fig. 8.6*b*), a pi circuit which comprises three capacitors and one inductor, the leakage inductance L_L.

If this network is connected to a source of fixed resistance and to a load, we can proceed to a general solution of the response at high frequencies. The solution of the relative transfer ratio E_H/E_M—that is, for the response at high frequencies—is the quotient of two polynomials.* Unfortunately, a general solution is too complicated to be of practical use. We are usually compelled, although at some risk, to select the simplest expression which seems adequate for the design problem in hand.

The simpler solutions, those which represent the response by means of one polynomial in the denominator, are the most widely used. In the more general case (considered in Sec. 8.5*c*), one or two zeros occur in the polynomial of the numerator. The choice of an equivalent circuit, which determines whether to use a simpler or more general solution, will depend on the significance which we ascribe to the capacitors. This, in turn, generally depends on the impedance level of the circuit, on the frequency range, and on whether we are designing a step-up or step-down transformer.

b. Polynomial Solutions[6]

In most circuits, the bridging capacitor, being small, is neglected. As a consequence, each of the equivalent circuits in Fig. 8.7 may have either one, two, or three reactive elements.

---

* E_H and E_M are the transfer ratios in the high- and low-frequency bands, respectively.

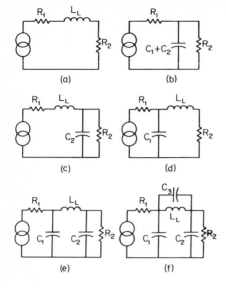

Fig. 8.7 High-frequency circuits: low-pass models. (a) One-pole LR (lag) network; audio step down. (b) One-pole RC network; audio step up or high impedance. (c) Two-pole step-up circuit $n < 1$; large road resistance. (d) Two-pole step-down circuit $n > 1$; large source resistance. (e) Three-pole pi circuit at high frequency. (f) Three-pole, two-zero, m-derived circuit.

The relative transfer ratio is:

$$\frac{E_H}{E_M} = \frac{E_2}{aE_1} = \frac{ne_L}{RI_G} = \frac{1}{H} \tag{8.27}$$

where the polynomial H, the transmission factor, is of the first, second, or third order.* The order, or degree, of the polynomial is the same as the number of poles and corresponds to the number of reactive elements in the circuits of Fig. 8.7.

The high-frequency attenuation A_H is, in decibels:

$$A_H = 10 \log |H|^2 \tag{8.28}$$

The circuit engineer is often particularly interested in how the rate of attenuation and the phase shift θ vary with the frequency. In the following pages we shall stress the relation of these variables to the order of the polynomial.

(1) First-Order Response. The one-pole response may be determined from the transmission factors.†

$$H_a = 1 + j\omega \frac{L_L}{R_S} \tag{8.29}$$

$$H_b = 1 + j\omega C_T R \tag{8.30}$$

where $R_S = R_1 + R_2$ and $C_T = C_1 + C_2$.

---

* The attenuation factor a is defined by Eq. (8.4).

† The subscripts a and b of the transmission factor refer, respectively, to Fig. 8.7a and b.

The *LR* circuit of Fig. 8.7*a* is applicable to the audio-frequency low-impedance step-down transformer. The *RC* circuit of Fig. 8.7*b* is applicable to the high-impedance step-up transformer when circuit capacitance predominates over leakage inductance, as it does when cable capacitance is unusually high or when the secondary winding feeds a capacitive load comprising many power transistors connected in parallel.

Attenuation is:

$$A = 10 \log \left[1 + \left(\frac{\omega}{\omega_2} \right)^2 \right] \tag{8.31}$$

where the 3-dB down frequency is $\omega_2 = 2\pi f_2$.

Phase shift may be calculated from the following relationships:

$$\left. \begin{array}{ll} \omega_2 = \dfrac{R_S}{L_L} & \tan \theta = -\dfrac{\omega L_L}{R_S} \\[3mm] \omega_2 = \dfrac{1}{C_T R} & \tan \theta = -\omega C_T R \end{array} \right\} \tag{8.32}$$

Attenuation is at the rate of 6 dB per octave. Phase shift at the output approaches $-90°$ as ω increases indefinitely.

(2) Second-Order Response. Beyond the range of audio frequencies it becomes necessary to include capacitance in the equivalent circuit. Capacitance must be reckoned with even at audio frequencies if circuit impedances are large. In Fig. 8.7*c* and *d*, a capacitance C_1 or C_2 has been added to either the input or output branch. The frequency range for which the second-order solutions (H_c and H_d) are used is the band in which ω is less than either the reciprocal of the input time constant $R_1 C_1$ or less than the reciprocal of the output time constant $R_2 C_2$. In most analyses, capacitance is placed at the output if $n < 1$ (step-up), or at the input if $n > 1$ (step-down) and the source resistance is high.

The two-pole response may be determined by substituting the following values of H into Eq. (8.28):

$$\left. \begin{array}{l} H_c = 1 + j\omega \left(C_2 R + \dfrac{L_L}{R_S} \right) - \omega^2 L_L C_2 a_2 \\[3mm] H_d = 1 + j\omega \left(C_1 R + \dfrac{L_L}{R_S} \right) - \omega^2 L_L C_1 a_1 \end{array} \right\} \tag{8.33}$$

Here $a_2 = R_2/R_S$ and $a_1 = R_1/R_S$.

A simplifying connection between the step-up and step-down circuits can be recognized by employing the reciprocity theorem. If the generator at the input of the step-up circuit is removed and then placed in the output branch in series with the load, we obtain the step-down circuit. The responses of the two circuits are *identical*, provided we change the subscripts of Eq. (8.33) from R_2, C_2, and a_2 to R_1, C_1, and a_1.

In the solution for the two-pole attenuation, Eq. (8.33) must be squared, leading to a quartic equation of the form:

$$|H|^2 = 1 + \alpha\omega^2 + \beta\omega^4 \tag{8.34}$$

where

and

$$\left.\begin{array}{c} \alpha = \left(CR + \dfrac{L_L}{R_S}\right)^2 - 2LCa \\[3mm] \beta = (L_L Ca)^2 \end{array}\right\} \tag{8.35}$$

In Eq. (8.35), a equals R_2/R_S or R_1/R_S, depending on whether C is the output capacitance (step-up) or input capacitance (step-down).

The concept of the *characteristic impedance* Z facilitates analysis:

$$Z = \sqrt{\frac{L_L}{C_D}} \tag{8.36}$$

where C_D is the distributed capacitance (i.e., the sum of the effective self-capacitances), which equals C_1 in the step-down and C_2 in the step-up transformer.

The transmission factor (and the relative magnitude of the output voltage) can now be stated as a function of the characteristic impedance:

$$H = \sqrt{(-a\Omega^2 + 1)^2 + \left[a\Omega\left(\frac{R_1}{Z} + \frac{Z}{R_2}\right)\right]^2} \tag{8.37}$$

where $\Omega = \omega/\omega_r$ and $\omega_r = 1/\sqrt{L_L C}$, the frequency of resonance. The term in brackets is a minimum when:

$$Z = \sqrt{R_1 R_2} \tag{8.38}$$

When the response is maximally flat (see Sec. 9.2a), and when source and load resistances are matched, then:

$$Z = \sqrt{\frac{L_L}{C_D}} = R_2 \tag{8.39}$$

and the high-frequency response does not peak. Substantial peaking does occur if the circuit has high Q, i.e., when $\omega L_L/R_1 > 1$ and $R_2 \gg R_1$.

The high-frequency response in the maximally flat case has a phase shift of 90° at its 3-dB down frequency ω_2:

$$\omega_2 = \frac{1}{\sqrt{L_L C_2 a_2}} \tag{8.40}$$

and falls off at the rate of 12 dB per octave. The phase shift approaches $-180°$ as ω increases. The second-order response, unlike the first-order response, which is monotonic, *can peak inside the pass band* before falling in the stop band.

One- and two-pole equivalent circuits are adequate representations of most designs. Hence the simple LR circuit (Fig. 8.7a) and the LRC half-T circuit (Fig. 8.7c) are those most used by design engineers.*

(3) **Third-Order Response.**[7]   In some circuits, both input and output capacitance must be reckoned with. Both C_1 and C_2 must be taken into account if the resistance of the source is high, the turns ratio n is close to unity, and the transformer is phase-inverting.

The appropriate pi circuit is depicted in Fig. 8.7e and the transmission factor H_e is:

$$H_e = 1 + j\omega\left(C_T R + \frac{L_L}{R_S}\right) - \omega^2 \frac{L_L}{R_S}(C_1 R_1 + C_2 R_2) - j\omega^3 L_L C_1 C_2 R \tag{8.41}$$

In order to compute the attenuation predicted by Eq. (8.27), we square Eq. (8.40) and obtain a sixth-order equation. The responses, relating such key parameters as $R_1/R_2 = B$, $C_2/C_1 = A$, $R_2/\omega_r L_L = D$, and ω/ω_r, have been plotted for a variety of conditions. The frequency ω_r is defined as occurring when $\omega L_L = A/(A+1)\omega C_2$, or (by some writers) as equal to $1/\sqrt{L_2 C_2}$.

A salient feature of the third-order response is that both a valley and a peak may occur inside the pass band. This behavior, called *ripple,* is usually undesirable.

A high-frequency response with ripple not exceeding 0.5 dB is assured when D is between 1 and 1.6. An extended high-frequency response is obtained when $A = B = 1$, which corresponds to the case of matched input and output time constants:

$$C_1 R_1 = C_2 R_2 \tag{8.42}$$

If $D = \sqrt{2}$, a ω/ω_r of 2, with a small ripple (not exceeding 0.25

---

* The class B circuit, a special but important case in which a pair of two-pole networks is used, is discussed in Sec. 9.4a. In this circuit, the source resistance used to predict high-frequency response is the active value of the resistance in series with each half of the primary.

dB) becomes possible. When $D = 1$, the ripple vanishes and the response is 3 dB down at $\omega/\omega_r = 2$. The response can fall as much as 18 dB per octave. The shift in phase can exceed 180° at about $2\omega_r$ and can approach 270° as ω increases beyond $10\omega_r$.

c. General Solutions

There are times when the bridging capacitance C_B cannot be ignored, even if it is small (see Fig. 8.7f). Its presence alters the effective input and output capacitances, so that the values of the three capacitances (referred to the primary) are:[8]

$$\left. \begin{aligned} C_1 &= C_p + C_B\left(1 - \frac{1}{n}\right) \\ C_2 &= \frac{C_s}{n^2} + C_B\frac{1-n}{n^2} \\ C_3 &= \frac{C_B}{n} \end{aligned} \right\} \tag{8.43}$$

Interestingly, there are several situations in which C_1 or C_2 can be negative.

As a consequence of the bridging capacitance, the general solution for the high-frequency response becomes the quotient of two polynomials: the denominator has three poles and the numerator has two zeros. Analysis is exceedingly complicated if we are dealing with a general case in which the input and output time constants are unequal.

The frequency response may dip, or be notched, outside the pass band before the final descent at infinite frequency (see Fig. 8.8). The first dip at ω_M occurs when:

$$\omega_M = \frac{1}{\sqrt{L_L C_B/n}} \tag{8.44}$$

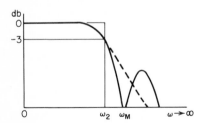

Fig. 8.8 High-frequency response: *m*-derived case. Solid curve, dip or notch at ω_M. Dotted curve, notch eliminated (see text).

Ordinarily, the notch is sufficiently far out in the stop band so as not to be troublesome. There are, however, times when ω_M will occur only slightly further out than the normal cutoff ω_2. This can, for example, happen, in a class B coupling transformer where $C_B \cong C_p/2$ because both the primary and secondary windings have few layers and the turns ratio is close to unity. The dip in frequency can also occur in a multiwinding transformer in which one of the windings is inactive and leakage inductance is high.

On occasion, measurements reveal anomolous high-frequency behavior (e.g., multiple resonances), usually in the stop band. Such behavior reminds us that the lumped-parameter model is artificial, especially when the magnitude of the wavelengths is comparable to the dimensions of the coil. O'Meara has examined the conditions under which a distributed-parameter circuit is more appropriate.[9]

The presence of bridging capacitance, as well as parasitic capacitance in a transformer circuit, can be exasperating to the designer. Because these capacitances are difficult to eliminate or even control, we must consider alternatives to the conventional coil and its concentric layers of magnet wire. One such alternative, used in the transmission-line transformer, is discussed in Sec. 9.5.

d. Phase Response and Distortion

A major characteristic of a network or filter is its phase response, which is related to the distortion or fidelity of the signal's waveform. The phase response of a network, being intimately connected with the delay time, also links analysis in the frequency domain to analysis in the time domain.

A suitable criterion of *fidelity,* one which is both general and stringent, is that the shape of the output wave should resemble the input, but be delayed by a constant interval. Such a requirement is satisfied if the phase shift of each frequency component of the Fourier spectrum of the waveform is a linear function of the frequency. This criterion, uniform or *flat delay,* it is interesting to note, characterizes the ideal transmission line. Such a criterion is particularly important in standard systems of modulation, e.g., amplitude modulation.

We shall digress for a moment in order to trace the evolution of the concept of group delay. In an amplitude-modulation system, the phase velocity of the carrier v_p is:

$$v_p = \frac{d\omega}{d\theta} = \frac{\omega_c}{\theta_c} \tag{8.45}$$

and the velocity of the modulated envelope is ω_m/θ_m.

The group velocity v_g is defined as the relationship of the difference in frequency between the carrier and the sideband $(\Delta\omega_m)$ and the difference in phase constant between the carrier and sideband $(\Delta\theta_m)$:

$$v_g = \frac{\Delta\omega_m}{\Delta\theta_m} = \frac{d\omega}{d\theta} \tag{8.46}$$

If ω is proportional to θ, then both the carrier velocity and the group velocity are equal. Thus, the information embodied in the modulated envelope will arrive at its destination undistorted in phase.

The relevant concepts in network analysis are the reciprocals of the phase and group velocities. If θ/ω is called the *phase delay* of a network, then its derivative, the slope of the phase, is called the *time delay* t_g:

$$t_g = \frac{d\theta}{d\omega} \tag{8.47}$$

Because t_g is also a measure of the delay time of the components in the envelope of a modulated carrier, it is also called *envelope delay*. In a system whose signals consist of a group of pulses, or a pulse train, the integrity of the information may depend on the constancy of the time delay. In this context, t_g is called *group delay*, and its variation is important to the designer.

The criterion of linear phase response is, in the strict sense, not fulfilled in a conventional network containing discrete lumped-constant elements of inductors and capacitors.*

8.6. BANDWIDTH

Before we consider (in the next chapter) the various design techniques at our disposal, it is well to obtain an overall view of the wideband transformer by considering the basic relationship involved in a specified bandwidth.

The bandwidth is ordinarily defined as the difference $f_2 - f_1$. The ratio spanned by the cutoff frequencies is $\omega_2/\omega_1 = f_2/f_1$. The distinction enables us to observe that identical bandwidths (for example, 100 to 10,100 and 1,000 to 11,000) can have markedly different span ratios (101 and 11).

---

* Feasible approximations of linear phase response are discussed in Secs. 9.2 and 10.4. Moreover, linear phase response can be achieved, in principle, in a distributed network—the transmission line. Because of its close connection with the subject of coil geometry, which is considered in Sec. 9.4, we defer most of our discussion of the transmission-line transformer to Sec. 9.5, although its use in hybrid circuits is described in Sec. 8.8*b* of this chapter.

The span ratio of the *first-order model,* the model which is appropriate for the step-down audio transformer, is:

$$\frac{f_2}{f_1} = \frac{R_S/L_L}{R/L_p} = \frac{L_p}{L_L}\left(\frac{1}{m} + m + 2\right) \tag{8.48}$$

where m is the matching factor, $n^2 r_L/R_1$.

This equation may also be written in terms of the geometry and core permeability μ_e:

$$\frac{f_2}{f_1} = g_a\left(\frac{n^2 r_L}{R_1} + \frac{R_1}{n^2 r_L} + 2\right) \tag{8.49}$$

where g_a is a shape factor which takes into account the window proportions and the degree of interleaving of windings (see Sec. 9.4a).

From these equations we can infer that both a large core permeability and a mismatch of generator and load resistances are desirable. Very small resistance of the generator *or* very small resistance of the load will increase the span ratio. The generator which feeds the step-down audio transformer should be a constant-voltage generator (one with small source resistance) or a constant-current generator (one with large source resistance). Note that neither equation explicitly includes size. Size, as we noted earlier, is a function of the insertion loss (which is related to the efficiency of the copper), the voltage level (which is related to the flux density), and the allowable harmonic distortion.

In the important and frequently used *second-order model* of the transformer, the upper cutoff frequency depends on the effective capacitance and the turns ratio. To keep complications to a minimum, we choose for f_2 the value obtained for the matched case of maximally flat response. The equation is derived from Eq. (8.40), when $a_2 = 1/2$.

$$2\pi f_2 = \sqrt{\frac{2}{L_L C_2}} \tag{8.50}$$

The span ratio is then:

$$\frac{f_2}{f_1} = \frac{L_p}{R}\sqrt{\frac{2}{L_L C_2}} = \frac{L_p}{L_L}\left(\frac{\sqrt{2Z}}{a_1 R_2}\right) \tag{8.51}$$

where $a_1 = R_1/(R_1 + R_2)$.

O'Meara has derived an expression which relates all the key variables of the circuit and the transformer and which thus is particularly useful to the transformer designer:[10]

$$\frac{f_2}{f_1} = \frac{M_1 g_3}{\sqrt{2\pi\epsilon_o}}\sqrt{\frac{\mu_e}{\epsilon}\frac{1}{f_1 K_L \sqrt{R_1 r_L}}} \tag{8.52}$$

where $\epsilon_o = 8.854\,(10)^{-12}$.

Equation (8.52) reveals that the band of frequencies spanned is a function of seven major parameters:

1. A winding factor, the turns-ratio function M_1
2. The geometry (i.e., winding configuration) g_b and configuration or shape factor g_c*
3. The lowest operating frequency f_1
4. The mean circuit impedance $\sqrt{R_1 r_L}$
5. The low-frequency impedance ratio K_L†
6. The effective permeability of the core μ_e
7. The dielectric constant of the insulation ϵ

In the following paragraphs, each of these parameters is described briefly so that we can gain perspective on the problem of designing a transformer with a large bandwidth.

(1) **The Winding Factor.** The size of M_1 depends on the turns ratio; it is maximum when $n = 1$. A high step-up turns ratio (i.e., voltage gain) entails a reduction in bandwidth.

(2) **The Geometry.** The factors g_b and g_c are variables which depend on the geometry of the coil and core [see Eqs. (9.22) and (9.23)]. The significant factor g_c is defined as:

$$g_c \propto \left(\frac{A}{\lambda^2 l}\right)^{1/2} \tag{8.53}$$

where A is the net area of the core and λ and l are, respectively, the mean lengths of coil and core. It becomes clear that the band-span ratio is inversely proportional to the square root of the characteristic linear dimension; that is:

$$\frac{f_2}{f_1} \propto \frac{1}{\sqrt{\text{centimeters}}} \tag{8.54}$$

Consequently, in the presence of capacitance, a large band-span ratio is obtainable only if we make the transformer small.

(3) **The Lowest Operating Frequency.** Since f_1 is in the denominator of Eq. (8.52), we expect it to be increasingly difficult to obtain wide bandwidths as the lower cutoff frequency is increased. And, since permeability generally decreases with frequency, we expect, and find, that a large span ratio is increasingly difficult to achieve as we ascend the spectrum from audio to radio frequencies.

(4) **The Mean Circuit Impedance.** Since M_1 is a maximum when $n = 1$, then $r_L = R_2$ and the source is matched to the load. A low

---

* Discussed further in Sec. 9.4c.
† Described in Sec. 8.4b.

operating circuit impedance $\sqrt{R_1 R_2}$ is clearly desirable. It is often advantageous to match source to load when the transformer is physically separated from either the load or generator. This makes it easier to match the cable impedance to the transformer's characteristic impedance. Standard cable impedances, which are low (50, 75, and 300 Ω), favor the attainment of a wide bandwidth when the transformer is used in long-line circuits.

(5) The Low-Frequency Impedance Ratio. Low levels of distortion and high levels of efficiency can be achieved by making source resistance small, K_L high, and size large. Mismatching the load and source may thus become desirable or even necessary. However, the improvement in low-frequency performance may adversely affect the high-frequency response. This is most likely when the voltage levels of the transformer are high. Then a reduction in flux density means a concomitant increase in the size of the transformer, at the expense of the high-frequency performance.

Note that when the impedance factor K_L is increased (in order to reduce both the loading and the low-frequency distortion of an amplifier) the span ratio is reduced. On the other hand, the flux density is small when voltage and power levels are low. Then K_L can also be low, and it is easy to keep distortion small. Under these conditions, it is logical to shift the circuit criterion from maximum efficiency to maximum transfer of energy. Since this occurs when source and load are matched ($R_1 = R_2$), the maximum bandwidth is achieved when circuit impedance is minimum and $n = 1$. Low values of K_L, around 0.5, are possible.

(6) The Effective Permeability of the Core. The choice of core material, as we noted in Secs. 5.3c and 6.2, depends not only on the lowest operating frequency but also on the power level, which determines the maximum flux density.

(7) The Dielectric Constant of the Insulation. Insulation that has a low dielectric constant is clearly desirable.

These considerations permit us to make a quick assessment of the difficulty of designing a transformer for a specified bandwidth. That a design is feasible does not mean it is simple; indeed, a successful physical model requires considerable care, and attention to certain geometric factors.[11,12]

We can now conclude that a transformer designed for one set of impedances and bandwidth may also be operated with another set of impedances but at a different set of high and low cutoff frequencies. The lowest practical limit of circuit impedances equals the total reflected winding resistance; the highest practical limit of circuit impedance equals the power loss attributable to the resistance of the core in an open circuit. The principles we have described make it possible to characterize the performance of a transformer by means of a closed contour graph, a method which permits the widest use of the transformer.[13,14]

8.7. AMPLIFIER TRANSFORMERS

Transformers are often used in power amplifier circuits to perform several functions: impedance matching; polarity inversion; combining, splitting, or isolating of signals;* and conversion from single-ended to push-pull drive. The transformer is an essential component of linear amplifiers to achieve an efficiency commensurate with a specified amount of waveform distortion.[15]

A listing of the various levels of efficiency that are feasible with sinusoidal and nonsinusoidal amplification or processing of signal energy is useful. Table 8.2 includes, in addition to the traditional classes of amplifiers (A, B, C), the less familiar designations which have been assigned to switching circuits.[16]

T A B L E 8 . 2   Classes of Amplifiers

Class	Conduction angle, rad	Percentage of efficiency	Type of load
A	2π	50 max	R
B	π	78.5 max	R
C	$<\pi$	80–90	LC
D	—	85–90	Tuned
E	—	100 max	Tuned
F	—	85–88	Several resonators
S	π	100 max	LCR

The letters A, B, and C describe classes of operation which differ with respect to the angles of conduction of current and the resulting efficiency. In the class C power amplifier, a pulse of current at the input results in amplified power in a tuned LCR circuit at the output. The level of efficiency is high.

In class D amplifier circuits, the drive and coupling resemble the class B circuits, but there are important differences in the waveforms.[16,17] Because the transistors act like bipolar switches, the drive waveforms are square. The load has either a tuned series or shunt LC circuit, depending on whether the drive is a square waveform of voltage or of current.

In Fig. 8.9a, we depict the push-pull voltage switching circuit in

---

* This is accomplished with a hybrid connection, discussed in Sec. 8.8.

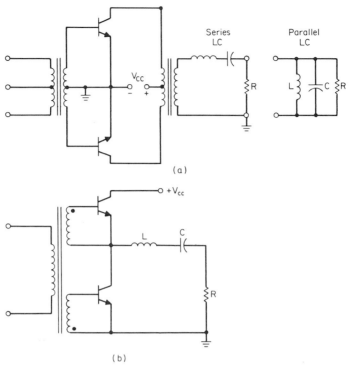

Fig. 8.9 The class D amplifier. (*a*) Push-pull coupling; shunt *LC* is used with current switching. (*b*) Complementary drive.

which both a drive transformer and an output transformer are used. In Fig. 8.9*b*, we see the simpler complementary circuit with a single drive transformer.

A common type of drive is *pulse width modulation* (PWM), in which the pulse rate is constant but the duration (width) of each pulse varies with the amplitude of a reference, or input, signal. The bandwidth of a class D system cannot exceed one-half the rate at which the input signal is sampled. The behavior of class D amplifiers is analogous, in certain respects, to the width-modulation of power that occurs in the switch-mode converter circuits (see Secs. 2.10 and 5.7).

In class E and F amplifiers, a single transistor is used. The load contains either a single resonant circuit (class E) or several resonant circuits (class F).*

In the class S circuit with an *LCR* load (see Fig. 8.9*a*), the shape of the amplified wave depends on whether the drive is provided by voltage-switching or current-switching transistors. The class S designa-

---

* Other designations, such as classes G and H, are described in ref. 16.

tion is used by some writers to define the circuit operation of those inverters in which a square wave of voltage is transformed (by means of pulse modulation) into a sine wave (see Sec. 2.9).

8.8. THE HYBRID TRANSFORMER[18]

The wide-band transformer is generally used to perform basic functions, such as matching a generator to its load. A more specialized application occurs, however, when two or more signal sources are to be coupled between ports and isolation is desired between specific ports.*

In the three-winding hybrid transformer, signals are combined or split under specified conditions of balance.[19,20] In the hybrid circuit shown in Fig. 8.10a, we see four ports, each of which contains a generator and an impedance connected to a winding. Each port is designated by the letters A, B, C, and D. The ability to combine and split signals with the hybrid transformer accounts for the terminology—*combiner*

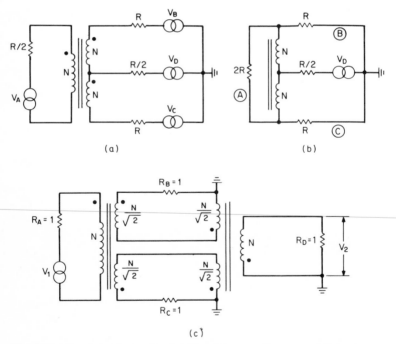

(a)

(b)

(c)

Fig. 8.10 The hybrid circuit. (*a*) Equal turns ratio, unequal impedances. (*b*) Autotransformer. (*c*) Equal impedances, modified turns ratio.

---

* A *port* is defined as any pair of terminals in a network from which signals enter or exit.

(or coupler) and *splitter*—encountered in radio-frequency (RF) circuits. Before undertaking a discussion of the general case, we will describe some simple applications.

The prototype of the function involves three ports: A, B, and C. We wish to couple energy from A to B and from A to C, but not from B to C. The transfer of energy is *bilateral* (i.e., the direction can be from B to A as well as A to B).

Television antenna circuits are often connected to a three-port hybrid transformer. Radio-frequency energy from a VHF antenna (port B) and a UHF antenna (port C) combine at port A, from which a transmission line is connected to the television receiver. There, the same type of transformer splits the signal, furnishing RF energy to both a VHF port and a UHF port on the chassis. As a combiner, the hybrid transformer combines the output of several low-power amplifiers to obtain kilowatts of RF power at frequencies to and beyond 100 MHz. The hybrid circuit is attractive to the designer for several reasons. An important one is that small differences in phase at ports B and C result in only a moderate dissipation of power at port D and in low distortion at the destination, port A.

a. Analysis

The behavior of the general hybrid circuit of Fig. 8.10*a* is perhaps more readily understood when it is viewed as a bridge circuit. Visualize each of the secondary windings N as the arm of a bridge. Port A contains the oscillator and is connected to the primary winding; port D is connected to the null detector.* Port B contains the standard, or known, impedance, and port C contains the unknown impedance. Balance occurs when the impedance of B equals the impedance of C.

Assume that the impedances in A and B are matched and that the bridge formed by the secondary circuit is balanced. Under these conditions, analysis reveals four basic properties of the hybrid circuit:[18]

1. Ports A and D are not coupled to each other. This lack of coupling, called *conjugacy*, means that a signal at A produces no output at D and that a signal at D produces no output at A. Ports B and C are similarly conjugate. The implications are important. If, because of a malfunction, one port of a conjugate pair is open-circuited or shorted, the transmission of a signal between the coupled ports is not impaired.

2. Assume that the impedances of conjugate ports are adjusted for

---

* In a bridge circuit a sensitive meter is connected in series with $R/2$ (Fig. 8.10*a*).

maximum transfer of power to their respective destinations. If the impedances have been adjusted to produce noncoupling (conjugacy), and if a match exists at *one* port, then a match exists at *all* ports.

3. The hybrid transformer may be designed to produce a desired distribution or ratio of power between coupled ports. With a suitable turns ratio, the power is divided as follows:

$$\frac{P_B}{P_A} = \frac{r}{r+1} \qquad \frac{P_C}{P_A} = \frac{1}{r+1} \tag{8.55}$$

Here the subscripts denote the power in the respective ports and the power ratio r equals P_B/P_C.

The condition $r = 1$ is the common case (Fig. 8.10a) where half the power of A is sent to B and half to C. (Half of D's power also arrives at B, and half at C.) Further, if conjugacy exists, the power lost in transmission from A to B is identical to the loss in the transmission from C to D. (The loss from A to C is also identical to the loss from B to D.)

4. Consider four transmission paths, e.g., D to B, D to C; and A to B, and A to C. The phase shift between ports is zero in three of the paths; but in one of the four paths, the phase shift is 180° with respect to all other paths; that is, one of the four signals is inverted in polarity.

The hybrid circuit of Fig. 8.10a can be developed into several levels of complexity. In the simplest case, that is, an autotransformer, if winding N of port A is omitted, then the nonisolated load at A is connected across $2N$, as shown in Fig. 8.10b. A proper match now results when the impedance at A is four times that at D and twice that at B and C. Altering both the normal 1:1 turns ratio and the impedance ratio to a ratio of $k:1$ results in the delivery of power to the ports in a prescribed proportion. For example, power from D can be divided so that P_C is k times P_B, provided that R_B is k times R_C.*

The use of two cores (each with three windings as in Fig. 8.10c), permits all four ports to have the same terminal impedance when the ratio of one winding to the other two is changed to $1:\sqrt{2}$. If each of the two cores has four windings of equal turns N, then a four-way split of power is achieved with eight ports.

Its bilateral behavior is of major importance and has led to the use of the hybrid transformer in bidirectional amplifier circuits (called *repeaters* in the telephone system). The absence of an output (at A)

---

* Also, $R_D/R_C = k/(1 + k)$ and $R_A/R_C = 1 + k$.

when a signal (at D) is placed in series with the center tap provides the basis for carrier suppression in the balanced modulator circuit which is used in single-sideband systems.

The degree of balance depends on the symmetry of leakage inductance, self-capacitance, stray capacitance, and dc winding resistance. Consequently, the hybrid transformer succeeds to the extent that thorough-going symmetry and precision of coil geometry can be maintained.* In certain long-line communication circuits, a large number of transmission transformers is cascaded. Performance will then depend on the *uniformity* of values of L and C as well as on the symmetry of stray capacitances. Extraordinary measures are sometimes taken to ensure the precision of the coils' construction. It should be noted that while close coupling between halves of the balanced winding is important, the exact matching of impedance at each port is not required.†

b. Phase Inversion and Grounding

In Sec. 9.5 we will discuss a specific type of coil construction known as the *transmission-line geometry*. Here, for the sake of completeness, we describe several types of transmission-line connections which make use of the hybrid characteristic (that is, conjugacy) and permit a desired type of phasing and grounding at a specified port.

An important characteristic of amplifier circuits is the ability to provide coupling between push-pull (bipolar) and single-ended (unipolar) circuits, with or without phase inversion.[21] In a hybrid circuit this usually requires two transformer cores. Port A of Fig. 8.10a and b cannot be grounded without altering (usually markedly) the balance and isolation characteristics of the circuit. In Fig. 8.11a, we have redrawn Fig. 8.10b in the format of the transmission-line connection. In this single-core circuit, *either* port A or D, but not both, can be grounded.

Greater flexibility in grounding and phasing results from the use of two cores, as in Fig. 8.11b and c.[22] In the two-core combiner (Fig. 8.11b), a signal (+E) from an amplifier at port B and a signal (−E) at port C combine in port A. The failure of one amplifier has no adverse effect on the other. In the same circuit, there is *phase inversion* between ports B and C. All ports (B, C, D, and A) can be grounded. In the circuit shown in Fig. 8.11c, however, ports B and C are *in*

---

* If a dc polarizing bias is present, the symmetry of the flux paths of the core becomes important.

† The subject of coil design is discussed further in Sec. 9.4.

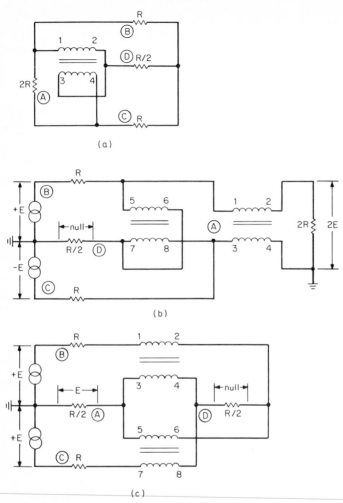

Fig. 8.11 Hybrid transmission-line connections. Conjugate ports are A-D and B-C. (*a*) Single-core construction. (*b*) Two-core combiner, with all ports grounded; note the phase inversion between B and C. (*c*) Two-core combiner, with three ports grounded; note that ports B and C are in phase.

phase, but only three ports (B, C, and A) are grounded. Port D, which is conjugate to A (and therefore receives no power), is not grounded.

In Chap. 9 we shift our perspective from analysis to synthesis. We, therefore, discuss a range of topics which are conducive to the objective of matching the specifications of a black box (the wide-band transformer) to its performance.

REFERENCES

1. E. Snelling, *Soft Ferrites* (London: Iliffe Books, 1969), pp. 238–40.
2. Reuben Lee, *Electronic Transformers and Circuits*, 2d ed. (New York: John Wiley & Sons, 1959), pp. 155–58.
3. H. Lord, "Design of Broad-Band Transformers," *Transactions AIEE*, Vol. 69 (1950): 1,005–10.
4. T. O'Meara, "Analysis and Synthesis with the Complete Equivalent Circuit of the Wide-Band Transformer," *Transactions AIEE*, Vol. 81 (March 1962): 55–62.
5. ———, "Wide-Band Transformers and Associated Coupling Networks," *Electro-Technology*, Vol. 70, No. 3 (September 1962).
6. A. Zverev, *Handbook of Filter Synthesis* (New York: John Wiley & Sons, 1967).
7. J. Howe et al., *High Power, High Voltage, Audio Frequency Transformer Design Manual: Final Report* (Holyoke, Mass.: General Electric Company, 1964), pp. 36, 102, 103. ASTIA Document 607, 774.
8. MIT Electrical Engineering Staff, *Magnetic Circuits and Transformers* (New York: John Wiley & Sons, 1943), p. 490.
9. T. O'Meara, "A Distributed Parameter Approach to the High Frequency Network Representation of Wide-Band Transformers," *IRE Transactions on Component Parts*, Vol. CP-8, No. 1 (March 1961): 23–30.
10. ———, "A Comparison of Thin Tape and Wire Windings for Lumped-Parameter, Wide-Band, High-Frequency Transformers," *IRE Transactions on Component Parts*, Vol. CP-6, No. 2 (June 1959): 49–57.
11. G. Gray, "Toroidal Transformers Pass Video Bandwidths," *Electronics*, Vol. 29 (May 1956): 150–53.
12. L. Symons, "Small-Signal Opto-Electronic Transformer," *Electronic Engineering* (November 1969): 35–39.
13. F. Kasper, "Characterizing Transmission Transformers for Multiple Use," *IEEE Transactions on Component Parts*, Vol. CP-11, No. 2 (June 1964): 322–29.
14. *Contour Characterization of Electronic Transformers*, IEEE Standard P669 (New York: Institute of Electrical and Electronic Engineers, 1977).
15. L. Giacoletto, *Electronic Designers' Handbook* (New York: McGraw-Hill Book Company, 1977), Sec. 15.
16. H. Krauss, C. Bostian, F. Raab, *Solid-State Radio Engineering* (New York: John Wiley & Sons, 1980), Chaps. 12 and 14.
17. J. Murray, G. Oleszek, "Design Considerations in Class D MOS Power Amplifiers," *IEEE Transactions on Industrial Electronics*, Vol. IE-CI 26, No. 4 (November 1979): 211–18.
18. E. Sartori, "Hybrid Transformers," *IEEE Transactions on Parts, Materials and Packaging*, Vol. PMP-4, No. 3 (September 1968): 59–66.
19. E. Bartelink, *Telephone Transmission Theory*, 2d ed. (Concord, N.H.: Bartelink Enterprises Corp., 1974).
20. T. Gross, "Hybrid Transformers Prove Versatile in High Frequency Applications," *Electronics*, Vol. 50, No. 5 (March 3, 1977): 113–15.

21. H. Granberg, "Combine Power without Compromising Performance," *Electronic Design* (July 19, 1980): 181–87.

22. O. Pitzalis, T. Couse, "Broadband Transformer Design for RF Transistor Power Amplifiers," *Proceedings National Electronics Conference, 1968,* pp. 207–16.

CHAPTER 9

The Wide-Band Transformer: Synthesis

The task of the transformer designer is to synthesize, from a set of specifications, a satisfactory design. The behavior of a wide-band transformer was described in Chap. 8, and we know, from Eq. (8.52), that seven major parameters must be taken into account if we are to achieve a desired band span. Not infrequently, however, it is necessary to expand the bandwidth at both ends in order to achieve the desired fidelity and compensate for the limitations imposed by feedback. The performance desired may be the limit of the art. This raises the question of feasibility; consequently, compromises, tradeoffs, and optimization become important design considerations.

It is at this juncture that the analytic approach used in Chap. 8 can prove awkward and even disadvantageous. There we assumed that one or another model (i.e., an equivalent circuit) would be appropriate and then asked: What is the response? When we use the synthetic approach to design, we first ask: What performance do we want? And then: Which model is most appropriate to the desired performance? How do we shape the transformer elements into that model? This is the creative approach. Synthesis is the approach we use to achieve a dominating objective of present-day design of electronic equipment—limit-of-the-art performance. We have the specifications; our job is to synthesize the elements inside the black box, the wide-band transformer.

In Sec. 9.1, after making a comprehensive checklist of specifications, we discuss feasibility, fidelity, and feedback, principally in order to explain why it is necessary to expand the bandwidth beyond the specified limits. In Sec. 9.2, we describe how optimum performance can be obtained (or approached) by choosing certain filter models and by using techniques borrowed from the theory of network synthesis. In Sec. 9.3, we describe considerations in the design of the core, principally its geometry and size, and the value of multiple cores. An optimum network model has repercussions on the design of the coil, and these are discussed in Sec. 9.4. In Sec. 9.5, we describe the transmission-line transformer. Sec. 9.6 is a summary of the rudiments of a rational design procedure.

9.1. EVALUATING THE SPECIFICATIONS

Because of the many factors to be weighed in the design of a wide-band transformer, it is desirable that specifications be complete.

a. Specifications

Our checklist of electrical specifications begins with the basic *mid-band* information (items 1 through 5). These are essentially the specifications required for the design of the narrow-band power transformer discussed in Chap. 5:

1. Power level and voltage level
2. Maximum flat loss, or minimum efficiency
3. Size
4. Source and load impedances
5. Turns ratio or the desired ratio of impedances

For the wide-band transformer we must, of course, have a specified *bandwidth:*

6. The lowest and highest operating frequency, and the number of decibels (dB) down with respect to the mid-frequency

For certain applications, the *shape* of the amplitude response is specified:

7. Either the permissible ripple, i.e., the fluctuation of amplitude in the pass band, or whether the response is to be flat

8. The shape of the skirt in the stop band, i.e., the slope, or rate of attenuation

9. The minimum attenuation allowable in the stop band, and the amount of ripple, if any, beyond cutoff

When *fidelity* of wave shape is important, it may be necessary to have specified:

10. The nature of the generator (e.g., transistor), its power capability or level, the class of operation (e.g., A, AB, B), and the degree of unbalanced direct current

11. Variation in input impedance of the loaded transformer at low and high frequencies

12. Maximum phase shift in the pass band

13. Maximum harmonic distortion contributed by the transformer core

14. Whether feedback is employed; if so, the amount of feedback, whether the transformer is inside or outside the feedback loop, and the slope of the phase shift in the stop band

15. The circuit schematic, especially if the source or load is nonlinear or the circuit is complicated

b. Feasibility

A careful study of the specifications will then lead to one of several conclusions about feasibility:

1. The specifications are practical.

2. The specifications are feasible but not necessarily practical. Sometimes certain simple optimization techniques can be employed; other, less practical, designs will require auxiliary networks.

3. The specifications are not feasible. A reevaluation often results in a compromise or a tradeoff of incompatible requirements.

c. Fidelity and Feedback

To meet stringent requirements of fidelity, we have to consider one or more of the following design objectives:

1. Reasonably uniform impedance in the loaded transformer over the pass band

2. Negligible harmonic distortion resulting from core hysteresis

3. A small, preferably linear, phase shift in the pass band

4. The use of stable negative feedback

(1) Uniform Impedance. As we saw in Secs. 8.4*a* and 8.5*d*, when the generator is nonlinear a reactive load produces, at both high and low frequencies, an elliptical load line whose phase angle does not ordinarily exceed 30°. Distortion is slight, but it may become appreciable if the impedance of the load departs substantially from its midband value. These considerations, which may influence our choice of the characteristic impedance of the transformer or the network model, are discussed later.

(2) Harmonic Distortion. This is often substantially reduced if we lower the low-frequency cutoff, the flux density, or the resistance of the generator.

(3) Linear Phase Shift. Traditional filter theory demonstrates that ideal transmission through a network occurs when it has a linear phase characteristic and produces a constant amplitude of waveform at the output. As noted in Sec. 8.4*e*, a substantial departure from linearity causes nonuniform time delay. While this is not of special significance in audio and many other sinusoidal applications,* it can have adverse effects on the transmission of certain nonsinusoidal waveforms and pulse trains.[1] This topic is discussed further in Secs. 10.4*b* and 11.3*b* because of its special relevance to pulse response and fidelity.

(4) Negative Feedback. When a transformer is used in a feedback amplifier, additional factors come into play. The performance and stability of the amplifier will depend on: (1) whether the transformer is inside or outside the feedback loop, (2) the *rate* at which the amplitude and phase shift change with frequency, (3) the maximum phase-shift change at frequencies far beyond cutoff, and (4) the amount of feedback. In Sec. 8.5*b* we noted that at high frequencies a transformer can behave like a one-, two-, or three-pole network and, therefore, produce phase shifts approaching 90°, 180°, or 270° in the stop band.

If a coupling transformer is inside the feedback loop (as it would be if feedback is obtained from the secondary winding), then the possibility of oscillation is very real. This may be inferred from:

$$A_f = \frac{A}{1 - A\beta} \tag{9.1}$$

where A_f is the gain with feedback, A is the gain without feedback, and β is the fraction of the output fed back to the input. β is negative with inverse feedback, but it becomes positive when phase shift reaches 180°. Instability begins when $A\beta = 1$.

Since the problem of instability is of great importance, we shall try to present some guidelines for the specification of transformers to be

---

* It is important in long-time transmission, however, in order to minimize echo.

used in a feedback system. Consider, for example, 2 one-pole networks (e.g., *RC* networks) preceding an audio transformer which is equivalent to a one-pole *LR* network. Assume that each network has the same time constant. It is desired to use feedback from the secondary winding via a nonreactive network to the input. The ensemble of networks reaches 180° when each network reaches 60°. This occurs when amplification of each network falls by a factor of 2, and *A* of Eq. (9.1) changes by 2^3, or 8. Stability is assured only if $A\beta \leqq 8$. The amount of feedback, at mid-band frequencies, must not exceed 20 log (1 + 8) = 19 dB. If, in this three-stage amplifier, the transformer behaves like a two- or three-pole network, either the amount of feedback or the number of stages traversed by the loop must be reduced.

It is axiomatic for designers of feedback amplifiers to provide one additional octave of bandwidth for each 10 dB of feedback, plus one extra octave for safety. A transformer inside a feedback loop, for the audio range of 40 to 10,000 hertz (Hz), may therefore be required to pass the expanded range of 10 to 40,000 or even 5 to 80,000 Hz. In the expanded bandwidth, phase shift is reduced at 40 to 10,000 Hz.* When a large amount of feedback over many stages is necessary, special steps must be taken to shape the amplitude and phase characteristics of the amplifier and of the feedback loop itself.[2] As a rule, it is simpler to reduce the number of stages traversed by feedback. Many times, instability can be avoided simply by obtaining the feedback from the primary rather than the secondary side of the transformer, especially if its behavior reveals that its high-frequency response is that of a three-pole or *m*-derived model (see Fig. 8.7*f*).

The effective source resistance depends on $A\beta$ (the amount of feedback), which will in turn depend on whether the loop is connected to the primary or secondary of the transformer. If there is a large amount of voltage feedback, the low-frequency response can be extended. As a result, audio amplifiers can have exceptionally wide bandwidth and little distortion with transformers not much larger than conventional power transformers.[3] The effective source resistance also influences the phase shift at high frequencies. It should be known or calculated, so that the leakage inductance and capacitances can be properly proportioned. Thus the effective source resistance has a bearing on our choice of both the low- and high-frequency constants of the transformer.

We conclude that fidelity, with or without feedback, is enhanced

---

* An expanded high-frequency response may be necessary even when there is no feedback because group delay [defined in Eq. (8.47)] may be constant over only the first two-thirds or three-fourths of the bandwidth. See Table 10.2.

by a bandwidth in excess of the operating range. Further, if the trans-former is part of a feedback loop, ripple and peaking in the pass band are undesirable and a gradual rate of cutoff in the stop band is desirable.

It is important to appreciate the fact that the design will also be affected by the character of the signal. If the signal has discontinuities (e.g., tone bursts), then the transformer's response should be flat, exhibit no ripple, and fall off gradually in the stop band. A one- or two-pole model is preferable to a three-pole model. If it is not possible to avoid using a three-pole model, or if its use is indicated for some other reason, the bandwidth must be substantially increased in order to keep the phase shift low within the operating range.

It is clear, then, that the specifications and design of a transformer will be influenced by the amount of feedback, the location of the trans-former with respect to the feedback loop, and the type of signal to be passed.

9.2. OPTIMUM PERFORMANCE AND NETWORKS

Since the transformer behaves like a band-pass filter, it is susceptible to *network analysis.* Indeed, this is the point of view used in Chap. 8. In designing the transformer, however, our objective is to make sure it conforms to the specifications. Since we must exercise complete control over the various physical constants of the transformer in order to synthesize the specifications, it is quite appropriate to apply the methods of *network synthesis* to the design of a wide-band transformer.

a. Filter Models[4,5]

The transformer is a band-pass filter, and the relevant branch of modern network synthesis is the theory of wave filters. When a model based on this approach is constructed, it exhibits some *optimum* type of performance.[6]

In designing a wide-band transformer, we desire one of three basic types of performance:

1. If *fidelity* is central, then a flat, smooth response, with gradual attenuation in the stop band, is desired. When constant time delay is also required, phase shift in the pass band should be a linear function of frequency. For this type of performance, the available models are: (*a*) the *m*aximally *f*lat *a*mplitude (MFA) or Butterworth filter; (*b*) the *m*aximally *f*lat *e*nvelope *d*elay (MFED) or flat time delay filter, the Thomson, Bessel filter; and (*c*) the transmission line, to be discussed in Sec. 9.5.

A carefully designed wide-band transformer (with isolated, multi-layer windings), when properly matched to source and load, can be expected to behave like a Butterworth filter. But analysis of the three basic models (poles of the first, second, and third orders) reveals that phase shift is not linear in any, nor is delay constant (see Table 10.2).

When a group of complex signals passes through a transformer, the individual components of the Fourier spectrum are delayed in unequal amounts. Some degree of distortion of the wave shape is inevitable. This distortion is most conspicuous when the applied waveforms are discontinuous or change abruptly in magnitude. When the fidelity of the wave shape is to be stringently constrained, the designer must strive for linear variation of the phase angle as the frequency changes.

The simplest procedure is to extend the high-frequency response beyond that which suffices to assure uniform magnitude of the response. In the absence of a specification of phase-shift variation, the designer sensibly chooses an upper frequency limit which produces an insertion loss of say 0.1 dB. This limiting frequency will occur at about six to ten times the nominal cutoff frequency.*

2. If *maximum bandwidth* is desired, then the rate of cutoff in the stop band will be high and there will be ripple in the pass band, stop band, or both. The uniformity of amplitude in the pass band is gauged by the amount of ripple permissible. For this type of performance, we have available steep-cutoff, maximum-bandwidth filters: (*a*) the Tchebyshev filter produces coequal ripples (also called equal ripples and equiripple characteristic) in the pass band and (*b*) a maximally flat response in the stop band.

3. *Optimum performance* of some favored parameter (e.g., rise time) may be desired. This is accomplished by deliberate distortion of certain performance characteristics (see Sec. 11.3).

Of the various possible filter responses, two are particularly relevant to transformer design: the Butterworth (Fig. 9.1*a*) and the Tchebyshev (Fig. 9.1*b*). In Table 9.1 we have listed the normalized values of the elements of these two filter models, using for the Tchebyshev filter only the three-pole values. The corresponding one-, two-, and three-pole circuit models were shown schematically in Fig. 8.7. The values of elements C_1, L_L, and C_2 are normalized by making R_1, R_2, and the cutoff angular frequency all equal to unity. (It is very convenient to use normalized values of the elements, and we shall come back to the topic later in this section.)

In the Butterworth filter, the cutoff frequency is 3 dB down from the mid-band frequency. In the Tchebyshev filter, the cutoff fre-

---

* Filter models and phase response are discussed further in Secs. 10.3, 10.4, and 11.3.

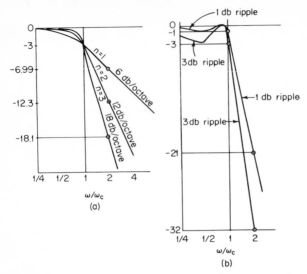

Fig. 9.1 Filter responses. (a) Butterworth responses: the number of poles $n = 1, 2, 3$. (b) Tchebyshev responses: $n = 3$; 1-dB and 3-dB ripple responses.

quency ω_c is defined as being $\frac{1}{10}$, $\frac{1}{4}$, $\frac{1}{2}$, 1, 2, or 3 dB down from the mid-band frequency, in each case equaling the ripple, in decibels. Table 9.2 is a list, for each Butterworth model, of the normalized frequency and of the phase shift at which attentuation is $\frac{1}{10}$, $\frac{1}{4}$, $\frac{1}{2}$, 1, 2, and 3 dB.

Table 9.1 also contains the normalized values of the characteristic impedance $\sqrt{L_L/C_D}$ and of the $L_L C_D$ product (where $C_D = C_1 + C_2$).

TABLE 9.1 Values of Elements of the Butterworth and Tchebyshev Filters*

Figure	Butterworth			Tchebyshev					
	9.6a	9.6c	9.6e			9.6e			
Number of poles, n	1	2	3	3	3	3	3	3	3
Ripple, dB	0	0	0	$\frac{1}{10}$	$\frac{1}{4}$	$\frac{1}{2}$	1	2	3
C_1	—	—	1	1.032	1.303	1.596	2.024	2.711	3.349
L_L	2	$\sqrt{2}$	2	1.147	1.146	1.097	0.994	0.833	0.712
C_2	—	$\sqrt{2}$	1	1.032	1.303	1.596	2.024	2.711	3.349
$Z^2 = L_L/C_D$	—	1	1	0.556	—	—	0.246	0.154	0.107
$L_L C_D$, 3 dB	—	2	4	2.36	—	—	4	4.51	4.77
$L_L C_D$, 1 dB	—	1.18	2						

* Element values are normalized with respect to $\omega_c = 2$ and $R_1 = R_2 = 1$. $C_D = C_1 + C_2$ (see Fig. 8.7).

TABLE 9.2   The Butterworth Filter: Attenuation and Phase Shift vs. Normalized Frequency (ω/ω_c)

Order, n	Attenuation, dB	$\frac{1}{10}$	$\frac{1}{4}$	$\frac{1}{2}$	1	2	3	6.93	12.30	18.18
1	ω/ω_c	0.152	0.243	0.350	0.51	0.763	1	2	3.9	8
	degrees θ	8.66	13.7	19.3	27.1	37.4	45			
2	ω/ω_c	0.39	0.493	0.591	0.713	0.875	1	1.45	2	2.85
	degrees θ	—	—	—	—	—	90			
3	ω/ω_c	0.535	0.624	0.705	0.798	0.914	1	1.25	1.6	2
	degrees θ	—	—	—	—	—	135			

Note that the product $\sqrt{L_L C_D} = 2$ is the same for the Butterworth three-pole model as for the Tchebyshev three-pole 1-dB ripple model. We may write, for either three-pole model:

$$\omega_2 = \omega_c = \frac{2}{\sqrt{L_L C_D}} \tag{9.2}$$

For the two-pole Butterworth model we may write:

$$\omega_2 = \frac{\sqrt{2}}{\sqrt{L_L C_D}} \tag{9.3}*$$

By choosing the 1-dB ripple model we obtain, inside the pass band, more uniformity (not flatness) of amplitude, than we obtain with the Butterworth model. This may be seen in Fig. 9.1. Since the product $L_L C_D$ is approximately constant in a given coil, it should be possible to extend the high-frequency response by choosing the Tchebyshev 1-dB ripple model rather than any of the Butterworth models. However, the extension of the high-frequency response is not obtained without cost: the rate of attenuation sharpens in the neighborhood of the cutoff frequency. The Tchebyshev model is therefore ordinarily shunned if the transformer is to be *inside* a feedback loop.

Note that only in the Butterworth models is the characteristic impedance Z matched to the load:

$$Z = \sqrt{\frac{L_L}{C_D}} = R_2 \tag{8.31}$$

It is possible now to confirm our earlier statement that it is desirable to match input and output time constants in the three-pole model:

$$C_1 R_1 = C_2 R_2 \tag{8.42}$$

---

* This statement is the same one made in Eq. (8.40) because matching requires that attenuation a equal $\frac{1}{2}$.

This equation has important repercussions in the design of a step-up transformer, which we shall now discuss.

b. Limitations of the Step-Up Ratio

It is of considerable practical importance to note that the high-frequency response and bandwidth of a matching transformer (with step-up ratio N) encounters a natural limitation imposed by the load time constant $R_2 C_2$ (see Fig. 8.7c). To illustrate, if our load consists of $R_L = 10$ kilohm (kΩ) shunted by a small value of $C_L = 10$ picofarads (pF) [a time constant of 0.1 microsecond (μsec)], then the cutoff frequency (when the source resistance is matched to the load resistance) cannot exceed:

$$\omega_c = \frac{1}{R_2 C_2} = \frac{1}{(R_L/N^2) C_L N^2} \tag{9.4}$$

If $C_1 = C_2$, the cutoff frequency can be between 5 and 10 MHz but cannot exceed 10 MHz regardless of how good the transformer is or how its elements are proportioned.*

We can also infer, from a theorem of Bode, that there is a basic limitation on the useful step-up ratio of an *ideal* wide-band transformer.[2] The theorem deals with an ideal matching network having a flat transmission characteristic. It can be shown that:

$$N_{max}^2 \leqq \frac{\pi}{2} \frac{1/\omega C_L}{R_L} \tag{9.5}$$

where N_{max}^2 is the maximum step-up impedance ratio of an ideal flat transformer and $1/\omega C_L$ is the reactance of the capacitance at some upper frequency ω at which response is still flat.

Example. Let us consider some typical circuit values, say, a load resistance of 10 kΩ, shunted by a stray capacitance of 10 pF. If response is to be flat to 100 kHz, the turns ratio may not exceed $\sqrt{(\pi/2)15.9} = 5$. Now if we want the response to remain flat out to 1 MHz, an additional decade, we must reduce either the load resistance or the ratio by the factor $\sqrt{10}$. In Sec. 9.4d we shall see that a large step-up ratio N is unfavorable to a wide band span. We now note that the natural upper bound of this ratio is set by the lumped-circuit input and output capacitances C_1 and C_2:

$$N_{max} = \sqrt{\frac{C_1}{C_2}} \tag{9.6}$$

---

* In theory, there is a value of leakage inductance which will raise the cutoff frequency to its maximum.

Practical values are in the range of 5 to 10, which often disappoints the circuit designer who is anxious to step voltages up from an impedance R_1 to some arbitrarily high value of R_2.

There are several ways of overcoming this limitation on the step-up ratio (provided that the bandwidth is narrow). One is to use the flyback transformer circuit described in Sec. 5.7a. Another is to make use of the fact that in an LC circuit, at resonance, the voltage across either reactive element is proportional to Q (quality factor). The voltage gain (i.e., the step up) can then be made proportional to the product of the Q of two resonant loops.[7] Gain is traded for bandwidth, since they are inversely related. Third, where high dc voltage is needed, a voltage multiplier circuit can be added to the output of the step-up transformer.

c. Network Techniques

The first step in adapting filter theory to transformer design is to select suitable low-pass and high-pass models. The next step is to translate the normalized model into a form which, subjected to certain techniques, will yield practical final values of the inductances and capacitances.

The procedure involves various techniques and theorems developed in network theory and employed by the filter design engineer.[5] These include: (1) scaling from normalized values of L and C to actual values, (2) using the partition theorem to obtain element values when source and load resistance are mismatched, (3) transforming the low-pass to the high-pass filter in order to obtain a preferred low-frequency as well as high-frequency characteristic, (4) applying the duality principle in order to convert a pi-section into a T-section filter, and (5) using the reciprocity theorem and Norton's theorem to predict the step-down response from the step-up response and to transform a voltage source into a current source.

The first four network techniques are illustrated in Fig. 9.2. The entire subject is of great interest and all five techniques are useful, but we shall limit our discussion to the topics of scaling, reciprocity, and wide-band matching networks.*

(1) **Scaling.** In Fig. 9.2, as in Table 9.1, the values of the elements [in farads (F) and henrys (H)] were reached by setting the cutoff radian frequency and the terminating resistances equal to unity. To obtain the actual values of L' and C' when the cutoff frequency ω_c and the

---

* A more complete discussion may be found in ref. 5.

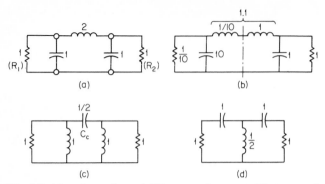

Fig. 9.2   Transformations. (*a*) Prototype low-pass pi Butterworth model, normalized; matched source and load resistance. (*b*) Partition of (*a*): $R_1 = R_2/10$. (*c*) Transformation of low-pass (*a*) into pi high-pass model. (*d*) Dual of high pass (*c*), converting pi network into a T network.

load resistance R are specified (in the following example, R is equal to the source resistance), we perform the following scaling operations:

$$L' = L\frac{R}{\omega_c} \qquad C' = C\frac{1}{R\omega_c} \tag{9.7}$$

Suppose we are to design a line transformer to match a 500-Ω source to a 500-Ω load with a maximally flat response and a cutoff at 20 kHz. We select for the 1:1 ratio transformer the Butterworth three-pole model (Fig. 9.1*a*) and obtain the required values, 7.93 millihenrys (mH) and 0.0159 μF (for the elements input and output capacitance), from Eq. (9.7).

(2) **Reciprocity.** In Fig. 9.2, both the input and output are terminated by shunt resistances. (The generator is omitted in order to emphasize the fact that either a voltage or a current generator may be used.) This practice facilitates our use of the reciprocity theorem, which, because of its great practical importance, is here stated in its several forms:

1. If a voltage in one branch (of a linear, bilateral, passive network) produces a current in the second branch, the same voltage in the second branch will produce the same current in the first branch. This may also be stated conversely, in terms of a current at a node producing a voltage at the output.

2. If a constant-voltage generator at the input is interchanged with an ammeter in the output branch, there will be no change in the reading. This may also be rephrased in terms of the interchange of a constant-current generator and an output voltmeter.

3. If the voltmeter and current source of a network are interchanged, then, except for a constant multiplier, the transfer impedance is unchanged.

With the aid of the reciprocity theorem, we can deduce the step-down response from the step-up response. In Fig. 9.3, in which the generators are interchanged, the output capacitor becomes the input capacitor. It is clear that the step-up and step-down circuits are made identical simply by interchanging the subscripts 1 and 2 (see also Sec. 8.5).

Norton's theorem enables us to transform a constant-voltage generator E in series with a resistance R into a constant-current generator $I = E/R$. As a consequence, we may equate the performance of constant-current circuits (Fig. 9.3) with their counterparts, the constant-voltage circuits (Fig. 8.6c and d).

(3) **Auxiliary Networks.** One method of extending the bandwidth is to use two complementary transformers in parallel, one behaving like a low-pass filter and the other like a high-pass filter. O'Meara[8] has demonstrated the feasibility of extending the bandwidth by a factor of 40 (about five octaves) as compared with the bandwidth of a single wide-band transformer.[9] But the drawbacks of such multiple-pole networks become evident in feedback amplifiers or when faithful pulse transmission is required. Fidelity is compromised by the excessive overshoot and nonuniform time delay.

In the megahertz (MHz) band, a simple way of moderately extending the high-frequency response is to add discrete capacitors across the input and output terminals. This build-out technique optimizes the LC elements in a filter model. Optimum values of C are determined on the laboratory bench.[10] More elaborate build-out networks (e.g., an additional coupling capacitor and shunt inductance) are also used.[11]

The transformation of impedance over a broad band can be achieved, of course, with a transformer having a conventional coil geometry (concentric, multiple-layer windings). At frequencies in the VHF (30 to 300 MHz) and the UHF range (300 to 3,000 MHz), departures from

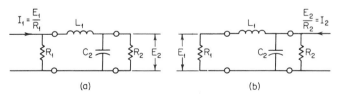

Fig. 9.3 Reciprocity: constant-current circuits. (*a*) Step-up case, $N > 1$ ($n < 1$) equivalent to Fig. 8.7c. (*b*) Step-down case, $n > 1$; similar to Fig. 8.7d.

low-frequency geometry are feasible. One example is the transmission-line transformer discussed in Sec. 9.5.

Another type of matching element, well known to antenna designers, is the *quarter-wave transformer*.[12] It is useful over only a very narrow band of frequencies, and we mention it because of its relevance to transmission at very short wavelengths. This component is simply a transmission line one-quarter wavelength long, whose characteristic impedance is chosen to equal the geometric mean of input and output terminating impedances. Insertion loss is a function of the ratio between the impedances and the bandwidth.

9.3. DESIGN OF THE CORE

The choice of a core involves considerations of geometry and of size, which is in turn affected by its magnetic characteristics.

a. Geometry

The subject of geometry is of great importance in the wide-band transformer and, therefore, warrants extended discussion. In this section we consider the choice of the shape of the core. Later (in Sec. 9.4) we deal with the geometry of the conventional coil and (in Sec. 9.5) the geometry of the transmission-line transformer.

The choice of a specific core *shape* is strongly influenced by: (1) the type of coupling (e.g., push-pull), (2) whether balance between windings is important, and (3) the desired high-frequency response. From experience we learn that the shape of the core evolves from the desired shape of the coil.

If the bandwidth is not greater than a decade, the selection of the core's shape is likely to be based on convenience or cost. When maximum bandwidth is mandatory, however, the toroid is the preferred shape.[13] But in circuits where unbalanced direct current is significant, a small series air gap is necessary. The E core or pot core is then a more suitable choice.

b. Size of the Core

The major factors which affect the low-frequency response were discussed in Sec. 8.4. We have noted that the size and Q of the core are closely related to its magnetic characteristics. Data for normalized inductance A_L, reactance, and shunt resistance R_P (to account for power loss in the core), when furnished by the manufacturer, are helpful in the initial phase of the design of the core.

The choice of the magnetic material is based on the minimum frequency to be transmitted. Size depends on the power level, efficiency (or insertion loss), and other factors which we will now consider.

Ferromagnetic distortion, as we noted in Sec. 6.5, varies inversely with the square root of the volume as well as with the impedance factor K_L [see Eq. (8.25)]. In the wide-band transformer, size depends on the allowable distortion (which influences the impedance factor K_L) and on the regulation, which depends on the mid-band efficiency η (a function of the flat loss). These factors may be related by the Q expression:

$$2Q \cong Q_{\mathrm{Cu}} \cong \frac{4\omega L_p}{2R_a} = \frac{4\omega L_p}{R_L \text{ reg}} = \frac{4K_L}{1 - \eta} \tag{9.8}$$

where L_p is the primary inductance, $2R_a \cong R_{ab}$ is the total winding resistance reflected to the primary, and $R_L = R_2$ is the reflected load resistance.

We infer that concomitant requirements of high inductance and low flat loss result in high Q and a large core. The size (or area product $A_{\mathrm{Fe}}A_{\mathrm{Cu}}$) corresponding to the Q may be obtained from graphs of Q vs. core size (e.g., Fig. 6.10) or inferred from charts of volt-ampere (VA) ratings of the power transformer (e.g., Tables 5.2 and 5.7). A convenient starting point for choosing the size of the core is the relation:

$$\mathrm{VA} = \omega L I^2 \tag{2.6}$$

where $\omega = 2\pi f_1$, f_1 is the lowest frequency at which power is to be transmitted, and I is the ac load current in the primary winding ($L = L_p$).

Since high Q can be obtained more easily at high frequencies, the size of the wide-band transformer (like the power transformer and the inductor) is inversely related to the low-frequency cutoff. Another way to state this relation is to express the minimum frequency as an inverse function of the geometric factor g_c:[10]

$$g_c \propto \left(\frac{A}{\lambda^2 l} \right)^{1/2} \tag{8.53}*$$

This factor will be discussed again, later, in connection with coil geometry [see Eq. (9.27)].

Obtaining a wide bandwidth and, concurrently, a wide dynamic

---

* Note the similarity between Eq. (8.53) and Eq. (4.10): $g_0 = (A_s A_{\mathrm{Fe}}/l_t\, l)^{1/2}$. The presence of the surface area A_s in the latter equation arises from the fact that temperature rise is a major factor in the power transformer but not in the wide-band transformer.

range can be difficult. If, for example, a transformer is to reproduce a Gaussian distribution of frequencies, it must transmit a lot of power at very low frequencies and at a high flux density B_m. Fortunately, a statistical method can sometimes be used to design a transformer whose size is not impractically large. This method involves choosing a flux density which corresponds to the geometric mean of the high and low cutoff frequencies.[14]

Since a polarizing dc bias can seriously reduce permeability and, therefore, inductance, the transformer designer must be alert to the effect of unbalanced direct current in push-pull circuits. In transistor circuits, particularly, seemingly small unbalanced currents can create a surprisingly substantial magnetic bias in a small transformer whose magnetic length must necessarily be short (see Sec. 7.8a). This effect must be compensated for by an air gap; moreover, the size and distortion (which includes the even harmonics) are greater in the polarized transformer. Balanced (as opposed to single-ended) circuitry is much to be desired and, if used, can make an impractical design practical.

We may conclude that distortion can be most easily minimized when the matching factors m and K_L are large and B_m is small. Since a small B_m implies a large core, the most economical way to keep the transformer small is to use a mismatched circuit. Negative feedback, which can drastically reduce generator resistance, helps us to achieve a miniature transformer. (The relationships between low-frequency distortion, phase shift, and feedback were discussed in Sec. 9.1b.)

c. Multiple-Core Transformers

In our analysis of the wide-band transformer, we have assumed that the shunt inductive and power-loss components (L_p, R_p) of the core remain constant and that high-frequency response is dependent only on the leakage inductance and capacitance elements of the circuit. There are, however, transformers in the megahertz band (typically with upper limits in the range of 8 to 200 MHz), in which R_p (shunt core power-loss resistance) falls sharply as frequency increases. When power is substantial, a band span in excess of two decades will be difficult to realize.

It is then worthwhile to consider a composite core geometry (discussed in Sec. 6.3c). The use of two contiguous cores (A and B in Fig. 6.9) enables the designer to select each with an R_p characteristic that is optimum for each portion of the frequency spectrum.[15] An example of a trial design for achieving a bandwidth of 0.5 to 710 MHz is as follows:

Core A   ferrite   1000 μ 0.5 to 250 MHz
Core B   iron powder   6 μ 250 to 710 MHz

9.4. DESIGN OF THE COIL

In this section we shall describe those physical aspects of coil design which determine or influence the high-frequency behavior of the transformer.

Traditionally, the designer endeavors to keep both the leakage inductance L_L and the capacitance C of the coil to a minimum; but when one parameter is reduced, the other is usually increased. The central problem becomes minimizing the product $L_L C$.

In network synthesis, however, once the circuit resistances and the network model have been chosen, the values of all the elements are determined. The specific values of L_L, C_1 and C_2 are optimum. If we define C_D as equal to $C_1 + C_2$, then the task of the coil designer is to synthesize (i.e., physically realize) *specific*, rather than minimum, values of $L_L C_D$ and L_L/C_c.* The problems encountered can be difficult, and sometimes frustrating, because we must control and proportion both leakage and capacitance.

In the discussion which follows, the topics of control of capacitance and coil geometry are especially important.

a. Leakage Inductance

The concept of leakage inductance is a convenient way to account for the imperfect magnetic coupling k between the primary and secondary windings, whose inductances are L_p and L_S, respectively. It may be related to mutual inductance M, the leakage coefficient σ, and the leakage factor m_L, by means of the following expressions:

$$\left.\begin{aligned} k &= \frac{M}{\sqrt{L_p L_S}} \qquad L_m \cong k L_1 \\ \sigma &= 1 + k^2 \cong \left(\frac{1}{k^2}\right) - 1 \cong 2(1+k) = 2m_L \end{aligned}\right\} \tag{9.9}$$

Total leakage inductance L_L, referred to a winding (e.g., a primary with N_p turns), is:

$$L_L = (1 - k^2)L_1 \tag{9.10}$$

In terms of geometry, it is expressed by the approximate formula:[16,17]

---

* C_c is the series capacitor element of the filter in Fig. 9.2c.

$$L_L = F_L \mu_o N_p{}^2 \frac{\lambda}{W_L} \left(d + \frac{d_{\text{Cu}}}{3} \right) \qquad (9.11)$$

where F_L is a distribution factor which denotes the extent to which leakage is reduced by subdividing, or sectionalizing, the coil windings. The constant $\mu_o = 4\pi(10)^{-9}$ when the dimensions are in centimeters (cm). The coil dimensions (shown in Fig. 9.4) are the coil mean-length turn λ, the winding length W_L, the interwinding spacing d, and the total copper thickness $d_{\text{Cu}} = d_1 + d_2$.

It is common practice to multiply Eq. (9.11) by an empirical factor to account for measured values of L_L which are lower than those calculated with the equation. The correction factor is about 0.9 when the window aspect ratio (length/width) is 3, its value in the scrapless E-I lamination.

There are various ways of reducing leakage. One is to make both windings the same length, since asymmetry increases leakage.[18] The geometry is also very important, for certain proportions favor small leakage. These proportions are obtained by placing a thin, long coil in a window with a large aspect ratio. In this respect, the geometry of the toroid is ideal because the winding length is spread over the entire circumference of the core. Still, small L_L may be obtained by using a more practical version of the toroid—the *two-coil construction* on an L, U, or C core (see Fig. 2.4). If the turns of the primary and secondary windings of the single coil (Fig. 9.4), are halved in order to form two coils in series (Fig. 9.5a), then $F_L = \frac{1}{2}$. L_L is reduced by somewhat more than a factor of 2 because the mean turn λ is also reduced.

In the layer-wound coil itself, a substantial reduction of L_L is obtained by subdividing one winding into sections and then sandwiching the other winding between them. This process is referred to as *sectional-*

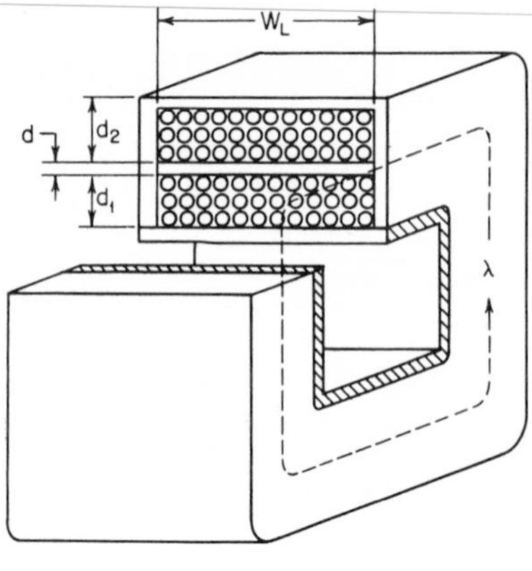

Fig. 9.4 Leakage inductance geometry. Cutaway view of layer-wound coil.

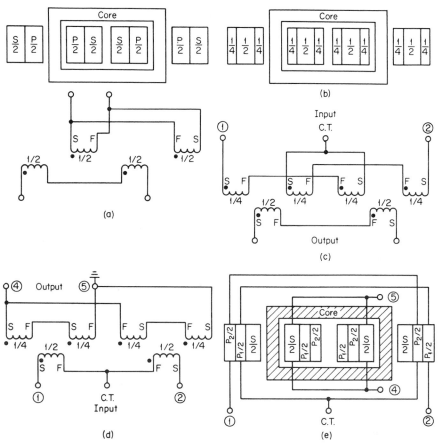

Fig. 9.5   Core-type interleaving. (a) Two-coils; primary and secondary are each split. The schematic shows the secondaries in parallel. (b) Two coils, each interleaved. (c) Schematic of (b) shows connection for balanced push-pull (P-P) primary to ungrounded secondary. (d) Schematic of (b) shows connections for P-P to single-ended (grounded) secondary. (e) Class AB and B geometry: P-P to single-ended secondary. Low leakage inductance between half-primaries. (Figure is discussed further in Sec. 9.4b; for shell-type geometry, see Fig. 9.7d.)

ization or *interleaving.*   There is an economical and, therefore, preferred method of interleaving, in which the maximum number of effective sections is obtained with the least number of windings N_w.[19] The winding arrangements which are most often used (see Fig. 9.6) result in a leakage distribution factor of:

$$F_L = \frac{1}{N_w{}^2} \tag{9.12}$$

When this definition of F_L is substituted into Eq. (9.11), the term d_{Cu} is redefined as the sum of all copper winding depths and d as the sum of all insulation spacings between sections. It is rare for more than seven windings (i.e., six sections) to be used to subdivide the primary and secondary because of the appreciable reduction in the copper space factor and the progressive increase in the capacitance of the coil.

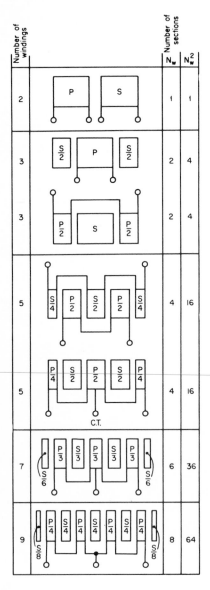

Fig. 9.6 Interleaving of windings. P denotes total primary turns; S denotes total secondary turns.

b. Push-Pull Coupling

Push-pull coupling can pose problems. Special considerations are entailed if symmetry of leakage and resistance is desired and if class B coupling is used. Several ways of interleaving a push-pull coil, using five windings, are shown in Fig. 9.7. If the secondary is a single-ended low-impedance winding, at ground potential, then the arrangement shown in Fig. 9.7a results in minimum capacitance between primary and ground. But it also results in unequal resistance in each half-primary and in unequal leakage inductance between each half-primary and its neighboring half-secondary. The configuration is usually satisfactory in the class A audio amplifier but is undesirable at higher regions of the frequency spectrum or in the class B amplifier. In Fig. 9.7b the interwinding connections have been altered to produce equal resistance in each half-primary and equal leakage to each half-secondary. The alternative arrangement of Fig. 9.7c, while similar to 9.7b, results in lower capacitance across one half-primary than across the other.

In the class B circuit, it is desirable that the active section of the

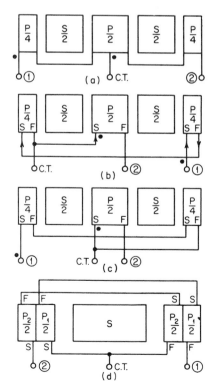

Fig. 9.7 Push-pull interleaving: shell geometry. (Split secondary windings may be connected in series or parallel.) (a) Minimum primary capacitance but unequal dc resistance, leakage inductance, and primary self-capacitance. Class A operation. (b) Equalized primary resistance, leakage inductance between windings, and primary self-capacitance. Class AB and B operation. (c) Capacitance lower across one half-primary than across other half-primary. Class AB and B operation. (d) Low leakage inductance between half-primaries. Class B operation.

primary be tightly coupled to the entire secondary. It is also desirable to couple the two half-primaries tightly. Lord has shown that high-frequency distortion is less than 5 percent when:[11]

$$\frac{L_L \text{ (half-primary to half-primary)}}{L_L \text{ (half-primary to secondary)}} \leqq 1.5 \qquad (9.13)$$

Windings arranged to conform to this objective are depicted in Figs. 9.7*d* and 9.5*e*. (For further comments on the class B transformer, see Sec. 9.4*c* and Fig. 9.17.)

When complete symmetry of resistance, leakage, and capacitance is necessary, the two-coil core-type configuration is preferred. Complete symmetry is also possible with sector, or pi-wound, coils. Such geometry results in lowered capacitance but, ordinarily, in increased leakage also.

Several coil arrangements are depicted in Fig. 9.5. Figure 9.5*a* shows the two-coil construction previously described, in which the primary and secondary are distributed between the two coils without interleaving. In Fig. 9.5*b*, each coil has been interleaved to reduce leakage even further. Connections for an ungrounded, or floating, secondary are shown in Fig. 9.5*c*, and connections for a grounded secondary are shown in Fig. 9.5*d* and *e*. When symmetry of capacitance must prevail, the arrangements of Fig. 9.5*c* and *d* are preferred. Geometry of this type is often chosen when balance is of great importance, as in the *hybrid çircuit* (discussed in Sec. 8.8).

Drastic reductions in leakage inductance are made possible by *bifilar* winding. A pair of insulated wires is wound simultaneously and contiguously (i.e., close enough to touch each other). Each wire constitutes a winding; their proximity reduces L_L by several orders of magnitude more than ordinary interleaving. But when the voltages are even moderately high, the problem of adequately insulating the winding becomes severe because the film on magnet wire is rarely rated at more than 80 volts per milli-inch (V/mil). The advantages of bifilar winding can, however, be great enough to warrant the use of magnetic wire insulated by triple and even quadruple thicknesses of film. The technique is limited to transformers in which the turns ratio is unity or a small integer. It is widely used in low-impedance class B transformers, in dc converter transformers, where it provides very tight coupling between the half-primaries, and in transmission-line transformers.

We shall find that a reduction in leakage inductance usually increases capacitance, but the relevant relationships will be discussed later (in Sec. 9.4*d*), after we consider the subject of capacitance in more detail.

c. Capacitance[20]

When the transformer is energized, different voltage gradients arise almost everywhere. In the pervasive electric field, a large variety of capacitances correspond to each of these gradients:

1. Between turns
2. Between layers
3. Between windings
4. Between terminals
5. Between the core and the end turn of a layer (i.e., the margin)
6. Between the core (i.e., the ground) and each terminal

In this chapter, as elsewhere in this book, we rely heavily on the concept of lumped rather than distributed parameters. In Chap. 8 we assumed that no more than three lumped capacitances (C_1, C_2, and C_3), defined by Eq. (8.43), are needed to represent the distributed capacitances inside and outside the transformer. Here it is convenient to express these lumped capacitances in terms of the step-up ratio $N = 1/n$:

$$\left. \begin{array}{l} C_1 = C_P + C_B(1 - N) \\ C_3 = C_B N \\ C_2 = C_S N^2 + C_B(N^2 - N) \end{array} \right\} \tag{9.14}$$

The extent to which each of the six categories of lumped capacitances affects the assumption that three are adequate requires some discussion.

Turn-to-turn capacitance is so small that its effect is usually ignored below 5 MHz. It is the layer and interwinding capacitances which are most important. The self-capacitance of the primary coil C_P and the interwinding and self-capacitance of the secondary coil C_S may be computed from a consideration of the energy stored between each pair of layers. The basic procedure begins with a calculation of the interface capacitance C_1 between one pair of layers (see Fig. 9.8a). This capacitance is a function of the dc, or static, capacitance C_o (see Fig. 9.8b) between layers which are treated like plates of a capacitor:

$$C_o = \epsilon_o \epsilon \frac{W_L \lambda}{d_1} \qquad C_o = 0.225 \epsilon \frac{W_L \lambda}{d_1} \quad \text{(in pF)} \tag{9.15}$$

Here ϵ is the average relative dielectric constant, or permittivity, of the insulation; $\epsilon_o = 8.85(10)^{-12}$ in the first expression if the dimensions are in meters; and d_1 is the spacing between layers. When dimensions are in inches the second expression is used and capacitance is expressed in picofarads.

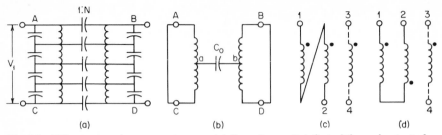

Fig. 9.8 Effective interlayer capacitance. (*a*) Capacitance distributed throughout windings (stray capacitance to the core is not shown.) (*b*) Static capacitance C_o. (*c*) Z or flyback winding traverse. (*d*) U (conventional) winding traverse.

Any capacitance C in a coil is computed with the aid of the formula:

$$C = C_o \frac{(V_{AB}^2 + V_{AB} V_{CD} + V_{CD}^2)}{3V^2} \tag{9.16}$$

Here the value of C depends on which pair of terminals the capacitance is referred to and on the voltage V which appears across those terminals. The other voltages are identified in Fig. 9.8*a*.*

If $V = V_1$ is the voltage across one layer, then the effective capacitance referred to that layer is:

$$C = C_1 = \tfrac{1}{3} C_o \tag{9.17}$$

The value of $\tfrac{1}{3}$ is also obtained when we compute the capacitance between a helical layer of wire and an equipotential surface such as ground or a screen (i.e., electrostatic shield).

The primary and secondary self-capacitances (C_P and C_S) depend on the number of effective layers N_{L1} (of the primary) and N_{L2} (of the secondary):

$$C_P = \frac{4}{3} C_o \frac{N_{L1} - 1}{N_{L1}^2} \qquad C_S = \frac{4}{3} C_o \frac{N_{L2} - 1}{N_{L2}^2} \tag{9.18}$$

The numerical coefficient in this formula will be $(\tfrac{4}{3})0.255$, or 0.30, when it is combined with the coefficient of C_o in Eq. (9.15). If the layer thickness d_1 is different in each winding, then C_o is different in each of the expressions in Eq. (9.18).

(1) The Two-Layer Transformer. When we can reduce both the primary and secondary layers to one layer each, C_P and C_S both vanish. This is of great import in the design of wide-band transformers because it favors the upward extension of the high-frequency cutoff. We are still left, of course, with interwinding and stray capacitances.

---

*In the figure, V_1 is the voltage applied to terminals A-C; and V_{AB} and V_{CD} are the voltages between terminals A-B and C-D, respectively.

To compute the interwinding capacitance, we introduce the concept of a dynamic or effective capacitance C'. We associate this capacitance with the energy stored between primary and secondary, and we apply Eq. (9.16), C' being taken as equal to C. This capacitance can then conveniently be referred either to the primary $(V = V_1)$ or to the secondary $(V = V_2)$:

$$C_1 = C'N^2 \qquad C_2 = \frac{C'}{N^2} \qquad (9.19)$$

where $N = V_2/V_1 = N_S/N_P$.*

For the important step-up transformer $(N > 1$ and $V_{AB} \neq V_{CD})$, we can infer from Eq. (9.14) that $C_2 \gg C_3$. This important inequality justifies the common practice of neglecting the bridging capacitance. We now identify C' with the effective secondary distributed capacitance. When this is expressed as a function of the turns ratio, we obtain:

$$C_2 = C' = (N-1)^2 \frac{C_o}{3} \qquad (9.20)$$

Note that when $N = 1$, C' vanishes.

When either the angular rotation or the direction of traverse of one winding layer is reversed, $(N - 1)^2$ in Eq. (9.20) becomes $(N + 1)^2$. It is considered good practice to maintain the same direction of rotation and traverse throughout a coil containing few layers. Interconnections or splices are made external to the coil. The Z (also called flyback) winding traverse shown in Fig. 9.8c is preferred to the U traverse (Fig. 9.8d) even though it is more expensive to wind.

If $V_{AB} = V_{CD}$, as it does when the volts per layer are equal and $N = 1$, then no electrostatic energy is stored between the layers. The bridging capacitance C_B then consists only of a net unsymmetrical leakage or stray capacitance between the input and output terminals. Now, when the transformer is connected for phase *inversion* (defined as polarity reversal, i.e., the primary start and the secondary finish are at the same potential), then:

$$C_2 = C' + (N^2 - N + 1) \frac{C_o}{3} \qquad (9.21)$$

Note that $C' = C_o$ when $N = -1$ (unity turns ratio, but inverted).

---

* Such a procedure is used to compute bridging capacitance $C = C_B$ in a multilayer coil. We proceed by using the relations: $V = E_1 - E_2$, where $E_1 = N_{L1}V_1$, $E_2 = N_{L2}V_2$, and $V_2 = V_{BD}$. If, as a result of sectionalizing, there are i interfaces, then the sum $\sum_1^i C_o/3$ is used in place of $C_o/3$ in Eq. (9.20).

When N differs little from unity, it may prove unrealistic to ignore the bridging or leakage capacitance. For example, if $N = \frac{3}{2}$ and we neglect the external shunt capacitances associated with the source and load, then $C_1 = -\frac{1}{2}C_B$; $C_3 = \frac{2}{3}C_B$; and $C_2 = \frac{3}{4}C_B$. When $N = 1$ and the external shunt capacitances are neglected, we are left only with:

$$C_3 = C_B$$

We conclude that the deliberate neglect of the bridging capacitance is most warranted when N differs substantially from unity.

The many capacitances are a nuisance to deal with, even though it is possible (often by indirect methods) to compute and measure them. A simplifying approach has great appeal, so we posit one inclusive distributed capacitance C_D, which we identify with the stored electrostatic energy throughout the coil. Suppose, therefore, that we define C_D as equal to $C_1 + C_2 + C_3$. When we refer C_D to the primary, it becomes:

$$C_D = C_P + C_S N^2 + C_B(N-1)^2 \cong C' + N^2(C_2 + C_L) \qquad (9.22)^*$$

This lumped capacitance can be usefully identified either with the input capacitance $(C_1 = C_P)$ or the output capacitance $(C_2 = C_S N^2)$ of the second-order model (Fig. 8.7c and d) when n or N is large and C_B is small.

(2) **Stray Capacitance.** Stray capacitance increases the total distributed capacitance, but only slightly. More important, it leads to asymmetry and unbalances the winding capacitances of the coil. Consider, as an illustration, the line input transformer, which is operated at low voltages—microvolts (μV) or millivolts (mV). If the designer has not striven for symmetry, then, when the primary is connected to a long line, asymmetrical ground currents containing noise will degrade the signal-to-noise (S/N) ratio.†

When we have taken pains to select a geometry which favors balanced leakage inductance and balanced dc resistance, it is disconcerting to find that subsequent measurements reveal electric unbalance which is traceable to neglected stray capacitances. It is not wise to neglect these capacitances, though we can often safely do so. Because they can be important, we depict in Fig. 9.9 the three major types of windings in which such capacitances are wont to manifest themselves: (1) the grounded single-ended winding (Fig. 9.9a), (2) the

---

* It is this value which is used to represent the distributed capacitance in a pulse transformer [see Eq. (10.14)].

† The degree of symmetry is determined by tests for *longitudinal balance,* which is particularly important in the hybrid transformer circuit.

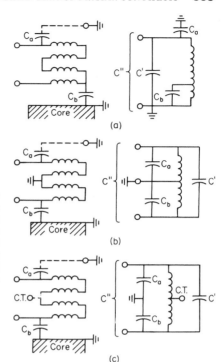

Fig. 9.9 Stray winding capacitances. In general $C_a \neq C_b$.
Equivalent schematic shown at
right. See Table 9.3 for values of C_a, C_b and the effective
stray capacitance C'' across
winding. (*a*) Single ended. (*b*)
Grounded center tap (CT).
(*c*) Ungrounded.

grounded center-tap winding (Fig. 9.9*b*), and (3) the ungrounded center-tap winding (Fig. 9.9*c*). Shown next to each physical schematic
is an equivalent circuit containing the lumped stray capacitance C_a
(with respect to a screen) and C_b (with respect to the core, which is
the other equipotential surface).

Values of the stray capacitances C_a and C_b and of resulting total
capacitance $C'' > C'$ are presented in Table 9.3. In all cases, C_a and
C_b can be made equal only if the geometry—the spacing of windings
with respect to the screen and core, and the areas ($W_L \lambda$) of screen
and core—are meticulously proportioned. The importance of the
screen in this context transcends its customary purpose of providing
electrostatic isolation between windings.

(3) **The Control of Capacitance.** As we have indicated, the task
of controlling the various capacitances in a transformer can be difficult.
The designer has several objectives:

1. To reduce capacitance to a minimum
2. To synthesize specific values of distributed capacitance (in order,
for example, to obtain a specific value of $\sqrt{L/C}$ or $\sqrt{LC}$)
3. To obtain balanced stray and self-capacitances

TABLE 9.3   Effect of Stray Capacitance on Total Coil Capacitance

| Figure | Stray capacitance | | Total capacitance* |
	C_a	C_b	C''
9.9a	$\dfrac{C_o(N_L - 0.5)}{N_L}$	$\dfrac{C_o}{3}$	$C' + C_a + \dfrac{C_b}{N_L{}^2}$
9.9b	$C_o \dfrac{N_L - 1}{N_L}$	$C_o \dfrac{N_L - 1}{N_L}$	$C' + \dfrac{C_a}{4} + \dfrac{C_b}{4}$
9.9c	$C_o \dfrac{N_L - 0.5}{N_L}$	$C_o \dfrac{N_L - 0.5}{N_L}$	$C' + \dfrac{C_a C_b}{C_a + C_b}$

* $C' = \tfrac{1}{3} C_o(N_L - 1)/N_L{}^2 =$ self-capacitance; $N_L =$ number of layers in winding; and $C_o =$ static (dc) capacitance between layers.

4. To eliminate, on occasion, certain undesired capacitative coupling between windings by means of a screen

There are times when capacitance must be reduced to a minimum even though leakage inductance is thereby increased.   In limit-of-the-art applications, in which the toroid and single-layer windings are often utilized, capacitance between the inner winding and ground is kept small by winding on a thick former (spacer) with low ϵ.   Self-capacitance, which is thus reduced to the very low turn-to-turn capacitance, can be reduced further by symmetrical winding with fewer than the normal turns per inch.

In multilayer coils, self-capacitance may be reduced by vertical sectionalization (see Fig. 9.10).*   When a coil is split into two adjacent halves, the surface area of each half and, therefore, the capacitance is halved.   When the halves are connected in series, overall capacitance is reduced by one-fourth.   In Fig. 9.10a, the secondary is split into two sectors in order to reduce its capacitance, but stray capacitance is unequal.   In Fig. 9.10b, symmetrical stray and self-capacitances are obtained by reversing the winding direction of one of the sectors and introducing a screen between the primary (unbalanced) winding and the secondary sectors.   This type of geometry makes it possible to reverse the polarity of the secondary (either terminal may be grounded) without altering the capacitance and, thereby, the high-frequency cutoff.   In Fig. 9.10c, both primary and secondary are split into sectors and then criss-cross connected in order to obtain symmetry of capaci-

---

* Two allied techniques are the *universal* winding, which is used to provide a self-supporting sector, and the *bank* winding, which is a kind of continuous vertical sectionalization in which adjacent turns are as nearly as possible consecutive.   Bank winding is sometimes used on toroids to reduce self-capacitance when there are few layers.

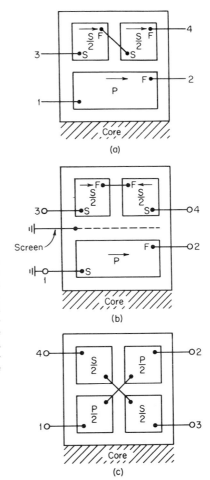

Fig. 9.10 Sector windings. (*a*) Secondary capacitance reduced, but stray capacitances cause asymmetry. (*b*) Symmetrical secondary capacitances and dc resistance, obtained by reversing winding direction of one sector and using screen between windings. (*c*) Symmetrical capacitance, resistance, and leakage of both windings.

tance and dc resistance within a shell-type geometry. The capacitance between the windings has also been reduced, without increasing the leakage inductance.

The need for symmetry of stray capacitances can be very stringent in the hybrid transformer and in its equivalent, the bridge transformer, which is used to isolate a single-ended generator from an ungrounded bridge circuit. A high degree of symmetry, as well as low capacitance, is sometimes achieved in the bridge transformer by separately and completely enclosing both the primary and secondary windings in metallic foil screens. This ensures the greatest degree of capacitative isolation between the coils (about 0.3 pF in the General Radio type 578 transformer) as well as uniform and symmetrical stray capacitances

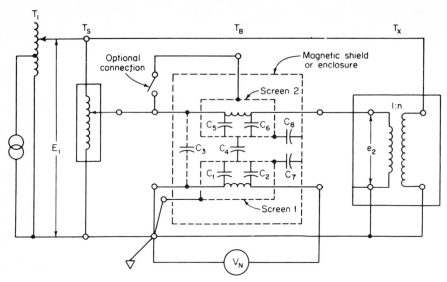

Fig. 9.11 Turns-ratio test circuit. T_1, transformer to set input voltage E_1. T_S, standard radio transformer. T_X, transformer under test. T_B, bridge transformer, 1:1 turns ratio, high impedance (typical values of capacitance: C_1, C_2, C_5, $C_6 = 200$ pF; C_7, $C_8 = 70$ pF; $C_4 = 30$ pF; $C_3 = 0.3$ pF). V_N, null meter (high impedance). Note: Phase shift $\gamma = \arctan V_N/e_2$ when e_2 is the secondary voltage.

of known value (see Fig. 9.11). The need for exact symmetry can even dictate that the physical splice to the screen be exactly equidistant, on the circumference of the screen, from the insulating strip between ends of the foil. We ground the output shield and the outside enclosure (which is the third shield) to establish identical stray capacitances from each input terminal to ground. The shield enclosing the unbalanced winding is usually left floating (i.e., not connected).

When leakage inductance has to be reduced and capacitance kept low, the designer tries to restrict the number of interfaces. One possibility, which is mechanically awkward and therefore uneconomical, is to sandwich full layers between sector-wound layers (which might consist of two halves of a high-impedance winding).

The use of more than one screen is an effective technique for providing equipotential surfaces, low interface capacitance, and symmetry.* The numerals next to certain terminals in Fig. 9.12 refer to the relative capacitance factor:

---

* Screens are also used in the high-voltage transformer to shape the electric field and to control the variation of voltage gradients and thereby eliminate corona (see Sec. 3.10).

$$\frac{\text{Capacitance from that terminal to screen}}{\text{Capacitance of an end layer to screen}} \qquad (9.23)$$

The relative capacitance factor of the conventional single-ended primary separated from a secondary winding by a screen (Fig. 9.12a) serves as a reference.   In Fig. 9.12b, the primary is sandwiched between two screens and half-secondary (push-pull) windings.   In Fig. 9.12c, one of the half-secondaries is wound in reverse.   Between the single-ended output which this permits and the core, there is little capacitance.   If the winding is to be center-tapped, the dc resistances will be equal.   The technique is readily adapted to the toroid geometry.

Although only a few winding schemes are depicted in Fig. 9.12, it should be clear that in the most effective ones the windings are so arranged as to minimize the voltage gradient at the interface and provide a means for maintaining capacitive symmetry.   Screens, when interface areas and spacings are carefully proportioned, make it possi-

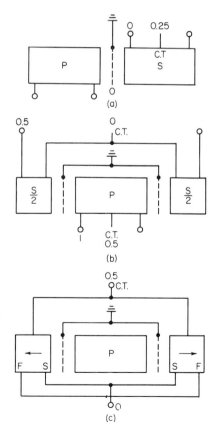

Fig. 9.12   Winding-to-screen capacitance.   (a) Single-ended primary.   (b) Push-pull secondary.   (c)   Single-ended   secondary.   One of shunt-connected windings is wound in reverse direction.

ble for the high-frequency response to be independent of the polarity of one (or even both) of the windings.

Screens have also been used in the high-impedance class B transformer in order to eliminate the bridging capacitance and thereby the frequency notch at ω_M. Double screens are used to minimize common-mode currents (see Fig. 8.8).

As a rule, it is easier to obtain both capacitive symmetry and low leakage inductance with the core-type geometry (e.g., L, U, C) than with the shell-type (e.g., E-I, E-E). In either type, the machine layer-wound coil is preferred to the random-wound coil.

d. Coil Geometry

The rational choice of a particular coil geometry requires that due consideration be given to various factors:

1. The particular region of the frequency spectrum in which the transformer is to operate
2. The impedance level
3. The voltage level
4. The degree of balance desired
5. The tolerable amount of externally induced hum or noise[21]
6. The cost, a consideration which often has a decisive effect on the geometry

The basic geometric desiderata in coil design can be phrased in terms of the basic objectives:

1. An optimum geometry for the required high-frequency response and impedance level
2. A geometry suitable for balanced circuitry
3. A geometry not conducive to the flow of common-mode currents
4. A geometry not vulnerable to stray magnetic fields

(1) The Characteristic Impedance. When we examine the elementary geometric expressions for leakage inductance and capacitance, it becomes evident that they may be expressed in dual form:

$$L_L = F\left(\frac{d}{W_L}\right) \qquad C_o = F\left(\frac{W_L}{d}\right) \tag{9.24}$$

that is, a reduction in one is attended by an increase in the other. This is as it should be, since the sum of the energy in the magnetic and electric fields is constant.

O'Meara[9] has shown that the high-frequency response and band-span ratio depend on geometric factors which can be expressed as the product of a winding factor M_1 and a coil factor g_b or coil-and-core factor g_c. The winding factor of Eq. (8.52):

$$M_1 = \frac{1}{\sqrt{1 + [(1/n) - n]^2/2}} \qquad (9.25)$$

is maximum when the turns ratio is unity ($n = 1$). This formula applies to the two-pole, two-zero model (see Sec. 8.5c).

The factor g_b is relevant to the high-frequency response. It is defined by:

$$g_b = \sqrt{\frac{F_c}{F_L} \frac{1}{(N_s \lambda)^2}} \qquad (9.26)$$

where F_c is a factor equal to 3 (the reciprocal of the coefficient ⅓ for the two-layer transformer), which depends on the influence of the winding geometry and polarity on effective capacitance; F_L depends on the extent of interleaving of the windings; and $N_s \lambda$ is the length of the coil wire.

An important inference from this equation is the desirability of *reducing wire length* to a minimum in order to obtain maximum high-frequency response. This implies that the optimum practical core cross section is circular (or square) and that the optimum coil is cylindrical.

The dependance of the band-span ratio on the geometry of the coil and core is expressed by:

$$g_c = \sqrt{\frac{F_c}{F_L} \frac{A}{\lambda^2 l}} \qquad (9.27)$$

This leads to the important conclusion that if high-frequency cutoff and band-span ratio are to be extended, the linear dimension—that is, the perimeter and therefore the size of the transformer—must be reduced:

$$\frac{f_2}{f_1} \propto \frac{1}{\sqrt{\text{centimeters}}} \qquad (8.54)$$

The small cylindrical transformer shown in Fig. 9.13 has an extended high-frequency performance and, concurrently, only moderate low-frequency distortion. Microminiature wide-band transformers (⅛-in cube) are also realizable.[22]

The importance of extreme size reduction becomes evident when the size of the transformer is to be made compatible with thick-film circuits. It is inevitable, therefore, that the geometry of a transformer be reduced in size to a printed spiral.[23]

As we noted earlier when we discussed the class B circuit, it is possible

Fig. 9.13 Miniature cylindrical transformer. Model DO-T: $\frac{5}{16}$ in diam. $\times$ $^{13}\!\!/_{32}$, $\frac{1}{10}$ ounce. (*By permission, United Transformer Company.*)

to have high-frequency distortion well inside the pass band. It is, therefore, desirable to place the high-frequency cutoff at least an octave higher than the highest operating frequency. If leakage inductance between the half-primaries is kept small by using bifilar windings, then the self-capacitance in the primary coil will be high and the characteristic impedance will be low. But if circuit impedances are high and the transformer is large, it becomes very difficult indeed to extend the high-frequency response and still keep the distortion small. The class B transformer performs best when the circuit impedance is low. In high-power audio circuits, this problem is often solved by using the emitter-follower circuit.

In coils containing few layers of magnet wire, once a geometry has been selected, the product $L_L C_D$ is restricted within a rather narrow range. For example, when a C core is chosen and layer insulation throughout the coil is designed for a *constant-voltage gradient*, $L_L C_D$ varies by only 50 percent even though a large variety of winding configurations can be used.[24]

If we endeavor to match the characteristic impedance $Z = \sqrt{L_L/C_D}$ to the load impedance, then the geometry can be important. High impedance implies a high L/C ratio and low impedance implies a low L/C ratio.

One practical application of the seesaw effect between L_L and C_D is a high-power (90-W) toroid transformer which is sector-wound so that either high or low $\sqrt{L_L/C_D}$ can be obtained by changing the winding interconnections.[25]

We can gain some idea of how Z changes with the geometry by considering the simple but important two-layer coil. If we assume that $d_{\text{Cu}}/3 < d$ (i.e., that the thickness of the copper is small compared with the interwinding insulation), then the characteristic impedance of the secondary winding is:

$$Z = 377 \frac{N_S}{(W_L/d)} \sqrt{\frac{F_c}{\epsilon}} \qquad (9.28)$$

where dimensions are in meters. Note that the term in parentheses is proportional to the aspect ratio of the window.

We conclude that the core of a high-impedance transformer should have a window with a low aspect ratio. If the voltage level of this transformer is high, sector winding (which favors low capacitance) is also indicated. The corresponding geometries—the wide-window shell-type E-I structure or the L structure with the coil on one limb—are the most suitable for high-impedance high-voltage transformers.

If impedance is low, then a long winding length (narrow window) is desirable and the concurrently large capacitance need not be disadvantageous. Appropriate structures are the toroid and the two-coil core-type geometry (i.e., L, U, D-U, and C cores). The toroid, a shape which is capable of providing the highest L_P/L_L ratio, is ideal for obtaining maximum bandwidth at low impedance levels. Unfortunately, even the slightest amount of unbalanced direct current is sufficient to reduce L_P drastically and thereby nullify the toroid's most important merit.

The geometry of the pot core is similar to the shell E-E or F-F shape. It also may be viewed as the geometrical dual of the toroid, in the sense that the role of core and coil are interchanged: that is, the core envelops the annular-shaped coil (see Fig. 6.12b). Although the L/C ratio tends to be greater than in the toroid, the other merits of the pot core—its hum-free structure and the low cost of winding its coil—often lead designers to favor it over the toroid.

(2) Foil and Bifilar Windings. It is often assumed that a screen is to be avoided because it increases the interwinding spacing and thus the leakage inductance. We have noted, however, that the screen introduces an equipotential surface into the coil and that certain advantages are thereby gained. These considerations have led some designers to the view that winding the turns spirally with thin foil, rather than with circular magnet wire, may result in a better high-frequency response than can be obtained with a conventional screen (consisting of one wrap of foil). Such tape is used in place of magnet wire when circuit impedance is low and operating frequencies extend into the radio-frequency region of the spectrum.[26,27]

An alternative to the concentric spiral geometry is the parallel helix geometry: the bifilar winding. The bifilar geometry is used in the transmission-line transformer, to be discussed in Sec. 9.5.

(3) **Common-Mode Coupling.**   In low-level circuits, signals must be distinguishable from spurious currents which flow in the ground or common circuit.   It is possible to reduce such currents by using an electrostatic shield (ES) between primary and secondary windings. The shield or screen can be in the shape of an insulated single-turn coil, wound concentrically over the primary coil.

A substantial reduction of spurious currents, that is, common-mode coupling, is obtained by the use of *two* screens between the windings. One screen, connected to the high side of the primary, serves to short-circuit a ground current which might otherwise flow via stray capacitance through the primary.   The other screen, connected to the grounded secondary terminal, serves to bypass current which might flow through the bridging capacitance between windings.   The measure of efficacy used is called *common-mode rejection.*[28]

It is also possible to achieve a drastic reduction of interwinding capacitance and leakage current if input and output windings are wound on opposite legs of an L lamination or opposite sectors of a toroid.[29] The poor inductive coupling may be of little consequence when the signals are rectangular pulses.

9.5. THE TRANSMISSION-LINE TRANSFORMER

We have seen that the response at the high end of the frequency band is restricted or impaired because of certain basic features of the transformer circuit.   These are:

1. The magnitude of distributed capacitance and the product $L_L C_D$
2. The step-up turns ratio n
3. The magnitude of the circuit impedances
4. Stray capacitances (which account for dips or peaks in the frequency domain and parasitic oscillations in the time domain)
5. Nonlinearity of the phase response

We have arrived at several conclusions about how to deal with these limitations.   These are:

1. Capacitance can be reduced to a minimum by confining each winding to a single layer.
2. Turns ratio should be a small integer.
3. Circuit impedances should be low, and the characteristic impedance Z_0 of the transformer should be matched to source R_g and load R_L:

$$Z_0 = \sqrt{R_g R_L} \tag{9.29}$$

4. Stray capacitance can be kept small but cannot be entirely eliminated. Bridging capacitance can be minimized by means of an electrostatic shield.

5. Nonlinearity of the phase response can be improved by selecting an appropriate filter model (e.g., the Bessel filter, Sec. 10.3c) and suitable element values of L_L and C_D.

In complying with these solutions, we arrive at a coil design in which primary and secondary are wound either (1) on two concentric layers or (2) on a single layer in the bifilar format in which both windings are contiguous, that is, wound side by side, with the starts at one end of the layer and the finishes at the other end. The bifilar format is shown in Fig. 9.14a. Compared with the concentric winding technique, bifilar wires result in a minimum of leakage flux, in exchange for an increase in capacitance. If the transformer is connected as a phase inverter in the normal manner (Fig. 9.14b), that is, primary (1,2) to the source, and secondary (3,4) to the load, then the product $L_L C_D$ sets a limit value to the upper end of the spectrum of the bandwidth.

At this juncture, we will step back and look at the problem in a

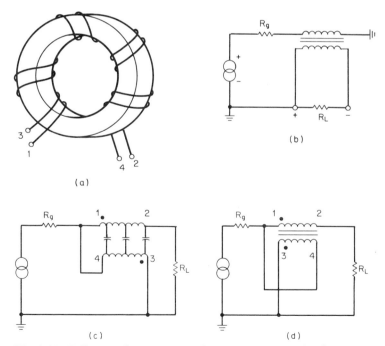

Fig. 9.14 Bifilar transformer. (*a*) Bifilar winding on toroid. (*b*) Conventional isolation connection. (*c*) Connected as transmission line. (*d*) Transmission line schematic, 1:4 impedance ratio.

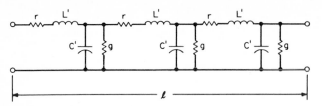

Fig. 9.15   Section of a uniform transmission line.

different way.   Bifilar windings are a pair of parallel wires wound as
a helix.   If this coil is stretched out into a straight path, we see that
the bifilar pair of coils is, topologically, a transmission line wound as
a helix on the core.   When we view it as a transmission line, we should
find it possible to transcend problems associated with stray capacitance
by making the distributed capacitance into an integral element of the
windings, as a transmission line[30] (see Fig. 9.14c).   We are now prepared to make a basic change in our design strategy.   We will first
ascertain how to achieve the optimum characteristics of the transformer, and then we will *design the circuit to conform to the optimum
transformer.* *

Some of the principles of transmission-line design have a bearing
on transformer design.[31,32]   First, any pair of wires connecting a source
to its load can be regarded as a transmission line.   Second, when the
line has uniform characteristics, it is analyzed as a cascade of LC sections, each having uniform proportions of series inductance, series resistance, shunt capacitance, and shunt conductance per unit length (see
Fig. 9.15).   Third, the output response of such a transmission line can
extend uniformly out to frequencies well into the gigahertz (10^9) band.

Our objective, then, becomes to design the coil as a transmission
line.   We note that, in an ideal transmission line, the transmitted waveform suffers minimum distortion and is also subjected to a *constant
time delay* because phase is a linear function of frequency (see Sec.
8.5d).   The equations relating the variables are as follows:

$$f\lambda = v = pc \tag{9.30}$$

$$v = \frac{\omega}{\beta} = \frac{1}{\sqrt{\mu\epsilon}} \tag{9.31}$$

$$\frac{1}{v} = \frac{t_d}{l} = \frac{\beta}{\omega} = \sqrt{L'C'} \tag{9.32}$$

$$Z_o = \sqrt{\frac{L}{C}} = \sqrt{R_g R_L} \tag{9.33}$$

---

* This is not a new point of view.   The circuit designer is, more often than not, in
the habit of using standard or optimum components as elements of a new circuit design.

where λ = wavelength of the frequency
v = phase velocity
p = relative phase velocity = v/c
c = velocity of light, $3(10)^{-8}$ m/sec
$\omega = 2\pi f$
$\beta \cong \tan \beta$, the phase constant
$\mu = \mu_o \mu_r$ $\mu_o = 0.4\pi(10)^{-8}$
$\epsilon = \epsilon_o \epsilon_r$ $\epsilon_o = 8.85(10)^{-12}$
t_d = time delay (time to reach 50 percent of final amplitude)
l = length of transmission line
L',C' = inductance and capacitance per unit length of line
$L = lL'$
$C = lC'$
R_g = source resistance
R_L = load resistance

Response is uniform out to a frequency whose wavelength is a fraction of the wavelength λ corresponding to the cutoff frequency, where $l = \lambda/2$. It can be shown that practical limits of the power ratio, insertion loss, and characteristic impedance are related to the effective (fractional) length l/λ of the line, that is, the length of the wires in the coil. In the analysis, the following definitions are employed:

$$\frac{l}{\lambda} = \frac{\beta l}{2\pi} \tag{9.34}$$

$$p_L = \frac{P_L \text{ (load power)}}{P_a \text{ (available power)}} \tag{9.35}$$

$$10 \log p_L = \text{insertion loss, dB} \tag{9.36}$$

$$z = \frac{(R_g R_L)^{1/2} \text{ (optimum impedance)}}{Z_o \text{ (actual impedance)}} = \frac{Z_{\text{opt}}}{Z_o} \tag{9.37}$$

Let us now consider a bifilar winding or transmission line which is connected as a step-up autotransformer when $n = 2$ and $R_L/R_G = 4$ (see Fig. 9.14d). For the design of the coil, we select a length of transmission line (i.e., the length of the pair of wires of the coil) that meets the constraints on the cutoff frequency and the insertion loss. The minimum length l of the coil should satisfy the inequality:

$$l < \frac{\lambda}{4} = \frac{v}{4f_2} \tag{9.38}$$

Here, f_2 is the upper limit of transmission.

The cross-sectional area of the wire chosen must result in an acceptable insertion loss, defined by Eq. (9.36).

The power output at high frequencies is a function of three impedances (R_g, R_L, Z_0) and of the length of line. When they are properly matched, the ratio of available power P_a to output power P_L is:

$$\frac{P_a}{P_L} = \frac{(1 + 3 \cos \theta)^2 + 4 \sin^2 \theta}{4 (1 + \cos \theta)^2} \tag{9.39}$$

where $\theta = \beta l = 2\pi l/\lambda$ is the length of the line, in radians (rad).[33] Maximum power is delivered when $P_a/P_L = 1$ and $\theta = 0$. The practical meaning of zero line length is that the line is much smaller than the wavelength of the highest frequency to be transmitted.

In practical designs, it is not mandatory that Z_0 be optimum or that the load and source be exactly matched. W. Blocker[34] has pointed out that, in the general case (where $R_L \neq 4R_g$), it is still possible for P_a/P_L to equal unity and, therefore, to obtain an optimum match when:

$$\sec \beta l = \frac{R_L}{2R_g} - 1 \tag{9.40}$$

The equation is satisfied when $R_L/R_g \geqq 4$. We can infer, therefore, that maximum power transfer is possible even when $\beta l > 0$. This finding becomes important when we want a substantial impedance step-up ratio, say 1:9, but have decided to use the standard 1:4 transmission-line configuration of Fig. 9.14d. In this situation, we can substitute $R_L/R_g = 4$ into Eq. (9.40) and solve for the length of the line:

$$\frac{l}{\lambda} = \frac{\beta l}{2\pi} = \frac{0.959}{2\pi} = 0.153 \tag{9.41}$$

Table 9.4 has been constructed so as to facilitate a rational choice of l/λ based on an impedance ratio of 1:4. Suppose we intend to use a transmission line as the coil and that we match it to source and load. Since $Z_0 = \sqrt{R_g R_L}$, then, as indicated in Eq. (9.33), $Z_{opt}/$

T A B L E   9.4   Transmission-Line Insertion Loss When Impedance Is Not Optimum*

Fractional line length l/λ	Z_{opt}/Z_0					
	1		3/2 or 2/3		2 or 1/2	
	P_L	dB	P_L	dB	P_L	dB
0.000	1.000	0.00	1.000	0.00	1.000	0.00
0.134	1.010	0.04	1.045	0.19	1.122	0.50
0.167	1.028	0.12	1.086	0.36	1.215	0.85
0.204	1.077	0.32	1.173	0.69	1.389	1.43
0.227	1.141	0.57	1.271	1.04	1.563	1.94
0.250	1.250	1.00				

* Load is matched to source ($R_L = R_G$).

$Z = 1$. From a perusal of the table, we see that any fraction less than one-quarter of a wavelength is suitable. For example, if we choose $l/\lambda = 0.204$, we can expect 0.32 dB insertion loss. If we use a bifilar winding and anticipate a nonoptimum design, say $Z_o = 2Z_{opt}$, then we cut a smaller line, say $l/\lambda = 0.134$ and settle for the higher loss of 0.5 dB. If a very low standing wave ratio is required, we stipulate a loss of 0.2 dB, for which $l/\lambda = 0.09$.[35]

a. Band Span

It will be useful to arrive at an estimate of the band span of a transmission-line transformer. We start, therefore, with the minimum practicable length of coil wire.

With the coil wire cut to a length $l_N = \lambda/8$ (in centimeters), the upper cutoff frequency f_2 can be calculated from:

$$f_2 = \frac{pc}{8l_N} = \frac{0.79(3)10^{10}}{8l_N} = \frac{2962}{l_N} \text{ MHz} \qquad (9.42)$$

We have selected a standard coaxial cable (RG-58U) whose dielectric results in a relative phase velocity of 0.79. The choice of a coaxial cable is deliberate. It has uniform and known characteristics: p (the relative phase velocity), p_L (the insertion loss), and a small ϵ_r (the dielectric constant).

The low-frequency cutoff f_1 is a function of five variables:[36]

d_m = the mean diameter of the toroid
l_N = the length of the pair of wires
$R_g = R_L/4$ = the source impedance
μ = the effective permeability of the core
K_L = the distortion factor

In the toroid, the inductance L_p of the coil is:

$$L_p = \frac{\mu_0 \mu N^2 r^2}{d_m} \qquad (9.43)$$

where r = the radius of the mean turn.

We shall assume that the proportions of the typical toroid are such that $l_N = 2\pi r N$ and that $d_m = l_N/2$. This results [after substitution into Eq. (9.43)] in the following formula:

$$L_p = \frac{\mu_0 \mu l_N}{2\pi^2} \qquad (9.44)$$

To satisfy any constraints on phase shift or distortion, we introduce a factor K (similar to K_L, described in Sec. 9.3):

$$K = \frac{\omega_1 L_p}{R_g} = \frac{2\pi f_1 N^2 A_L}{R_g} \qquad (9.45)$$

where

$$f_1 = \frac{\pi K R_g}{\mu_0 \mu l} = \frac{250 \, R_g}{\mu l} \text{ MHz} \qquad (9.46)$$

The band span is the ratio of Eqs. (9.42) and (9.46):

$$\frac{f_2}{f_1} = 11.85 \frac{\mu}{R_g} = 47.4 \frac{\mu}{R_L} \qquad (9.47)$$

The utility of Eq. (9.47) is illustrated by substituting values of permeability that are feasible in different regions of the frequency spectrum. In Table 9.5 we have used the following standard conditions: $R_g = 25$, $R_L = 100$, $Z_0 = 50$, $K = 2$, and $l_N = \lambda/8$ (cut from RG-58U cable). Some comments are in order:

1. The bandwidth $(f_2 - f_1)$ decreases as f_1 increases.
2. The band-span ratio changes with the power level. At higher levels of power, flux density B increases and permeability μ decreases. We have depicted this condition for a specific ferrite material in columns 2 and 3 of Table 9.5. Band span drops by a factor of 0.39 (= 1,050/2,700) when B increases from 10 to 3,000 gauss (G).
3. The band span is greater when we use ferrite than when we use iron powder.
4. When environmental stresses are severe (e.g., vibration or thermal shock) and cyclical polarizing dc bias and stability of magnetics are mandatory, many designers will choose iron powder instead of ferrite.

TABLE 9.5 Bandwidth of Transmission-Line Transformer at Low Power*

1	2	3	4	5	6	7	8	9	10	11
f_2/f_1	1,279	500	355	1,042	59.2	71	9.48	4.74	3.79	2.84
f_2	1,279	500	782	625	296	710	47.40	142.00	341	438
f_1	1	1	2.2	0.6	5.0	10	5.00	30.00	90	150
μ, 10 G	2,700	—	750.0	2,200	125.0	150	20.00	10.00	8.00	6.00
μ, 3,000 G	—	1,050								
$\tan \delta/\mu$ (in ppm)	105	—	100	100	52.0	90				
Ferrite grade†	3C8	3C8	3D3	3E2A	4C4	4C4				
Iron powder grade‡	—	—	—	—	—	—	C1	E2	SF	W

* Calculations are based on low power levels. Values are based on: impedance ratio 1:4, $Z_0 = 50$, and $K = 2$ (see text).

† Grades and data from *Ferroxcube Division Catalog*, 1977. Comparable data are available from other suppliers (see Table 6.3).

‡ Arnold Engineering Co., *Catalog PC 109-B*, and Micrometals, Inc., *Catalog 3*.

b. BALUN Coupling and Limitations

The transmission-line transformer is widely used, especially in RF circuits, where it provides coupling between balanced and unbalanced circuits. When used to couple a *bal*anced line to an *un*balanced line, it is called a *BALUN* (see Fig. 9.16).

To provide coupling between balanced circuits, a two-core construction (Fig. 9.17*a*) is used. Winding 3–4 on one core is joined to winding 5–6 on the other core, resulting in a symmetrical geometry. This construction has several desirable features. One is the ability to function in a push-pull (bipolar) drive circuit when the junction of terminals 4 and 6 (marked C.T. in Fig. 9.17*a*) is grounded. Another is that the outer windings can be of either braid or copper tubing, which results in very low dc resistance and is, therefore, desirable in the low-impedance branch of a high-power RF transformer circuit. The two cores can be integrated into a single core having two holes (Fig. 9.17*b*), a shape called the *BALUN core.*

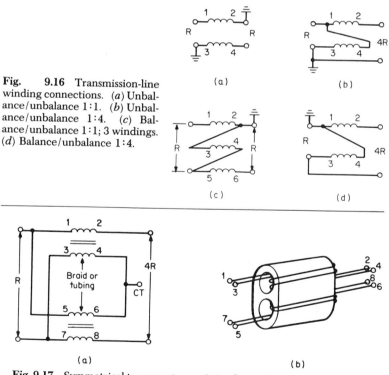

Fig.   9.16 Transmission-line winding connections. (*a*) Unbalance/unbalance 1:1. (*b*) Unbalance/unbalance 1:4. (*c*) Balance/unbalance 1:1; 3 windings. (*d*) Balance/unbalance 1:4.

Fig. 9.17 Symmetrical two-core transmission line. (*a*) Balance/balance, 1:4 impedance ratio. Center tap junction permits push-pull drive. (*b*) Integrated core geometry (BALUN core).

When the drive is push-pull class B or bipolar class D and the load is unbalanced (single-ended), two or more cores are used, depending on the impedance ratio. In Fig. 9.18a, where the impedance ratio is 1:4, two cores are used; in Fig. 9.18b, where the impedance ratio is 1:9, three cores are employed. Note that the characteristic impedances (in parentheses) are different in each set of coils.

The transmission-line transformer is a natural choice in the design of the hybrid circuit operating at high frequencies. As we noted in Sec. 8.8, the hybrid transformer is capable of transmitting energy between designated ports while providing virtual isolation between selected ports (see Fig. 8.10).

The limitations of the transmission-line transformer are regarded as a challenge, judging from the close scrutiny it has received in the literature. However, the desire to extend the bandwidth of a transmission line can be satisfied by cascading (connecting together) lines of different characteristic impedances or by designing a line with a gradual taper between adjacent sections.[32,37,38] Another objective has been to provide the isolation function of a conventional transformer without

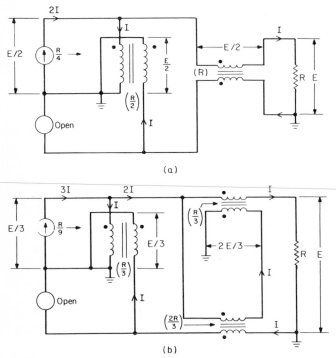

Fig. 9.18   Class B or D transmission line: push-pull drive to unbalanced load. (a) 1:4 impedance ratio. (b) 1:9 impedance ratio.

compromising the wide bandwidth associated with the transmission-line geometry.[39]

c. Some Practical Considerations[40-43]

The physical transmission line can be constructed in any of several ways: (1) as a standard braided coaxial cable (25, 50, 75 Ω), (2) in the shape of copper or brass tubing which encloses the coil's lead wires, (3) as a stripline cable,[32] or (4) as a pair of magnet wires (i.e., a bifilar winding).

Bifilar magnet wire deserves serious consideration because it permits, in principle, the maximum utilization of copper in the window area of the core. Or, viewed in another way, the bifilar technique offers the possibility of achieving minimum size and, therefore, maximum high-frequency response. Some discussion is in order.

The bifilar winding can be analyzed as a two-wire transmission line whose characteristic impedance is given by:*

$$Z_0 = 120 \sqrt{\frac{\mu_r}{\epsilon_r}} \cosh^{-1} \frac{D}{d} \qquad (9.48)$$

where $\mu_r = 1$, ϵ_r is the relative permittivity (dielectric constant) of the dielectric separating the wires, and D is the spacing between wires of diameter d. When two enameled wires are contiguous and we can assume that $D/d \cong 1.1$ and $\epsilon_r \cong 1$, we calculate that a bifilar pair would result in a characteristic impedance of 53.2 Ω. When the wires are twisted, ϵ_r increases and Z_0 decreases in the order of 10 to 25 percent.

A more exact analysis shows that, in practical wire sizes [21 to 38 American wire gage (AWG)], Z_0 varies with the film's thickness and the angle (or pitch) of the wire's twist.[44] For example:

$$Z_0 \cong 40\text{--}60 \qquad (5°)$$
$$Z_0 \cong 27\text{--}40 \qquad (45°)$$

Example. We will make some of the foregoing ideas more concrete. Assume that we are to design a low-power transformer to match a 25-Ω source to a 100-Ω load. Suppose that the maximum frequency to be transmitted is 500 MHz and that the minimum is 1 MHz.

We begin at the high end of the spectrum. First, we select a standard 50-Ω cable such as RG-58U, whose dielectric loss is low (0.071 dB/ft at 800 MHz) and whose diameter d is small (5 mm or 0.195 in).

---

* For computations with a calculator, we use the identity $\cosh^{-1} y = ln\,(y + \sqrt{y^2 - 1})$, where $y = D/d$.

Small power losses and reflections are desired, so we use a cable length of $\lambda/8$.* We compute λ from Eq. (9.30):

$$\lambda = [0.79 \ (3) \ 10^{-10}]/[500 \ (10)^6] = 47.4 \ \text{cm}$$

The coaxial wire will be cut to be 0.125×47.4 cm, or 59.3-mm long.

There are several cores to choose among. If we choose a toroid, then its *internal diameter* (ID) must be adequate to permit N turns to be wound. Preliminary calculations show that, in this example, we are restricted to a two- or three-turn design since we want the ID to be small. The toroid for this trial will, therefore, have an ID about twice the cable's diameter, which is 5 mm.

To satisfy the requirement for the low-frequency limit f_1, we need a core whose sectional area and diameter will satisfy Eq. (9.45). We noted earlier that K is an integer which is selected to satisfy the desired distortion or phase shift and that A_L is the inductance factor of the core defined in Eq. (6.60). In our example we make K equal to 2 and solve Eq. (9.45) for f_1 as a function of A_L:

$$f_1 = \frac{KR_g}{2\pi N^2 A_L} = \frac{2 \ (25)}{2\pi 3^2 A_L} = \frac{0.88}{A_L} \tag{9.49}$$

Thus, we plan to select a core with an ID of about 12 mm and, if $f_1 = 1$ MHz, to specify an inductance factor A_L of 0.88 μH per turn.

A perusal of catalog data on ferrite cores shows that a core with the following values will approximate the desired characteristics:†

OD d_o, mm	ID d_1, mm	Height h, mm	Core area A_e, sq mm	A_L, μH At 10 G	At 3,000 G
29	19	7.6	37.1	1.74	0.680

We infer, therefore, that a low-frequency cutoff of $0.88/1.74 \times 1 = 0.506$ MHz is feasible for low flux density. For a higher density of flux, 0.3T, we calculate a cutoff frequency of:

$$0.88/0.68 \times 1 = 1.29 \ \text{MHz}$$

The output power P_o can be calculated after determining the voltage levels E_s at several values of flux density and frequency (see Table 9.6). In this tabulation, we have restricted the flux density to 800 G at 1 MHz in order to limit the power loss W_{Fe} in the core.‡ Compara-

---

* Calculations show that if $l = \lambda/5$, the insertion loss is 0.4 dB and the standing wave ratio (a measure of reflections) is 2:1.

† One such core is Ferroxcube 3C8, part 502T300 (37.1 cu cm, $\mu = 2,700$; see Table 9.5). Other suppliers offer comparable materials.

‡ Power loss in the core has been calculated with the aid of Eq. (6.20), using as the reference level the value of 115 mW/cu cm at 25 kHz.

TABLE 9.6 Calculations of Output Power (P_o) and Power Loss in the Core (W_{Fe})

	Low frequency f, MHz			
	0.5	0.5	1.0	2.0
B, G	2,000	1,600	800	400
E_s, V	98	78.4	78.4	78.4
P_o, W	96	61.5	61.5	61.5
W_{Fe}, W	30.5	19.4	11.6	7.0

tively small toroids ($d_o = 51$ mm and $h = 14$ mm) can furnish kilowatts (kW) of RF power.[45]

In our design example we have implicitly ignored the consequences of power losses in the iron. However, if both high-power and extended-frequency performance are desired, we can investigate the feasibility of using a composite core structure, that is, two cores of different materials (see Sec. 9.3c).

d. Alternative Geometry

Now consider the alternatives to the conventional geometry we have just described. Instead of using a helix of three turns threaded through a toroid, we can make other choices.

We can abandon the thick cable and substitute bifilar magnet wire (or stripline) and then select an even smaller toroid, called a *bead*. (We have noted, in the literature,[33] that response out to 1,000 MHz can be achieved by using an 8-turn twisted pair of No. 37 Formex wire wound on a toroid bead whose ID is a mere 2 mm.)

Or we can use a one-turn design, and add the necessary number of cores to satisfy the requirements at the low-frequency end of the band. A fresh look at the basic transformer equation in Chap. 1, Eq. (1.8), reminds us that we can trade turns N for discrete cores of area A. For example:

$$4 \text{ turns} \times 1 \text{ core unit} = 1 \text{ turn} \times 4 \text{ core units}$$

With this in mind, we depict, in Fig. 9.19, three possible alternatives to the traditional single-core geometry. In Fig. 9.19a, the bifilar pair of wires, constituting one turn, passes through four toroids or beads.* In Fig. 9.19b, each set of four toroids is integrated into two sleeves. And in Fig. 9.19c, two sleeves are integrated into a single core with two parallel holes, which we recognize as the BALUN core. Having

---

* A UHF transformer in which beads (cores) are slipped over threads (wires) resembles, geometrically, a necklace of beads.

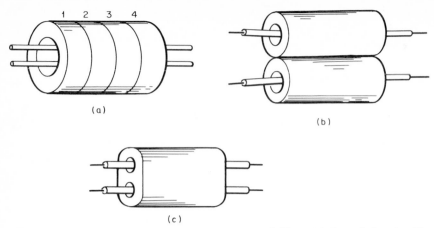

Fig. 9.19   Alternative core geometry.   (*a*) One turn (bifilar pair) through four toroids, equivalent to four turns on one toroid (see Fig. 9.14*a*).   (*b*) Eight toroids equal two sleeves (see Fig. 9.17*b*).   (*c*) Two sleeves integrated into one balun core.

traded turns for cores, we can now select a suitable number n of cores whose permeability μ and whose combined area nA will be adequate to satisfy the specification of the minimum frequency f_1 to be transmitted.

9.6. SUMMARY: THE DESIGN OF A
WIDE-BAND TRANSFORMER

We have seen that the analysis and synthesis of the wide-band transformer entails many factors.   We shall summarize them by outlining the steps followed in the design of such a transformer.   Our review is procedural rather than quantitative so that we can maintain our perspective and avoid becoming lost in a forest of details.   The techniques to which we refer are not limited to the concepts described in this chapter and in Chap. 8.   We also make use of the area product (Chap. 2), efficiency and size (Chap. 5), permeability and Q (Chap. 6), and polarizing bias (Chap. 7).

It is not wise to proceed without a complete set of specifications and a checklist.   Our first step is to determine the *feasibility* of the specifications.

We begin by examining the specified fidelity and band span, because these specifications tend to conflict with limitations on the transformer's size and cost.   Strict fidelity (less than 5 percent harmonic distortion) often requires expansion of the bandwidth (by two or more octaves) from the specified low- and high-frequency cutoffs.   Yet, if the band span is too great, we shall need auxiliary, or built-out, networks.

Seven major parameters affect the band span. We cannot ordinarily control the parameters which are established by circuit considerations (e.g., the circuit impedances, the turns ratio, and the lowest operating frequency), and too often we are unable to exploit the fact that maximum band span is achieved when circuit impedances are low and the turns ratio is unity. However, we can exercise control over some of the other variables:

1. The low-frequency impedance ratio $K_L = \omega L_P/R_L$, which is determined by the permissible low-frequency distortion
2. The core permeability μ_e
3. The insulation permittivity ϵ
4. The characteristic impedance, which influences the high-frequency response and has a bearing on high-frequency distortion
5. The coil geometry

A large band span is facilitated by *small size;* thus smallness is a prime and compelling motif in the design of wide-band transformers. An initial estimate of size can be made by adapting the low-frequency impedance factor K_L and the flat loss (a function of efficiency η) to the Q concept [Eq. (9.7)]. Q can be directly related to the area product $A_{Fe}A_{Cu}$. Size decreases as the low-frequency cutoff increases and as the efficiency decreases.

It is interesting to compare the limitations on the size of the wide-band transformer with those on the size of the narrow-band power transformer. In the power transformer, size reduction proceeds until it is limited by flux density, efficiency and regulation, dielectric considerations, or excessive temperature. In the wide-band transformer, reduction of the size of the core proceeds until one of the following factors sets a lower limit:

1. The allowable low-frequency flux density (a function of harmonic distortion, shunt inductance, and polarizing bias)
2. The copper efficiency and regulation (insertion flat loss) at the mid-band frequency
3. The dielectric considerations, i.e., the amount and type of insulation necessary for reliable performance, especially when voltages are high

So far as reduction in size is concerned, the distinction between the two types of transformers rests mainly on the magnitude of the flux density at the low cutoff frequency. In the power transformer, a high duty cycle and high flux density produce thermal problems. In the wide-band transformer, the low duty cycle of the signals and the low flux density account for the absence of a thermal problem.

Having decided how large, approximately, the area product and core must be, we now make a tentative choice of the core material and core shape. The core material is selected on the basis of the particular region of the frequency spectrum the transformer is to operate in. This will usually depend on the low-frequency cutoff ω_1. Calculations are then made to confirm that primary inductance L_p and flux density B_m satisfy the distortion allowable at the lowest operating frequency. Next we check to see that the size of the window of the core is adequate for the required flat loss at the mid-band frequency. Winding resistance is estimated. It must not be excessive.

If the required band span appears to be large after our first assessment, we take extra pains to reduce the size of the core and choose a filter model which will improve the prospects of obtaining superior high-frequency performance. Then we tackle the problem of geometry.

The choice of a particular geometry will be influenced by the circuit impedance, the voltage level, the required balance, and the constraints on the susceptibility to stray fields. The magnitude and proportions of leakage inductance and distributed capacitance will depend on the high-frequency cutoff, the network model (chosen to provide the specified character and shape of the response), and the characteristic impedance. If the voltage levels are high, the shape and relative size of the window become important. An appropriate geometry, one consistent with these considerations, is tentatively chosen.

It is now possible to compute the magnitudes and relative proportions of leakage inductance and distributed capacitance. Next, we decide whether the configuration shall be a core type having one or two coils, or a shell type having one coil. Certain balance and high-voltage problems may indicate the need for sector (pi or pancake) coils. The need for complete astaticism combined with an extremely wide band span may indicate that the core-type geometry be in the shape of a toroid. In the spectrum above a few kilohertz, good shielding may be economically obtained with the pot core.

The wide-band transformer may be regarded as a power transformer operating at low flux density and low duty cycle. Special attention must be paid to the coil geometry and to the core geometry. Consideration of the various factors, including cost (which has a way of becoming decisive), leads us to the conclusion that no one geometry is ideal.

REFERENCES

1. D. Preis, "Linear Distortion," *Journal of the Audio Engineering Society,* Vol. 24, No. 5 (June 1976): 346–67.

2. H. Bode, *Network Analysis and Feedback Amplifier Design* (Princeton, N.J.: D. Van Nostrand Company, 1945), pp. 286, 363, 471.
3. N. Grossner, "Transformer Design for Zero Impedance Amplifiers," *Audio*, Vol. 40, No. 3 (March 1956): 27–30, 69.
4. H. Blinchikoff, A. Zverev, *Filtering in the Time and Frequency Domains* (New York: John Wiley & Sons, 1976).
5. A. Williams, *Electronic Filter Design Handbook*, (New York: McGraw-Hill Book Company, 1981).
6. D. Jensen, "Speed Transformer and Op-Amp Design with a Fast-Working Desktop Computer," *Electronic Design* (July 19, 1979).
7. J. Harrison, "A New Resonance Transformer," *IEEE Transactions on Electron Devices*, Vol. ED-26, No. 10 (October 1979): 1,545–49.
8. T. O'Meara, "Very Wide-Band Impedance Matching Networks," *Transactions IRE PGCP*, Vol. CP-9, No. 1 (1962): 38–44.
9. ———, "Wide-Band Transformers and Associated Coupling Networks," *Electro-Technology*, Vol. 70, No. 3 (September 1962): 80–89.
10. H. Granberg, "Combine Power without Compromising Performance," *Electronic Design* (July 19, 1980): 181–87.
11. H. Lord, "The Design of Broad-Band Transformers for Linear Electronic Circuits," *Transactions AIEE*, Vol. 69 (1950): 1,005–10.
12. J. Krauss, K. Carver, *Electromagnetics*, 2d ed. (New York: McGraw-Hill Book Company, 1973), pp. 509–20.
13. G. Larson, "Critical Factors in Core Geometry for Improved Performance of Communication Transformers," *IEEE International Convention Record 1966*, Part 9, pp. 84–93.
14. M. Vore, "Design of Random Noise Transformers," *Transactions AIEE*, Vol. 78, Part 1 (March 1959): 59–63.
15. P. Allen, "Multi-Core Transformers Boost Bandwidth," *Electronic Design* (February 3, 1964): 29–32.
16. MIT Electrical Engineering Staff, *Magnetic Circuits and Transformers* (New York: John Wiley & Sons, 1943), pp. 357–62.
17. L. Blume (ed.), *Transformer Engineering*, 2d ed. (New York: John Wiley & Sons, 1951), pp. 72–78.
18. H. Stephens, "Transformer Reactance and Losses with Nonuniform Windings," *Electrical Engineering* (February 1934): 346–49.
19. N. Crowhurst, "Leakage Inductance," *Electronic Engineering*, Vol. 21, No. 254 (April 1949): 129.
20. K. Macfadyen, *Small Transformers and Inductors* (London: Chapman & Hall, 1953), pp. 57–60 and 222–23.
21. H. Ott, *Noise Reduction Techniques in Electronic Systems* (New York: John Wiley & Sons, 1976), Chap. 6.
22. A. Kusko, M. Caplan, "Microminiature Transformers," *IEEE Transactions on Magnetics*, Vol. MAG-10, No. 3 (September 1974): 698–700.
23. J. Casse, "Printed Transformers for High Frequency," *Electronic Engineering*, Vol. 41 (June 1969): 34–38.
24. G. Glasoe, J. Lebacqz, *Pulse Generators* (New York: McGraw-Hill Book Company, 1948), pp. 516–18.

25. H. Lamson, "A High-Power Toroidal Output Transformer," *General Radio Experimenter* (West Concord, Mass.), Vol. 26, No. 6 (November 1951).

26. T. O'Meara, "A Comparison of Thin Tape and Wire Windings for Lumped Parameter, Wide-Band High Frequency Transformers," *Transactions IRE PGCP*, Vol. CP-6, No. 2 (June 1959): 49–57.

27. R. Sette, "Search for a Better Transformer," *General Radio Experimenter*, Vol. 43, No. 11 (November 1969): 5–6.

28. B. Sommer, G. Plice, "Specification and Testing of Shielded Transformers," *Electro-Technology*, Vol. 71, No. 5 (May 1963): 102–105.

29. W. Olschewski, "Unique Transformer Design Shrinks Hybrid Isolation Amplifier's Size and Cost," *Electronics* (July 20, 1978): 105–12.

30. R. Matick, "Transmission Line Pulse Transformers," *Proceedings IEEE*, Vol. 56, No. 1 (January 1968).

31. ———, *Transmission Lines for Digital and Communication Networks* (New York: McGraw-Hill Book Company, 1969), Chaps. 1–3, 5, 8.

32. L. Dworsky, *Modern Transmission Line Theory and Applications* (New York: John Wiley & Sons, 1979).

33. C. Ruthroff, "Some Broadband Transformers," *Proceedings IRE*, Vol. 47 (August 1959): 1,337–42.

34. W. Blocker, "The Behavior of the Wide-Band Transmission-Line Transformer for Nonoptimum Line Impedance," *Proceedings IEEE*, Vol. 66, No. 4 (April 1978): 518–19.

35. O. Pitzalis, T. Couse, "Practical Design Information for Broadband Transmission Line Transformers," *Proceedings IEEE* (April 1968): 738–39.

36. ———, "Broadband Transformer Design for RF Transistor Power Amplifiers," *Proceedings National Electronic Components Conference, 1968*, pp. 207–16.

37. R. Irish, "Method of Bandwidth Extension for the Ruthroff Transformer," *Electronic Letters*, Vol. 15, No. 24 (November 22, 1979): 790–91.

38. S. Dutta Roy, "A Transmission Line Transformer Having Frequency Independent Properties," *Circuit Theory and Applications*, Vol. 8 (1980): 55–64.

39. W. Hilberg, "High Frequency Transformers with Separate Windings and Very Broad Transmission Bandwidth," *IEEE Transactions on Magnetics* (September 1970): 667–68.

40. O. Pitzalis, R. Horn, R. Baranello, "Broadband 60-W HF Linear Amplifier," *IEEE Journal of Solid State Circuits*, Vol. SC-6, No. 3 (June 1971): 93–103.

41. H. Krauss, C. Allen, "Designing Toroidal Transformers to Optimize Wideband Performance," *Electronics* (August 16, 1973): 113–16.

42. H. Granberg, "Broadband Transformers and Power Combining Techniques for RF," *Application Note*, No. 749 (Phoenix, Ariz.: Motorola Semiconductor Products, 1975).

43. M. De Mauw, *Ferromagnetic Core Design and Application Handbook* (Englewood Cliffs, N.J.: Prentice-Hall, 1981).
44. P. Lefferson, "Twisted Magnet Wire Transmission Line," *IEEE Transactions on Parts, Hybrids and Packaging*, Vol. PHP-7, No. 4 (December 1971): 148–54.
45. J. Sevick, "Broadband Matching Transformers Can Handle Many Kilowatts," *Electronics* (November 25, 1976): 123–28.

CHAPTER 10

The Pulse Transformer:
Analysis

In Chaps. 8 and 9, we described signal transmission in the frequency domain. We now consider the subject of signal transmission in the time domain. In this chapter we analyze the response of a linear network and of a transformer whose signals are pulse waveforms. In Chap. 11, we shall deal with the synthesis of the pulse transformer.

10.1. SCOPE OF THE DISCUSSION

When Faraday demonstrated electromagnetic induction in 1831, he had actually invented the pulse transformer. Closing a switch in series with a primary coil and a battery, he observed a pulse of current in the secondary coil connected to a galvanometer. Opening the switch a moment later, he observed a pulse of current in the opposite direction. The early telegraphic systems of the nineteenth century, in which a switch, a battery, and a coil were used, were pulse circuits. Faraday's experiment can be repeated with any transformer, and in this sense any transformer can be regarded as a pulse transformer. However, modern pulse transformers have been specifically designed either to transmit pulses of voltage or current with some reasonable

362

degree of fidelity to wave shape or to shape an input pulse into an arbitrary output waveform.   The term *pulse transformers* is usually restricted to such transformers.

Our survey comprises two chapters, analysis and synthesis.   In this chapter, we know what is in the black box (i.e., the constants of the network or transformer) and our approach is analytic.   Given a prescribed pulse waveform, we wish to determine the response at the output of the transformer.   In Chap. 11, we approach the subject from the point of view of synthesis.   Given a prescribed pulse waveform and the desired response at the output, we establish the constants inside the black box; that is, we design the transformer.   We have divided our present topic, analysis, into two parts:

Part A.   Linear Networks (Secs. 10.2 to 10.4)
Part B.   The Transformer Network (Secs. 10.6 to 10.10)

Part A is a summary of concepts in the theory of linear networks and filters.   We intend it to serve as a review and glossary of the terminology used by transmission engineers in the specification of pulse transformers.*   In Part B, we deal with the ferromagnetic pulse transformer.   The phenomena of saturation, remanence, and the nonlinear behavior of the core, which are not discussed in linear network analysis, have an important effect on pulse response, as we shall see.   We begin our analysis in Sec. 10.2, which follows, by describing the various types of pulse waveforms we are likely to encounter in electronic networks and systems.   In Sec. 10.3, we describe the network, or linear transformer circuit, in terms of certain concepts in the theory of networks and filters.   In Sec. 10.4, we describe the various types of pulse response that result at the output of the linear transformer when certain basic pulse waveforms are fed into the input of the transformer.   The basic waveforms we focus on are the impulse, rectangular, and step functions.   In Sec. 10.5 we describe the basic concepts of pulse code modulation and their relevance to the criteria of accuracy and bandwidth.

In Part B, Secs. 10.6 through 10.9, we focus on the response of the ferromagnetic transformer.   In Sec. 10.6, we discuss equivalent circuits and our rationale for using the step function in the piecewise analysis of the transformer circuit.   In Secs. 10.7, 10.8, and 10.9, we consider in some detail the response during the rise, top, and off periods of the last rectangular pulse.

We conclude our analysis in Sec. 10.10 by relating the pulse response of the transformer in the time domain to bandwidth in the frequency domain.

---

* Part A will be, for some readers, a digression from the main subject and can be skipped without loss of continuity.

PART A: LINEAR NETWORKS
10.2. PULSE WAVEFORMS

A variety of pulse waveforms is in use. The most popular shape is the rectangle, which is easy to generate with semiconductor switches and which produces minimal loss of power in the switch. Rectangular and square-wave circuits are efficient.

a. Preliminary Definitions

Our main concern is with the rectangular pulse as encountered in a periodic train of pulses (Fig. 10.1a). The repetition rate or *pulse repetition frequency* (PRF) $= 1/T$, is expressed in pulses per second (pps). The duty cycle, or duty ratio, is that fraction of an arbitrary time period T occupied by the pulse τ, that is, τ/T. If the on time of a periodic pulse train equals the off time, we have the square wave depicted in Fig. 10.1b, whose frequency is expressed as PRF $= 1/(2\tau)$.

b. Pulse Modulation

The rectangular pulse train is widely used in such systems as television, radar, power conversion, and *pulse code modulation* (PCM). We have, earlier in the book (in Secs. 2.10 and 5.7) described the power converter transformer, which employs *pulse width modulation* (PWM) as a means for the efficient transfer of regulated power. The width of the pulse τ changes in proportion to the amplitude or variation of the input or output voltage (or current). Hence the duty ratio τ/T, or modulation, contains the information about the variation of an amplitude function. There are many ways in which a pulse train can be modulated; for example, *pulse frequency modulation* (PFM), *pulse position modulation* (PPM), and *pulse code modulation* are in use.

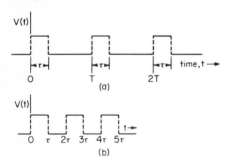

Fig. 10.1 Pulse trains. (*a*) Periodic; PRF $+ 1/T$, duty ratio $= \tau/T = \tau$PRF. (*b*) Square wave; PRF $= 1/2\tau$, duty ratio $= 1/2$.

In the PCM scheme, information about a signal (its amplitude or variation) is encoded into a sequence of pulses whose amplitude is constant but whose pattern (the code) varies. Pulse code modulation concepts, which are of great importance, are discussed later, in Sec. 10.5.

c. Basic Waveforms

A useful scheme for the classification of pulses is based on whether the pulse has sharp contours (that is, is discontinuous) or is smooth (continuous). On this basis, we sort the basic shapes as follows:

Discontinuity	Smooth
Rectangular	Cosine
Triangular	Raised Cosine
Exponential	Gaussian
Impulse	

Examples of both types of pulses are depicted in Fig. 10.2. In the left half of the figure we depict, in the time domain, four pulse shapes—rectangular, exponential, triangular, and Gaussian—each of which has a characteristic width of τ seconds. In the right half of the figure, we depict, in the frequency domain, the frequency spectrum corresponding to each pulse waveform. Note that in each case, the magnitude of the frequency components (harmonics) diminishes from its maximum value (when $\omega = 0$) to a value inversely proportional to the width τ of the pulse.

d. The Impulse Function[1]

The unit impulse function, also called the delta function $\delta(t)$, is a theoretical waveform of importance * We are to visualize a pulse of infinitesimal width and of infinite magnitude but with an enclosed finite area: unity. This seemingly fictitious concept can best be imagined as a pulse shape of finite area that progressively decreases in width a (on the time axis) as it increases in height to $1/a$, the inverse of the reduction in width; that is, its height is a function of its width. The evolution of the unit impulse is suggested for three shapes in Fig. 10.3. We see that as the ratio of height to width increases by a factor of 2, then 4, and so forth, the product remains unity. Suppose

---

* The delta function was introduced (ca. 1925) by Paul Dirac in his treatment of concepts in quantum mechanics.

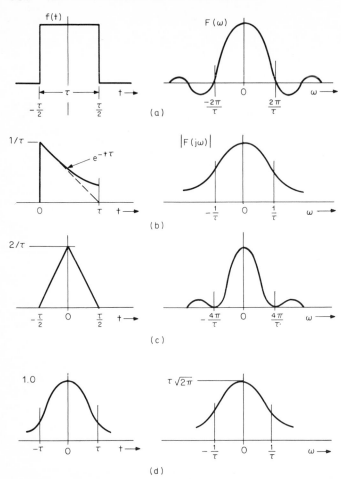

Fig. 10.2 Single pulses and frequency spectrums. (*a*) Rectangular and (sin *x*)/*x* response. (*b*) Exponential pulse and response. (*c*) Triangular pulse and response. (*d*) Gaussian pulse.

that the ordinate is a voltage V. Let $1/a = f(\tau) = V$ and let $a = t - \tau$. As $a \to 0$, $V \to \infty$, and $1/a \times 1 = 1$; that is:

$$V\tau = 1 \qquad (10.1)$$

Following the notation of Fig. 10.3*a*, we view the delta function $f(t)$ as equal to $\delta(t - b)$ when $t \to b$, where $b = 0$, on the time axis. Mathematically, the delta function is defined by an integral:

$$f(t) = \int_{-\infty}^{\infty} f(\tau)(t - \tau)\, d\tau \qquad (10.2)$$

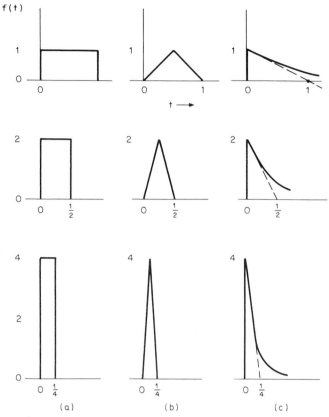

Fig. 10.3 Impulse function derived from three shapes. (*a*) Rectangular. (*b*) Triangular. (*c*) Exponential.

A practical approximation of the unit impulse is, therefore, a very narrow pulse of finite magnitude, such as the "spike" that the engineer encounters in switching circuits.

Trains of impulses are depicted in the left half of Fig. 10.4, and their spectrums (in the frequency domain) are shown in the right half of the figure. Three examples are given in order to illustrate the effect of increasing the interval T between impulses. Several significant characteristics can be inferred from the figure:

1. The harmonic frequencies, which are infinite in number, occur at discrete intervals $2\pi/T$, the period T being the reciprocal of the repetition frequency (Fig. 10.4*b* and *d*).

2. The magnitude of each frequency component is identical. This is not true of other pulse waveforms (such as those in Fig. 10.2), where the magnitude of the harmonics decreases as the frequency increases.

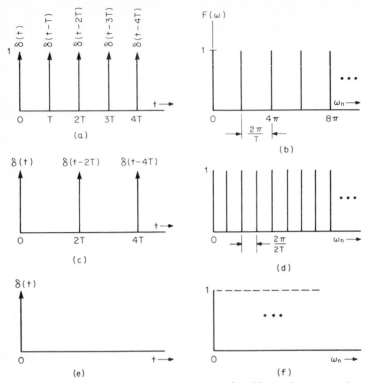

Fig. 10.4   Impulse train and spectrum (not band-limited). (*a*) High repetition rate, PRF = 1/T. (*b*) Wide spacing of discrete frequencies. (*c*) Slow repetition rate. (*d*) Narrow spacing of frequency components. (*e*) Single impulse. (*f*) Continuous spectrum of all frequencies.

3. In the spectrum of an impulse train, the spacing of the harmonic frequency components decreases as the time interval between impulses increases. We see this when we compare the train of pulses in Fig. 10.4*a* with the train in Fig. 10.4*c*. The train in Fig. 10.4*c* has a repetition rate which is half that of the train in Fig. 10.4*a*, but in the spectrum of train *c*, shown in Fig. 10.4*d*, the harmonic components are spaced more closely together. Specifically, in Fig. 10.4*d* there are twice as many harmonics in the interval 2π as there are in Fig. 10.4*b*.

4. In the limiting case (Fig. 10.4*e*) as the interval between pulses becomes unbounded, we arrive at the single impulse $\delta(t)$ and its spectrum (Fig. 10.4*f*). Now the frequency components are equal in amplitude but are so closely packed that $2\pi/T = 0$; that is, all *frequencies* (*zero to infinity*) *are present in the spectrum of the unit impulse*.

Because the unit impulse contains all frequencies, it can serve as a powerful tool in the analysis of linear networks. Further, it can be shown that if a network is defined, and if its response $h(t)$ to an impulse function is known, then the response of that network to any other arbitrary input waveform can be deduced.[2]

e. The Step Function

Another wave shape useful in the analysis of pulse circuits is the step function. In Fig. 10.5a we depict the unit step, $u(t)$ a function whose amplitude $f(t)$ instantaneously changes in magnitude from zero to 1. Mathematically:

$$\left.\begin{array}{ll} f(t) = 0 & t \leq 0 \\ f(t) = 1 & t > 0 \end{array}\right\} \tag{10.3}$$

The response to a unit step at the output of an "ideal" net network (Fig. 10.5b) is characterized by a delay in time and a tilt in the vertical axis, resulting in a finite interval called *rise time* t_R, during which the unit achieves its full amplitude.

There is a very important relation between the unit step depicted in Fig. 10.5a and the unit impulse shown in Fig. 10.4a; that is, the integral of the unit impulse is a unit step:

$$\left.\begin{array}{ll} \displaystyle\int_{-\infty}^{t} \delta(t - b)\, dt = 0 & t < b \\[2mm] \qquad\qquad\quad = 1 & t \geq b \end{array}\right\} \tag{10.4}$$

Since the inverse of an integral is a derivative, the derivative of the unit step is a unit impulse:

$$\left.\begin{array}{ll} \dfrac{d}{dt} u(t - b) = \delta(t - b) = \infty & t = b \\[2mm] \qquad\qquad\qquad\quad = 0 & t \neq b \end{array}\right\} \tag{10.5}$$

The implications of Eqs. (10.4) and (10.5) are of great utility in the analysis and design of pulse circuits. Later, in Part B of this chapter, we shall use the step function as an important tool (the forcing function) in our analysis of the pulse transformer.

10.3. NETWORK OF FILTERS

Pulse transmission is a major subject of analysis in the theory of networks and filters. It will be helpful to examine certain concepts used in these disciplines, especially since the circuit designer often

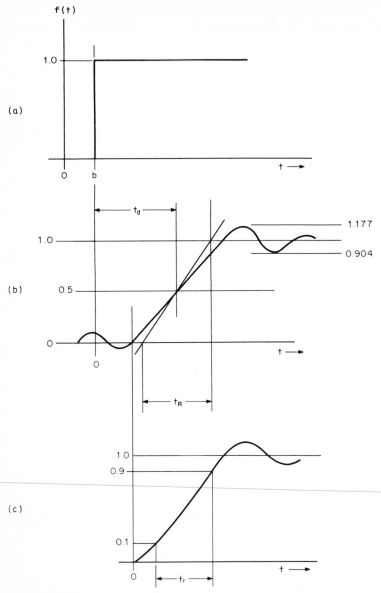

Fig. 10.5 Step function and response. (*a*) Unit step. (*b*) Delay time t_d and rise time t_R. (*c*) Rise time t_r defined by 10-to-90% response.

uses their special terminology, which we (the transformer specialists) must comprehend to understand not only the theory but also a set of specifications.

From the point of view of network theory, one can define four classes of networks and their attributes.

a. The Ideal Network (or Black Box)

The output is an exact replica of the input. Phase shift is a linear function of frequency. The output waveform differs from the input only by a scale factor in magnitude and a delay in time. The delay is constant for all frequencies contained in the input pulse (see Fig. 10.6*a* and *b*.)

b. The Ideal Filter

Response at the output of the black box is uniform up to a cutoff frequency ω_c beyond which response is zero; that is, the filter exhibits either a rectangular (flat) amplitude response (Fig. 10.6*c*) or a rectangular (flat) delay-time response (Fig. 10.6*d*).

c. Realizable Filters

The theory of filter networks provides the designer with practical procedures for approximating the ideal filter. One class of networks

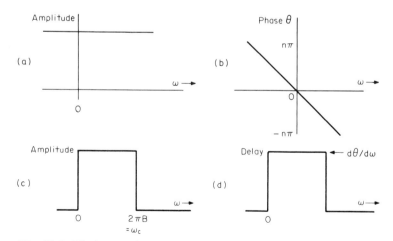

Fig. 10.6 Ideal network and filter characteristics. (*a*) Ideal network response, infinite bandwidth. (*b*) Linear phase shift θ; bandwidth B is finite. (*c*) Flat amplitude, band-limited ideal filter. (*d*) Constant flat time delay, band-limited case.

(e.g., the Butterworth filter) can be designed to approximate a flat amplitude out to a specified cutoff frequency ω_c and a specified slope [decibels (dB) per octave] in the spectrum beyond ω_c (see Fig. 10.7a). Another class of networks (e.g., the Bessel filter) can be designed to approximate a flat or constant delay out to a specified cutoff frequency ω_c (see Fig. 10.7b).

d. The Transmission Line

In Chap. 9, Sec. 9.5, we took note of the fact that the phase response of the ideal uniform transmission line is a linear function of frequency. Hence its response, in the time domain, results in a constant delay of a pulse waveform. Practical transmission lines come close in behavior to the ideal filter and delay line (see Figs. 10.6c and d). We shall discuss the application of the transmission line to the design of the coil of a pulse transformer in Chap. 11, Sec. 11.5.

10.4. NETWORK RESPONSE AND FIDELITY[2]

All realizable networks, filters, and transformers have a limited bandwidth and a characteristic phase response, linear or nonlinear. Apply an arbitrary waveform to the input of a nonideal black box, and the waveform at the output will be altered (distorted) in some manner, typically phase and shape.

a. Bandwidth

Nevertheless, the response has several basic characteristics:

1. The output pulse is delayed in time t_0.
2. The pulse retains most of its area. (See Figs. 10.8 and 10.9.) Hence, most of its energy is also preserved, occupying, so to speak, the spectrum limited to a radian cutoff frequency $\omega_c = 2\pi f_c$, which is inversely proportional to the width of the pulse τ. If we define bandwidth B of the network as equal to f_c (the cutoff frequency), then:

$$B = \frac{\text{constant}}{\tau} \tag{10.6}$$

3. When a step function is applied to a band-limited filter, rise time t_R, the time necessary to reach unit amplitude, is defined in terms of the maximum slope of a line intersecting 0 and 1, as in Fig. 10.5b.[3]

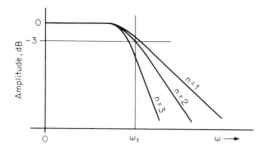

Fig. 10.7 Realizable filter characteristics. (*a*) The Butterworth response (maximal flatness). (*b*) The Bessel response (flat delay).

When so defined:

$$t_R = \frac{\pi}{\omega_c} \tag{10.7}$$

The bandwidth is, therefore, also inversely related to rise time:

$$B = \frac{1}{2t_R} \tag{10.8}$$

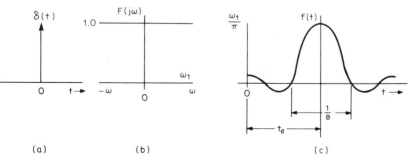

(a) (b) (c)

Fig. 10.8 Impulse response. (*a*) Unit impulse. (*b*) Continuous amplitude spectrum response to network of Fig. 10.6*a*. (*c*) Band-limited response to network of Fig. 10.6*c* (e.g., *LR* or *RC*).

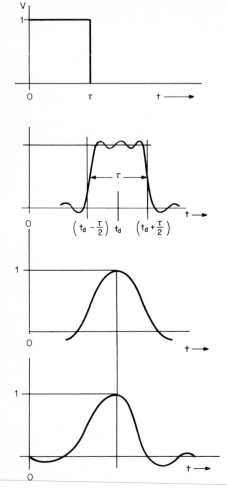

Fig. 10.9 Rectangular response. (*a*) Input rectangle. (*b*) Response, $B = 5/\tau$. (*c*) Response, $B = 1/\tau$. (*d*) Response, phase distortion.

An alternative definition of rise time (less elegant mathematically, but practical) is also shown in Fig. 10.5c. Circuit designers prefer this definition because it is easy to measure on an oscilloscope (also see Sec. 10.7).

4. Pulse waveforms can be characterized and assessed in terms of the approximate bandwidth of the altered shape of the pulse after transmission through the "ideal" filter. This can be seen, for example, when an impulse (Fig. 10.8) and rectangular pulse (Fig. 10.9) pass through an ideal filter (Fig. 10.6). In Table 10.1, we list the approximate bandwidths corresponding to each of the basic pulses shown in Fig. 10.2.

TABLE 10.1   Pulse Waveforms and Approximate Bandwidth*

Pulse waveform (Fig. 10.2)	Bandwidth × pulse width $B\tau$
Rectangular	1
Triangular	2
Cosine	1.5
Raised cosine	1
Gaussian	1
Gaussian†	$4/\pi$

* Based on the ideal filter (Fig. 10.6c).

† See text.

Analysis of the Gaussian pulse (Fig. 10.2d) shows that the normalized amplitude of its response can be expressed in terms of the relative amplitude $p < 1$. The value of p determines the area of the pulse.[4] Bandwidth is calculated from the formula:

$$B = \frac{2}{\pi\tau}\left(2\,ln\frac{1}{p}\right)^{1/2} \tag{10.9}$$

When $p = e^{-2}$ (or 0.135), the height of the pulse is reduced to 13.5 percent and $B = 4/\pi$. In this example, 95.5 percent* of the power of a Gaussian pulse is transmitted when $B = 1.26/\tau$.

We infer, from Eqs. (10.6) and (10.8), a relation between the width of a rectangular pulse and its rise time:

$$t_R = 0.5 \tag{10.10}$$

b. Group Delay Time[5]

Earlier, in discussing high-frequency response (in Chap. 8, Sec. 8.5d), we commented on the connection between the fidelity of a waveform and the linearity of phase shift as a function of frequency.

A linear variation of phase in the frequency domain results in a constant delay in the time domain. This can be inferred from the definition of group time delay t_d:

$$t_d = \frac{d\theta}{d\omega} \tag{10.11}$$

The high-frequency response of most transformers is comparable to that of a low-pass Butterworth filter. It is of interest, therefore, to compute the variation of delay time as a function of frequency.

* 95.5 percent corresponds to a standard deviation of 2 sigma.

TABLE 10.2 Group Delay with the Butterworth Filter

Frequency ω/ω_c	Group Delay, $t_d\omega_c$					
	0	0.5	0.7	1	1.5	2
$n=1$	1.0	0.80	0.67	0.50	0.31	0.20
$n=2$	1.41	1.66	1.70	1.41	0.75	0.42
$n=3$	2.0	2.34	2.66	2.50	1.16	0.58

This is done in Table 10.2 for the three cases: $n = 1, 2,$ and 3, depicted in Fig. 10.7. Each value of n corresponds to the order of the network model, as described in Secs. 9.2a and 10.3b. The delay and frequency response have been normalized by use of the relations:

$$\text{Group delay} = t_d\omega_c \tag{10.12}$$

$$\text{Relative frequency} = \frac{\omega}{\omega_c} \tag{10.13}$$

where $\pi = 2\pi f$ is the radian frequency, and ω_c is the cutoff frequency (the frequency at which response is down 3 dB with respect to the mid-band frequency). Examination of the table reveals the important fact that delay time is not constant, even in the pass band (where $\omega < 1$). Hence fidelity, in the strict or literal sense—that is, the exact replication of an applied pulse wave shape—should not be expected from a pulse transformer of conventional construction.

Consider, for example, the unit step (Fig. 10.5a) which passes through an ideal network (Fig. 10.6a and b). At the output in Fig. 10.5b we see that the step is tilted and that the unit value of the step has been exceeded by the overshoot of a damped oscillation. A sharp cutoff in the frequency domain has repercussions in the time domain. Thus, flatness of amplitude in the frequency domain does not guarantee flatness in the time domain, even granted an ideal filter.

An improvement in group delay response is obtainable with the Bessel filter, which produces a gradual rate of cutoff (as noted in Sec. 10.3c).

10.5. PULSE CODE TRANSMISSION AND ACCURACY[2]

A faithful replica of the waveform after it passes through a network is not always necessary to the faithful transmission of information. An important example is the system of pulse code modulation. In this scheme, the signal, of arbitrary shape in the time domain, is transformed into a sequence of pulses in accordance with a binary code.

Hence the information is conveyed by means of a code pattern rather than by means of an amplitude or shape characteristic peculiar to the original signal. Information can therefore be defined in terms of the time period T, the width of the pulse τ, and the number n of bits (*binary digits*) or levels used to approximate the original signal.

In PCM, the signal is sampled by a process called *quantization*. A set of binary pulses (off = 0, on = 1) corresponds, in binary arithmetic, to the sampled amplitude of the signal to be coded. For example, in Fig. 10.10 an amplitude level of six (one of seven quantized levels) corresponds to the code: $2^2 + 2^1 + 0 = 6$. It can be coded as three bits: 1, 1, 0 if the pulse stream is unipolar, or 1, 1, −1 if the pulses are bipolar.

In the theory of information transformation, Shannon uses, in his basic equation, the definition:[6]

$$\text{Information} = \frac{T}{\tau} \log_2 n \qquad (10.14)$$

The rate of transmission, or capacity C, of a channel, is:

$$C = \frac{\text{information}}{T} = \frac{1}{\tau} \log_2 n \qquad (10.15)$$

The fact that information is inversely related to the pulse width is a powerful incentive for using very narrow pulses in any and all types of transmission systems. The criterion of fidelity is changed from faithful reproduction of shape into accurate replication of digits. In a sequence of pulses, the fidelity of information is not degraded by a change in the shape of the pulses which represent the digits used to express the number n corresponding to the number of levels necessary to

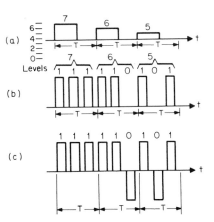

Fig. 10.10 Binary pulse coding. (*a*) Quantized samples. (*b*) On-off coded pulse. (*c*) Bipolar coded pulses. (*From D. Fink, ed., Electronics Engineers' Handbook, New York: McGraw-Hill Book Company, 1975, p. 14-41.*)

encode the instantaneous magnitude of a signal. Consider the code sequence of 6 bits (Fig. 10.11a):

$$101001 = 2^5 + 2^3 + 2^1 + 2^0$$

which corresponds to the integer $n = 41$. Fidelity now means accuracy. Replication would surely be poor if the above sequence were reproduced as 111001.

Digital systems of transmission make use of the Nyquist principle: all the information is retained provided that the sampling of the original signal occurs at a rate f_c corresponding to at *least twice the bandwidth B* of the signal to be encoded; that is:

$$f_c \geqq 2B \qquad (10.16)$$

Consider the task of transmitting analog voice signals as digital codes. Analog voice signals of a bandwidth of 4 kilohertz (kHz) are sampled

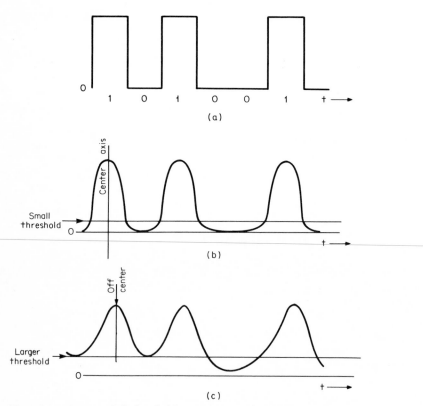

Fig. 10.11 Intersymbol threshold. (*a*) Input pulse train, 101001. (*b*) Linear phase shift response; small threshold between pulses. (*c*) Nonlinear phase shift; larger threshold between pulses.

at the rate of 8 kHz. If the number of quantized levels is $128 = 2^7$, then the coded transmission is at the rate of $8,000 \times 7 = 56,000$ bits per second (b/sec).

a. Distortion

In PCM, any disturbance to the integrity of the code is tantamount to distortion. Yet a strikingly important advantage of PCM derives from the fact that the coded signal, even after severe degradation of its waveform, can be regenerated by repeater amplifiers as many times as desired without the progressive distortion of the original signal.

The process of quantization introduces an error referred to as *quantization noise*. Its magnitude, which is less than one quantum step, can be treated as an additive noise with zero mean value but with a root-mean-square (rms) value of $1/\sqrt{12}$ times the height of the quantum step. Thus a measure of fidelity is the ratio of signal power to quantization power. *Quantizing distortion* can be reduced by increasing the number of levels in the sampling process. The resulting increase in the number of pulses in each group entails an increase in bandwidth. Hence the benefits of PCM arise from trading an increase in bandwidth for an improvement in the signal-to-noise (S/N) ratio.

When pulses are bipolar, as in Fig. 10.10c, excessive backswing* can result in the transmission of a negative digit ($-1 =$ off) when the correct digit is positive ($+1 =$ on). Moreover, a rectangular pulse may emerge from the output of a circuit with an extended fall time. Because the energy has spread (smeared out) into the interpulse period, the term *smearing distortion* is sometimes used to describe such behavior.

Fidelity of information requires simply that each digit be transmitted in its original position in a sequence. *Phase distortion is permissible,* but only to the extent that accuracy is not jeopardized. An error can result from a markedly nonlinear phase response if the result is an asymmetrical pulse shape (as in Fig. 10.9d). In Fig. 10.11b, the skirts of the symmetrical pulses in the train 101001 produce a small amplitude in the zero intervals between the 1's. However, when the pulse shapes are assymmetrical, as in Fig. 10.11c, a large amplitude in the zero (off) region can be interpreted as 1. The threshold level, above which an error can occur, is a measure of *intersymbol distortion*. We can infer, therefore, that the permissible threshold is a function of the noise level of the channel.

Transmission by means of pulse code modulation is widespread and

---

* Backswing is discussed later, in Sec. 10.9.

we will mention several of the basic concepts because they have a bearing on the analysis of pulse transformers. The basic considerations are the pulse shape and the bandwidth.

b. Shaped Pulses

Rectangular pulses are, we know, a natural choice of waveform in power-conversion circuits. They result in minimum heating losses in the active switch (such as the transistor and thyristor) and a high level of efficiency of the pulse generating circuit. We should note, however, one serious disadvantage: the discontinuous wave shape results in significant amounts of *radio-frequency* (RF) energy. This RF energy, in the form of *radio-frequency* *interference* (RFI) or *electromagnetic* *interference* (EMI), raises havoc in nearby circuits through either conduction or radiation. In those transformer circuits which are designed to operate with saturated core material, the output waveform resembles a trapezoid with rounded corners. This can be seen in the output of the ferroresonant regulating transformer (Sec. 5.5), which produces very little EMI because it provides a high degree of isolation from its source of power.

Various studies have been made of the merits of alternative wave shapes whose bandwidth B is limited. Waveforms such as the Gaussian, cosine, and raised cosine produce less energy in the RF band than the rectangular pulse and are, therefore, advantageous in search radar systems, where a pulse must be recognized but its shape is of secondary importance.* The transformer designer also finds limited bandwidth very desirable because it makes it easier to achieve other, possibly difficult, objectives in a set of specifications.

In PCM systems it is important to minimize intersymbol interference, as we mentioned earlier. And studies indicate the advantage of alternative shapes such as the $(\sin x)/x$ pulse (similar in shape to Fig. 10.2a) and the raised cosine. Such waveforms have the significant property of passing through the zero axis at multiples of the pulse width τ. Hence, in principle, intersymbol interference should be zero. The frequency spectrums of such pulses have components which fall to zero at discrete multiples of the cutoff frequency.

To illustrate this design concept, consider an important example, the single rectangular pulse (Fig. 10.2a) of amplitude V [volts (V)]

---

* A waveform with fast rise time is, however, of prime importance in tracking radar, loran, and certain pulse-modulation systems other than PCM. In PPM circuits, where this requirement is less exacting, a trapezoidal waveform is used.

and width τ [seconds (sec)]. The Fourier transform $F(\omega)$ of this pulse is:

$$F(\omega) = V\tau \frac{\sin (\omega\tau/2)}{\omega\tau/2} \tag{10.17}$$

Of mathematical import, therefore, is the quantity in parentheses, whose form represents the function:

$$f(x) = \frac{\sin x}{x} \tag{10.18}$$

This function is characterized by the fact that its undulations become zero at multiples of x. In the rectangular pulse, $(\sin x)/x = 0$ when:

$$\frac{\omega\tau}{2} = n\pi \qquad \tau = \frac{2n\pi}{\omega} = \frac{n}{f} \tag{10.19}$$

Here, n is the integral multiple harmonic of the fundamental frequency.

When n is equal to unity, most of the energy of the rectangular pulse is in the area bounded by π on the x axis. The implications are impressive. For it then becomes possible and practical to use a band-limited network in a PCM system without compromising fidelity. There are other pulses which exhibit the $(\sin x)/x$ characteristic, such as the triangular spectrum (Fig. 10.2b). The impulse (delta) function, which has a uniform spectrum at the output of an ideal network, also exhibits the $(\sin x)/x$ characteristic after passing through a nonideal network.

Notwithstanding its shortcomings, the rectangular pulse requires the same or smaller bandwidth than other common pulse shapes (see Table 10.2), is easily generated, and is the wave shape preferred for most schemes for the transmission of coded information.

c. Bandwidth

The accuracy of information transmitted by pulses is related to bandwidth and power in accordance with two equations:*

$$BT = \text{constant} \tag{10.20}$$

$$B \log_2 (1 + S/N) = \text{constant} \tag{10.21}$$

---

* Note the similarity in form of these equations to the important principle of amplifier design, which relates bandwidth B and circuit gain G: $BG = $ constant.

where B is the bandwidth, T is the time required for the transmission of a coded pattern, and S/N is the signal-to-noise power ratio of the channel. [The first of the two equations can be derived from Eqs. (10.6) and (10.14).] The second equation conveys the desirability of a high S/N ratio. Analysis shows that bipolar pulses, that is, pulses of alternating polarity ($+1$, -1, 0, $+1$, -1, as in Fig. 10.10c), result in a higher S/N ratio than unipolar pulses, that is, pulses of the same polarity (1, 0, 1).

There are other important conclusions to be drawn from these simple equations. We infer that it is possible to *trade bandwidth for time,* that is, the time necessary to convey the information. We can reduce the bandwidth without sacrificing accuracy if we are willing to accept a greater delay in receiving the information. If we equate fidelity with accuracy, an infinite (or very large) bandwidth is not necessary for the accurate replication of an input waveform. The technological implications are impressive. Consider, as an example, the fact that video signals, which normally require a bandwidth of 4 to 6 MHz, can be transmitted over a telephone channel limited to a bandwidth of 4 to 8 kHz. And, in the reproduction of music, the PCM system provides a degree of fidelity (that is, the faithful rendition of the original waveform) and dynamic range not practicable with analog methods of transmission. Moreover, Eq. (10.21) implies that it should be possible to trade bandwidth for signal power, without loss of information, in a noisy channel. (All channels have noise.)

PART B: THE TRANSFORMER NETWORK
10.6. EQUIVALENT CIRCUITS

Let us consider the relation between the pulse and wide-band transformers.

Every transformer is a band-pass filter in that it can pass only a limited band of frequencies without distortion of amplitude and phase. The exact representation of a periodic train of pulses, however, requires an infinite number of sine waves. A transformer which preserves the shape of an applied train of pulses must therefore be able to pass a very wide band of frequencies with very little distortion; that is, the pulse transformer designed for faithful transmission must be a wide-band transformer. Conversely, the wide-band transformer should be able to transmit a train of pulses faithfully. It follows that fidelity can be specified in terms of frequency response, for pulse as well as wide-band transformers. A brief examination of the appropriate equivalent circuits will bear this out.

The frequency response of wide-band transformers is usually analyzed by means of two or three equivalent circuits. In one circuit, low-frequency attenuation is associated with finite primary inductance. In another circuit, high-frequency attenuation is associated with leakage inductance. And a modified circuit is used for step-up and wide-band transformers, whose capacitance cannot be neglected, to account for the peak and dip in the high-frequency response.

Essentially the same restricted-band circuits discussed in Chap. 8 can be used for the piecewise analysis of pulse response. The droop at the top of a rectangular pulse (see Fig. 10.12) is associated with primary inductance and, therefore, with limited low-frequency response. Finite rise time in the output waveform is associated with leakage inductance and, therefore, with a limited high-frequency response. Oscillations in the neighborhood of the discontinuities are

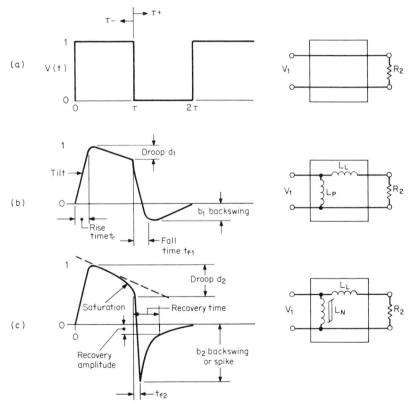

Fig. 10.12 Square-wave response. (a) Ideal network. (b) Linear transformer. (c) Nonlinear transformer.

accounted for by the resonance of capacitance and by leakage inductance.

a. Analytic Approaches[7]

The circuit engineer familiar with equivalent circuits can therefore specify a pulse transformer in terms of its frequency response, and the transformer engineer will then use steady-state analysis as the design technique. It is simpler and more desirable, however, to specify in terms of the actual shape of the pulse waveform at the output of the transformer (i.e., in the time domain). This is the more common requirement, and the transformer engineer uses time analysis, especially when the equivalent circuit changes during the pulse cycle.

In Part A of this chapter, we noted that a unit step (Fig. 10.5) or a square pulse (Fig. 10.9) undergoes a change in shape as it passes through a nonideal network such as a low-pass RC or LR filter. Consider the transformer network, and suppose that a square pulse $V(t)$ (Fig. 10.12) is applied to the input. The equivalent circuit in the black box of Fig. 10.12b can be represented by two inductances, series and shunt. We see that the output waveform is somewhat deformed but still recognizable. Four alterations of the shape are caused by the following:

1. A finite rise time t_r
2. A droop or tilt d at the top
3. A finite fall time t_f
4. A backswing b of the amplitude

And yet there is considerable symmetry in the response to a square wave:

$$t_r = t_f \quad \text{and} \quad d = b$$

However, in a transformer with a ferromagnetic core, saturation of the core (depicted by the symbol L_N in the black box of Fig. 10.12c) can distort the waveform even further. When the material used in the core is nonlinear, symmetry is lost because, in general:

$$t_r \neq t_f \quad \text{and} \quad d \neq b$$

Because of the possibility of magnetic saturation, we have chosen, in the sections which follow, to use the *unit step* (see Fig. 10.5a) as the driving force in our analysis of the ferromagnetic transformer.[8] An alternative driving function, the unit impulse $\delta(t)$, is worthy of consideration. In theory, if we know the response $h(t)$ of the transformer to $\delta(t)$, we can deduce its response to any applied waveform.[2,5]

We have chosen to use the step function because it provides us with a direct means of examining the consequences of saturation of the core.

We can employ a complete lumped-constant equivalent circuit (Fig. 10.13a) or a somewhat less complex one (Fig. 10.2b).[9] The lumped capacitance used in Fig. 10.13b is functionally equivalent to the distributed capacitances and can be very useful in piecewise analysis, but it requires discussion. For, while leakage and primary inductances can be regarded as lumped constants, distributed capacitance plays a major role in determining the rise and decay of pulses and poses special problems.

b. Capacitance

In lumped-constant circuit analysis, a real transformer is represented by an *LCR* network in which the unique isolation feature is achieved by adding an ideal transformer with a specified turns ratio. Physically, however (and this can be inferred from Fig. 9.8a), capacitance is not lumped but distributed throughout the windings.

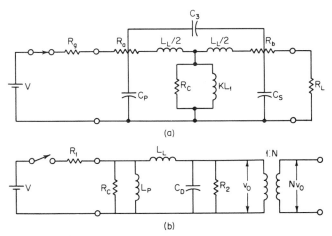

Fig. 10.13 Lumped-constant equivalent circuits. (a) "Complete" circuit based on unity turns ratio. (b) Simplified circuit, constants referred to primary; step-up case. R_g = generator source resistance; R_a = primary resistance; R_b = secondary resistance; R_L = load resistance; $R_1 = R_g + R_a$ = lumped source resistance; $R_2 = R_b + R_L$ = lumped load resistance; R_c = shunt core loss resistance; K = coupling coefficient; L_1 = primary inductance. $L_P \cong KL_1$ = primary shunt inductance; L_L = leakage inductance. C_P = primary distributed capacitance; C_D = secondary distributed capacitance; C_3 = bridging and direct leakage capacitance between input and output; N = step-up turns ratio.

In this chapter we use the simplifying concept of one equivalent distributed capacitance to account for the electrostatic energy stored in the windings when a voltage step $V(t)$ is applied to the primary winding and to provide mathematically simple, yet reasonably accurate, predictions of the output response. In Sec. 10.7a we shall provide further justification for this simplification.

Since the windings of a pulse transformer contain relatively few turns and layers, the effective capacitance C' associated with the energy stored between the primary and secondary windings is the most significant. The effective capacitance of the most common winding scheme, the single-layer primary and secondary winding, was summarized in Chap. 9 [see Eqs. (9.22) and (9.23)].

The effective capacitance of the transformer C' is shown in Fig. 10.14a. We know that it depends on many factors: the voltage gradient, the turns ratio, N, the polarity, the dielectric constant, and the geometry of the coil.

When C′ is lumped with the reflected load capacitance C_L (see Fig. 10.14b) and with the reflected self-capacitance of the secondary C_2 (if there are several layers), the result is the $L_L C_D$ combination shown in Fig. 10.14c. C_D, then, represents the total effective distributed capacitance:

$$C_D = C' + N^2(C_2 + C_L) \tag{10.22}$$

The above procedure makes it possible to arrive at the simple half-T rise-period equivalent circuit of Fig. 10.15a, and the mathematics becomes much more manageable. The simplicity of this figure should not, however, obscure the fact that the calculation of a number of capacitances is assumed in C_D. When the capacitances associated with the cable and load must be estimated, the error in C_D may be as great as 10 to 20 percent. This does not affect the utility of the LCR circuit, because the error in the predicted rise time is usually small. When the circuit engineer feels that the value of C_D will not be sufficiently accurate to justify its calculation, he or she may resort to performance measurements and ascertain C_D indirectly.

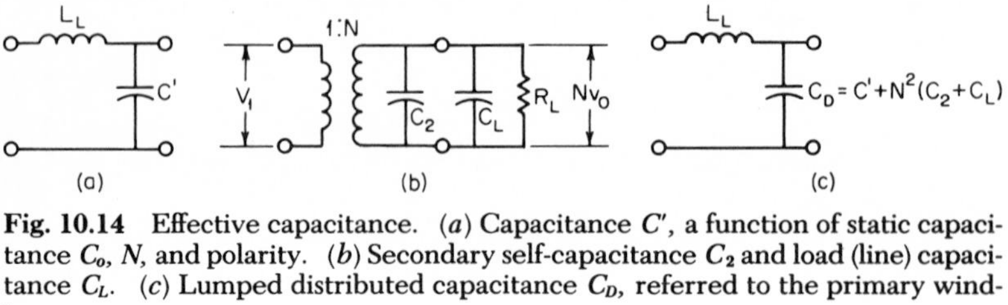

Fig. 10.14 Effective capacitance. (a) Capacitance C', a function of static capacitance C_o, N, and polarity. (b) Secondary self-capacitance C_2 and load (line) capacitance C_L. (c) Lumped distributed capacitance C_D, referred to the primary winding.

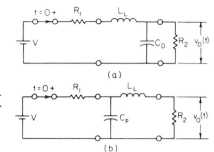

Fig. 10.15 Rise period equivalent circuits. (*a*) Step-up, $N >$
1. (*b*) Step-down, $N < 1$.

c. Top and Decay Circuits

Analysis of the equivalent circuit for the top period predicts the tilt or droop at the top of the pulse. Response during the decay period is predicted with the aid of the backswing and fall circuits (shown later in Figs. 10.20 and 10.25).

10.7. THE RISE PERIOD

The rise period is important for several reasons:

1. It sets a limit on the maximum *pulse repetition frequency* (PRF), which is high in many digital circuits. For example, since a square wave of 100 MHz has a pulse width τ of 5 nanoseconds (nsec), the rise time must be substantially less than this value if the pulse is to retain its identity as a rectangular or trapezoidal wave shape.

2. At still higher frequencies, the rise time sets an upper limit on the channel's capacity, i.e., the amount of digital information transmitted per unit of time [see Eq. (10.14)].

3. In timing and trigger circuits, rise time is a residual error which must be held to a minimum. A real pulse generator has a real rise time, and it is undesirable for the transformer to add appreciably to the total time of rise. The continuing reduction in the switching time of semiconductors, for example, creates the need for comparably fast transformer response.

4. In transformer circuits which drive switching transistors, slow rise and fall times can increase the temperature and dissipation of power at the junction to the point where the transistor will fail.

In consequence, substantial analysis and development has been devoted to reducing the rise time of transformers and other pulse networks.

a. The Solution

The choice of an appropriate equivalent circuit is a problem. We can be guided by wide-band transformer analysis and establish three lumped capacitances (primary, bridging, and secondary) in a pi configuration. From such a configuration, used in conjunction with the leakage inductance, we can predict multiple resonances in the wide-band transformer. However, when a similar circuit (the m-derived model) is used for the rise period of the pulse transformer, the mathematics becomes formidable. Even a solution of the third-order model, which neglects bridging capacitance, is so complex that it is rarely attempted.

If we apply the powerful techniques of network synthesis and use preferred filter models, we obtain *optimum* rather than completely general solutions. These are nevertheless of great value. Section 11.3 contains a discussion of the synthesis of pulse transformers which parallels the discussion of synthesis of the wide-band transformer in Sec. 9.2.

In the present chapter, we restrict our analysis to the simpler second-order model. The resulting *LCR* circuits (e.g., Fig. 10.15a) have the advantage of being familiar to the circuit and servodesign engineer, thus affording continuity with similar circuits studied in linear network analysis. In addition, they yield predictions which are accurate enough for most applications.

The equation for the normalized output voltage v_o is:

$$\frac{v_o(p)}{V} = \frac{1}{L_L C_D(p^2 + 2\alpha p + \omega_a{}^2)} \qquad (10.23)$$

Here the operator p stands for poles of the transfer function (the complex frequency of LaPlace transform analysis) and is used to yield an algebraic solution in the time domain. The denominator contains the characteristic equation which has been simplified by means of the following definitions:

$$\alpha = \frac{R_1}{2L_L} + \frac{1}{2R_2 C_D} \qquad \omega_a = \frac{1}{\sqrt{L_L C_D a}} \qquad \omega_n{}^2 = \omega_a{}^2 - \alpha^2 \quad (10.24)$$

where α is a rate-of-decay function, ω_a is the radian frequency (greater, by the factor $\sqrt{a}$, than the resonant frequency ω_r), a is the resistive attenuation due to series source resistance, $R_2/(R_1 + R_2)$, and ω_n is the natural frequency of the circuit.

We then obtain, in the time domain, the normalized solution:

$$\frac{v_o(t)}{aV} = 1 - \frac{p_1}{p_1 - p_2} e^{p_2 t} + \frac{p_2}{p_1 - p_2} e^{p_1 t} \qquad (10.25)$$

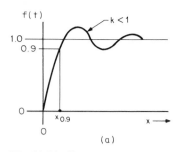 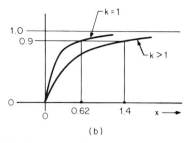

Fig. 10.16   Response to unit step. (*a*) Underdamping. (*b*) Critical and heavy damping.

in which the coefficients p_1 and p_2 represent the two roots of the complex frequency p.

A solution of this equation is facilitated by introducing the following definitions for the damping constant k, the time constant T, and the fractional time x:

$$k = \frac{\alpha}{\omega_a} = \frac{\alpha T}{2\pi} \qquad T = 2\pi\sqrt{L_L C_D} \qquad x = \frac{t}{T} \qquad (10.26)$$

We see, in Fig. 10.16, that the response in the rise interval can be either oscillatory or damped, depending on the magnitude of k. There are three distinct solutions:

1. Oscillatory, $k < 1$
2. Critically damped, $k = 1$
3. Damped, $k > 1$

The normalized rise period response for various values of k is plotted in Fig. 10.17. For the present, the rise time is considered to be $t_{0.9}$,* the time required for the pulse to reach 90 percent of its full value. Important values of the corresponding fractional rise time $x_{0.9}$, R_1, and the characteristic impedance Z are listed in Table 10.3, p. 391.

b. Criteria for Fast Rise

There are several criteria for fast rise:

1. Minimum damping
2. Moderate overshoot
3. Maximum transfer of energy
4. Critical damping
5. Optimum capacitance

---

* The subscript 0.9 means that the pulse reaches 90 percent of its unit amplitude in the interval $t_o = x_{0.9}T$ (see Fig. 10.16).

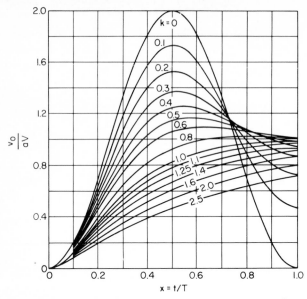

Fig. 10.17   Rise period response $C_D \neq 0$. (*From G. Glasoe and J. Lebacqz,* Pulse Generators, *New York: McGraw-Hill Book Company, 1948, by permission.*)

(1) Minimum Damping.   One readily infers from Fig. 10.17 that fast rise is associated with small values of damping.   Minimum rise time occurs when $k = 0$, but the condition for undamped oscillations can exist only when $R_1 = 0$ and $R_2 = \infty$, hardly a practical condition for fidelity to pulse wave shape.   In any case, as we also see from Fig. 10.17, overshoot cannot exceed 100 percent.

Since k plays a central role, Eq. (10.26) is rewritten in a more useful form:

$$k = \frac{\sqrt{a}}{2}\left(\frac{R_1}{Z} + \frac{Z}{R_2}\right) \tag{10.27}$$

where $Z = \sqrt{L_L/C_D}$, the characteristic impedance.

From previous encounters with equations of this form, we recognize that k is minimum when:

$$\frac{R_1}{Z} = \frac{Z}{R_2} \qquad Z = \sqrt{R_1 R_2} \qquad \frac{L_L}{R_1} = R_2 C \tag{10.28}$$

When source and load resistances are matched ($R_1 = R_2$), then $Z = R_2$, $k = \sqrt{\frac{1}{2}}$, and the overshoot is 4.3 percent.

The condition for minimum damping in the time domain corre-

T A B L E   1 0 . 3   Fractional Rise Time $x_{0.9}$ for Several Values of k and Z

k	$x_{0.9}$	R_1	Z	Comments
0	0.23	0	—	$R_2 = \infty$; pure ringing
$\sqrt{a_1}$	—	$\neq R_2$	$\sqrt{R_1 R_2}$	Minimum k
$\frac{1}{4}$	0.28	0	$R_2/2$	44.5% overshoot; $x_m = 2/\sqrt{15}$
$\frac{1}{2}$	0.33	0	R_2	17% overshoot: $x_m = \sqrt{\frac{1}{3}}$
0.59	0.37	$0.39 R_2$	R_2	10% overshoot; $x_m = 0.646$
$\sqrt{\frac{1}{2}}$	0.41	R_2	R_2	4.3% overshoot; $x_m = \sqrt{\frac{1}{2}}$
0.885	0.53	R_2	$R_2/2$; $2R_2$	Small overshoot
1	0.62	0	$2R_2$	No overshoot
1	—	$R_2/10$	$1.41 R_2$; $0.7 R_2$	Low source resistance
1	0.62	R_2	$(\sqrt{2} + 1)R_2$	No overshoot
1	0.62	R_2	$(\sqrt{2} = 1)R_2$	No overshoot
1	—	$10 R_2$	$3.16 R_2$	High source resistance
2	1.4	0	$4R_2$	Overdamped
—	2.3	R_1	—	$C_D = 0$; $T_o = L_L/R_S$*

* In the first-order solution $v_o/aV = 1 - e^{-t/T_o}$ and $t_o = 2.3 T_o$.

sponds to the condition of *maximal flatness* in the frequency domain. This can be seen when we compare Eq. (10.28) with its counterpart (see Sec. 8.5*b*):

$$Z = \sqrt{R_1 R_2} \qquad (8.38)$$

It should be noted that there is a lower bound on k when there is a finite load resistance. Assume, for example, that $R_2 = Z$. Using the definition $b = R_1/R_2$, we transpose Eq. (10.27) into $k = \sqrt{b+1}$, where $b = 4k^2 - 1$. Since R_1 must be real and positive, k must equal or exceed $\frac{1}{2}$.

(2) **Moderate Overshoot.** In digital circuits, moderate overshoot is often defined as 10 percent. We have already seen that overshoot is closely related to the damping constant. To find the value of k in terms of maximum overshoot, v_{om}, we solve:

$$\frac{v_{om}}{aV} = 1 + e^{-\pi k/\sqrt{1-k^2}} \qquad (10.29)$$

For example, when $v_{om}/aV = 1.10$, the damping factor is 0.59. The first overshoot occurs in the fractional time $x_m = 0.65$.

(3) **Maximum Transfer of Energy.** The condition for maximum transfer of energy (i.e., $Z = R_2$) results in a small damping constant. When source and load are matched, we obtain the Butterworth MFA model.* We find, from Eq. (10.27), that $k = \sqrt{\frac{1}{2}}$ and that a small

* See Secs. 9.2*a* and 10.3*c*.

overshoot (4.3 percent) results. The rise time, denoted by $t_{0.9}$ (from Fig. 10.17 or Table 10.1) is $0.41T$ or:

$$t_{0.9} = 0.82\pi \sqrt{L_L C_D a} \qquad (10.30)$$

However, when source resistance is much less than load resistance (mathematically expressed as $R_1 = 0$), we find that $k = \frac{1}{2}$, overshoot is 17 percent, and $t_{0.9} = 0.33T$.

It is interesting to compare the rise-time solutions we have discussed with the solution for the ideal filter, described in Sec. 10.3b. In that definition, the time to reach unit amplitude is evaluated at the interval t_d, the delay time after application of the unit step (see Fig. 10.5b). Rise time t_R is evaluated for the maximum slope at t_d:

$$t_R = \frac{\pi}{\omega_c} \qquad (10.31)$$

where ω_c is the cutoff frequency.

If we equate ω_c with the cutoff frequency of the MFA response, Eq. (8.40), we can write:

$$t_R = \pi\sqrt{L_L C_2 a_2} \qquad (10.32)*$$

A comparison of Eq. (10.32) with (10.31) leads to the conclusion that the optimum rise time for a matched realizable filter or transformer network corresponds to the optimum rise time of the ideal filter.

(4) **Critical Damping.** To obtain minimum rise time without oscillatory overshoot, we use the critical damping solution. The rise time $t_{0.9}$ is $0.62T$.

Rise time is not generally defined as $t_{0.9}$, as we have been doing, because in some pulse circuits the origin of the rise is obscured by initial oscillations or noise. A rise time t_r, defined as the elapsed time between 10 and 90 percent of the output, overcomes this difficulty. The 10 and 90 percent responses for the maximum transfer or Butterworth two-pole model correspond, respectively, to the values $x = 0.068$ and 0.41. When these are used, we obtain the important formula:

$$t_r = 0.342T = 2.15\sqrt{L_L C_D a} \qquad (10.33)$$

We can infer the advantage of a low turns ratio N. Since C_D varies approximately with N^2, t_r will increase if N increases [see Eq. (9.20)].

(5) **Optimum Capacitance.** Up to this point we have assumed that minimum rise time is achieved by making k as small as is consistent with an acceptable percentage of overshoot. This widely practiced

---

* $a_2 = R_2/R_s$; $R_s = R_1 + R_2$ (see Fig. 8.7).

procedure is used, for example, to adjust a servomechanism loop for optimum response.

Another approach to minimum rise time arises from a consideration of optimum capacitance. It is true that when a transformer has a given $Z = \sqrt{L_L/C_D}$, rise time decreases as k decreases, although overshoot increases. However, C_D can be made an independent parameter. The transformer engineer has some control over C_D, and the skillful circuit engineer can alter the external capacitance C_L as well as R_1 and R_2.

Another important possibility is to minimize the product $\sqrt{L_L C_D}$. This will also reduce rise time and has the further advantage of reducing the transformer's size. The reduction in size results from the fact that $\sqrt{L_L C_D}$ is proportional to $N_p \lambda$, where N_p is the number of turns on the primary coil and λ is the mean circumference of the coil. But $N\lambda$ is the length of magnet wire which, of course, will be minimum when the size of the coil is minimum.

(6) **Summary.** The foregoing discussion of the criteria for fast rise suggests the desirability of: (1) adjusting the circuit parameters so that k will be between ½ and 1, (2) employing a low turns ratio, and (3) using a load resistance as low as practicable. There are, however, circuits where the maximum transfer of energy approach ($Z = R_2$) is to be preferred, e.g., when a substantial physical separation between a pulse transformer and its load logically calls for a cable whose characteristic impedance is identical with that of the transformer and of the load.

It must be borne in mind that minimum rise time is not compatible with overall pulse fidelity. If, as in our last example, we reduce the rise time by increasing R_1 and R_2, k decreases but the top droop increases. If R_1 is increased to, say, $10R_2$ and Z is set equal to $3.16R_2$ to keep overshoot small, then the rise time decreases, but so do the output voltage and gain. If R_1 alone is reduced in order to decrease droop, backswing may increase. And if $Z = R_2$, overshoot will increase. The overshoot may be eliminated by increasing Z to $2R_2$, which results in unity damping.

Thus, a favorable change in one region is likely to have adverse repercussions elsewhere. Further consideration of the topic of fidelity is deferred to Sec. 11.3b.

c. The Step-Down Transformer

When N is moderately less than 1, the primary and secondary windings will each contain, as does the step-up transformer, a single layer.

The formulas for C' and C_D are the same as for the step-up transformer, but the net result is a low value of C_D.

When $N \ll 1$, the primary coil will contain many more layers than the secondary coil. Consequently, most of the electrostatic energy will be stored in the primary coil. The effective primary distributed capacitance C_P is, from Eq. (9.18):

$$C_P = \frac{4}{3}\left(\frac{N_{L1} - 1}{N_{L1}{}^2}\right)C_o \qquad (10.34)$$

where N_{L1} is the number of layers of the primary and C_o is the equipotential or "static" capacitance between layers.

Since the thickness of the insulation between the layers of the primary is ordinarily much less than between the windings, C_P will be greater than C'. When $N \ll 1$, the reflected capacitance $N^2(C_2 + C_L)$ will be small compared with C_P. Thus, $C_P \gg C_D$, and Fig. 10.15c, in which capacitance has been moved to the input, becomes the appropriate equivalent circuit.

The step-down pulse (as well as wide-band) transformer poses fewer problems than the step-up transformer. Consider, for example, the situation when the source resistance is smaller than the load resistance. In the limit, if $R_1 \ll R_2$, the effective primary capacitance C_P is shunted directly across the pulse source. The simplified equivalent circuit is the simple LR circuit and the rise period response (in terms of p) is given by a simple exponential equation.

When the source resistance is appreciable, the analytic procedure is the same as for the step-up transformer and results in the same basic solution.[10] However, the positions of R_1 and R_2 in Eqs. (10.24) and (10.26) are interchanged in the expressions for the attenuation factor α, the damping factor k, and the time constant T. Note also that C_D becomes C_P. Thus, the time constant T has, under the radical sign, the attenuation factor $a_1 = R_1/(R_1 + R_2)$ instead of $a = R_2/(R_1 + R_2)$:

$$\alpha = \frac{1}{2R_1 C_P} + \frac{R_2}{2L_L} \qquad (10.35)$$

$$k = \frac{\sqrt{a_1}}{2}\left(\frac{R_2}{Z} + \frac{Z}{R_1}\right) \qquad (10.36)$$

$$T = 2\pi\sqrt{L_L C_P a_1} \qquad (10.37)$$

When these definitions are used, all the step-up equations and graphs (which have been expressed in terms of the normalized output voltage

v_o/aV, normalized rise time $x = t/T$, and damping constant k) become applicable to the step-down transformer.*

10.8. THE TOP PERIOD

Very soon after the switch has been closed, leakage inductance may be neglected. The shunt (loss) resistance R_e is also usually neglected because it is much greater than the load resistance R_2. When it cannot be neglected, as in high-power circuits, it may be combined with R_2.

Primary inductance dominates the top period, and Fig. 10.18 becomes the appropriate equivalent circuit. Thévenin's theorem enables us to arrive at the simpler circuit of Fig. 10.18b, in which $R = R_1R_2/(R_1 + R_2)$, the parallel combination of load and source resistance.

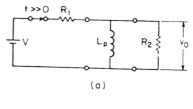

(a)

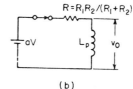

(b)

Fig. 10.18 Top period response. (a) Equivalent circuit. (b) Thévenin equivalent of (a), $T_d = L_p/R$. (c) Droop response, $v_o/aV = e^{-xd}$.

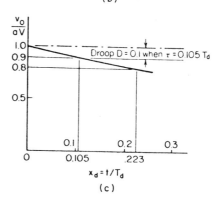

(c)

___

* We reached a similar conclusion about the wide-band step-down transformer and the interchangeability of symbols in Secs. 8.5b and 9.2c.

In the time domain, we obtain the simple exponential for an RL circuit:

$$\frac{v_o(t)}{aV} = e^{-\alpha_d t} = e^{-x_d} \tag{10.38}$$

where α_d is the reciprocal of the time constant T_d for the top period:

$$T_d = \frac{L_P}{R} \tag{10.39}$$

and $x_d = t/T_d$, the normalized time for the top period of the pulse. Equation (10.38) is plotted in Fig. 10.18c, which shows that the droop or tilt at the top of the pulse response is dependent on the relative duration of x_d.

This fractional droop D (denoted earlier by d) is conveniently expressed as:

$$D = 1 - e^{-x_d} = 1 - e^{-\tau/T_d} \tag{10.40}$$

where τ is the duration of the pulse.

The Taylor expansion* enables us to approximate small values of x_d in Eq. (10.40):

$$D \cong x_d = \frac{\tau}{T_d} = \tau \frac{R}{L_P} \tag{10.41}$$

When the droop is 10 percent or less, the approximation results in a maximum error of 5 percent.

When the exact time constant must be determined for a specified permissible droop and applied pulse width τ, Eq. (10.40) may be rewritten as $x_d = \ln [1/(1 - D)]$. The time constant will then be:

$$T_d = \frac{\tau}{\ln [1/(1-D)]} \tag{10.42}$$

Since the droop will be 10 percent when $x_d = \ln (1/0.9) = 0.1053$, the pulse width τ will be:

$$\tau = 0.105 T_d \cong 0.1 \frac{L_P}{R} \tag{10.43}$$

Important values of droop and corresponding fractional time values are given in Table 10.4.

In the elementary treatment of the top period we assumed that the primary inductance is constant. But inductance, defined as $10^{-8}N$ $d\phi/di_m$, is proportional to $\Delta B/\Delta H$ or to permeability μ. This is be-

---

* $e^{-x} = 1 - x + \frac{x^2}{2!} - \frac{x^3}{3!} + \cdots$ or $e^{-x} = 1 - x$ when x is small.

T A B L E 1 0 . 4 Droop vs. Normalized Pulse Duration*

Droop D	0	0.05	0.10	0.15	0.20	0.30	0.40	0.50	0.90	0.95	1.0
$G = 1 - D$	1	0.95	0.90	0.85	0.80	0.70	0.60	0.50	0.10	0.05	0
$x_d = \tau/T_d$	0	0.0513	0.1053	0.162	0.223	0.357	0.511	0.693	2.30	3.0	∞

* Computations are based on Eq. (10.40).

cause flux density is defined at $B = \phi/A$, and the magnetizing current, i_m is related to the magnetizing force H by $H = 0.4\pi Ni_m/l$. The assumption of constant or linear inductance, therefore, implies a core material with a B-H characteristic which could be graphically represented by a straight line in Fig. 10.19a.* We are already familiar (see Sec. 6.1) with the fact that a B-H plot of ferromagnetic cores is characterized by a hysteresis loop (Fig. 10.19b) and that B is therefore a multivalued function of H. For each core material there is a family of B-H loops whose shape and slope depend on a number of parameters.

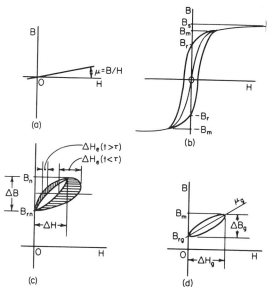

Fig. 10.19 B-H characteristics of the core. (a) Linear core material: constant but small permeability μ. (b) Ferromagnetic major hysteresis loop. (c) Minor hysteresis loop; loop widens asymmetrically as undirectional pulse width decreases. See Sec. 11.4b. (d) Minor loop with air gap reduces B_r to B_{rg}.

* The symbols in Fig. 10.19 and elsewhere in this section are based on the characteristics of a core discussed in Chaps. 6 and 7.

Primary inductance is actually a *nonlinear* function dependent on pulse duration, the magnitude of the applied pulse voltage, the thickness of the laminations or particles, the resistivity of the core, and temperature.

a. Pulse Permeability

If the applied pulses are unidirectional, the core becomes dc-polarized. The loop now begins and terminates in the remanent flux density B_r and is therefore smaller, or minor. The incremental flux density $\Delta B = B_m - B_r$ is also smaller (see Fig. 10.19c). When the pulse width is narrow and the duty cycle low, the waveforms of flux are sawtoothed. There is a greater eddy-current field H_e during $t < \tau$ than there is during the longer recovery interval $(t > \tau)$ between pulses, and the effect is a widening of the right flank of the minor hysteresis loop, shown as a striped area in Fig. 10.19c. The pulse permeability μ_p is likely to be less than the incremental permeability μ_Δ, which is associated with sine-wave excitation.*

The implications of the foregoing observations are that the top droop (a function of primary inductance) also depends on:

1. The magnitude of the applied voltage V_1, which must not exceed a value determined by the saturation flux density B_s, since the droop increases as B approaches B_s [see Eq. (10.45)].

2. Pulse duration τ (especially when it is less than 1 μsec), as compared with the time constant of the lamination, since the increment in the magnetizing force ΔH generally increases as τ decreases.[11]

3. Duty ratio, if the primary is in a single-ended circuit, since the value of B_r and the residual field H_r are determined by the average dc.

4. Duration of off time (time between pulses) in relation to the load time constant. When they are of comparable duration, $H_r > 0$; since the core is biased, μ_p is less than when $H_r = 0$.

5. The magnitude of an air gap (in series with the magnetic circuit), which reduces B_r and increases ΔH (see Fig. 10.19d).

Pulse permeability and resolution of pulses in high-PRF circuits are discussed further in Sec. 11.4d.

b. Magnetic Saturation

It is important that the primary inductance not decrease sharply as the voltage level reaches the specified maximum. If it does, the

---

* Incremental permeability is discussed in Sec. 7.1.

droop will increase excessively. The flux density B_m must therefore be held below the saturation density B_s. It is important to examine the significant relationships among these key parameters.

If we put the classic Faraday equation $V_1 = 10^{-8}d(N\phi)/dt$ into integral form:

$$N \int_{\phi r}^{\phi m} d\phi = \int_0^\tau V_1 \, dt \; 10^{-8} \qquad (10.44)$$

and solve the primary voltage $aV = V_1$, we obtain:

$$V_1 = \frac{N \Delta BA \; 10^{-8}}{\tau} \qquad (10.45)$$

where ΔB is in gauss (G), the core area A is in square centimeters (sq cm), and the pulse duration τ is in seconds (sec).

When droop is specified, the transformer designer will select an appropriate core material, establish a sufficient number of turns to obtain the necessary inductance, and take care not to exceed the safe maximum flux density for the particular core material. The circuit engineer, who is not ordinarily concerned with the details of the transformer, can safely satisfy the voltage saturation limitation simply by complying with the maximum volt-second product specified by the transformer designer. This product is a transposition of Eq. (10.45) into:

$$V_1\tau = N \Delta BA \; 10^{-8} \qquad (10.46)$$

where $V_1\tau$, or energy in joules (J), can be equated with flux in webers (Wb).

A rating of 100 volt-microseconds (V · μsec) on a transformer catalog data sheet means that the product of the pulse width (in microseconds) and the primary voltage should not exceed 100. The circuit engineer can use a 100-V pulse 1 μsec long, or a 1-V pulse 100 μsec long, without causing saturation.

10.9. THE OFF PERIOD

In this section, we discuss the shape and characteristics of the response during $t > r$, the fall response which follows the discontinuity, and the total decay response (see Fig. 10.12).

But first, some introductory comments. The response at the output of a linear network is, in general, symmetrical. In Figs. 10.8 and 10.9, the rise and fall intervals are equal. Similarly, the rise time t_r of the response of a linear transformer equals the fall time and the magnitude of the droop d should equal that of the backswing b, as shown in Fig. 10.12b. But, as we have previously noted, the pulse response in a transformer with a magnetic core is, in general, asymmetrical. Therefore:

$$t_r \neq t_f \qquad d \neq b \qquad\qquad (10.22)$$

Furthermore, as we shall see, the response during the off period can include oscillations.

We are about to embark on an extended discussion of the response during the off period. It warrants our close interest, especially if, as a result, we become better able to account for the seemingly anomalous or unexpected pulse response of a real (nonideal, nonlinear) transformer in a real (nonideal, nonlinear) circuit.

There is a variety of responses to be accounted for. We shall consider them in five distinct sections:

1. The backswing response. Here we deal with the possibility of tail oscillations and with the magnitude of the backswing. The character of the response depends closely on (1) the nature and behavior of the switch during the interval $\tau+$ immediately following the discontinuity and (2) the nature of the load, that is, whether it is linear or nonlinear.*

2. The fall response. Here we consider the response beginning with the discontinuity at $\tau+$, that is, the fall time t_f, and the subsequent shape of the response. The character of the response depends on whether the switch and the load remain connected to the transformer during the interval between $\tau-$ and $\tau+$.

3. A summary of the off-period response.

4. A statement about the equal-area theorem and the conservation of flux linkages.

5. The TV flyback transformer circuit, an interesting utilization of the off-period response.

a. The Decay Period: Backswing

The response during the off period is easy to analyze when $C_D = 0$, source and load resistances remain constant, and the primary inductance is linear. In practice, however, the physical source and load may be transistors or the load may include a series or shunt diode. In addition, the primary inductance during $t \gg \tau$ may differ from its value during $\tau+$.

A rigorous analysis requires a fairly accurate characterization of the

---

* We use the symbols $\tau-$ and $\tau+$ to denote the time intervals immediately preceding and following τ, the duration of the pulse (see Fig. 10.12a).

behavior of a transistor switch and appropriate circuit models of a bipolar junction transistor in both the normal and saturated modes. Such a model, described by J. Watson, portrays the saturation of the transistor as a function of both forward and reverse modes of bias.[12] It can then be shown that circuit behavior depends on whether the transistor, during the drive (on) interval, behaves as either a current source or a voltage source.

In our analysis, we deal with the nonlinear behavior of source and load by regarding them as switches whose resistance rises sharply during the off period. The time required for this increase in resistance is assumed to be less than the transformer's fall time. The term *source connected*, or *load connected*, refers to the constancy of source or load resistance during the on and off periods; the term *source disconnected*, or *load disconnected*, refers to the rapid increase in the resistance of the source or load during the off period.

(1) The Backswing Response. In the initial period $\tau-$, prior to opening the switch, the magnetizing current has reached a value of I_1, so that the primary voltage is $L_P I_1$. The distributed capacitance has accumulated a charge of q_o coulombs (C), so that the capacitor voltage is q_o/C_D. Thus, the initial conditions at time $\tau-$ are such that magnetic energy and electrostatic energy have been stored in the inductance L_P and the capacitance C_D.

In a large class of circuits, the pulse source is a switching transistor, or a thyristor biased so that when the switch is opened (at $\tau+$), the source resistance leaps. We approximate the condition by assuming that $R_1 = \infty$ at $\tau+$ (source disconnected) and that the load R_2 remains connected. The approximate equivalent circuit is Fig. 10.20a, which will be recognized as the *ringing* circuit.

The voltage equations in the time domain are:[10]

$$L_p I_1 - L_P \frac{di_1(t)}{dt} = \frac{q_o}{C_D} + \frac{1}{C_D}\int i_1\, dt - \frac{1}{C_D}\int i_2\, dt = R_2 i_2(t) \quad (10.47)$$

where zero time starts at $t = \tau$.

As we did during our analysis of the rise period, we define an attenuation factor α_1, a damping factor k_1, a time constant T_1, and a frequency ω_a:

$$\left.\begin{array}{cc} \alpha_1 = \dfrac{1}{2R_2 C} = \dfrac{2\pi k_1}{T_1} & k_1 = \alpha_1 \sqrt{L_P C_D} = \dfrac{\alpha_1 T_1}{2\pi} \\[3mm] T_1 = 2\pi \sqrt{L_P C_D} & \omega_o = \dfrac{1}{\sqrt{L_P C_D}} = \dfrac{2\pi}{T_1} \end{array}\right\} \quad (10.48)$$

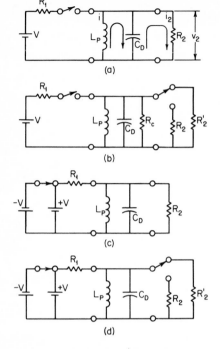

(a)

(b)

(c)

(d)

Fig. 10.20 Backswing circuits, $t > \tau$. (*a*) Source disconnected, load connected. (*b*) Source and load disconnected (R_c may be significant). (*c*) Source and load connected. (*d*) Source connected, load disconnected.

Here $I_1 = \Delta I_2 = \Delta V_2/R_2$, I_2 being the load current at $\tau-$ and Δ being the ratio of magnetizing to load current.

Normalized curves which employ informative values of Δ show the representative backswing shapes for the three cases: underdamping, critical damping, and overdamping (see Fig. 10.21*a*, *b*, and *c*). If we compare the elementary (first-order) backswing response of Fig. 10.12*b* with the magnified curves of Fig. 10.21, we see several differences. Each curve in Fig. 10.21 has a finite fall time and is rounded in the vicinity of the maximum backswing, thus assuming the appearance of a tail. When $k_1 < 1$, tail oscillation and return backswing will occur.

(2) **Tail Oscillation.** Although overdamping is frequently encountered, it should not be fortuitous. It is important to ensure that $k_1 \geq 1$, thus precluding the possibility of tail oscillation. The criterion can be expressed as a restatement of Eq. (10.48):

$$k_1 = \frac{Z_1}{2R_2} \geq 1 \qquad \sqrt{\frac{L_P}{C_D}} \geq 2R_2 \qquad (10.49)$$

where $Z_1 = \sqrt{L_P/C_D}$.

It will be even more useful to know when to expect that $k_1 \geq 1$.

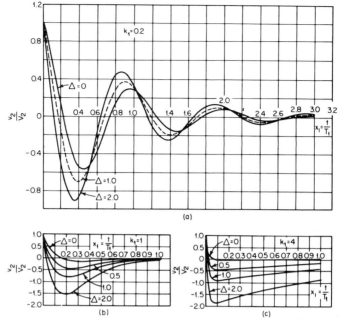

Fig. 10.21 The backswing response. (*a*) Tail oscillation, $k_1 = 0.2$. (*b*) Critical damping, $k_1 = 1$. (*c*) Heavy damping, $k_1 = 4$. (*From J. Millman* and *H. Taub*, Pulse and Digital Circuits, *New York: McGraw-Hill Book Company, 1956, by permission.*)

Let us recast Eq. (10.49) as a function of the characteristic impedance Z of the rise period:

$$k_1 = \frac{1}{2R_2}\sqrt{\frac{L_L}{C_D}\frac{L_P}{L_L}} = \frac{Z}{2R_2}\sqrt{\frac{L_P}{L_L}} \tag{10.50}$$

We see from Eq. (10.50) that the backswing is likely to be heavily damped (when R_2 is constant) because Z is typically of the order $2R_2$ or R_2, and L_P/L_L is invariably much greater than 4.

In an important group of circuits, the *load resistance is disconnected* (i.e., nonlinear, as in Fig. 10.20*b*), and the problem is the extent to which R_2 may increase before tail oscillations occur. Let us take the frequently encountered situation in which droop is 10 percent and rise is 90 percent. The relationship between L_P/L_L and $\tau_{0.1}/t_0$ is:

$$\frac{\tau_{0.1}}{t_0} = \frac{0.1 T_d}{2.3 T_0} = \frac{1}{23}\frac{R_S}{R}\frac{L_P}{L_L} \tag{10.51}$$

$$\frac{L_P}{L_L} = 23\frac{R}{R_S}\frac{\tau_{0.1}}{t_0} \tag{10.52}*$$

---

* $R = R_1 R_2/(R_1 + R_2)$ and $R_S = R_1 + R_2$.

When we substitute into Eq. (10.50) the value of L_P/L_L given in Eq. (10.52), we obtain:

$$k_1 = \frac{1}{2}\frac{Z}{hR_2}\sqrt{23\frac{R}{R_S}\frac{\tau_{0.1}}{t_o}} \tag{10.53}$$

where $h = R_2'/R_2$, the ratio between the higher and the normal load resistances. This equation permits us to arrive at the upper limit of R_2'.

For example, when *damping* is *critical* ($Z = 2.41R_2$, $R_1 = R_2$, and $R/R_S = \frac{1}{4}$), then:

$$k_1 = \frac{2.89}{h}\sqrt{\frac{\tau_{0.1}}{t_o}} \qquad h \le 2.89\sqrt{\frac{\tau_{0.1}}{t_o}} \tag{10.54}$$

If k_1 is to be greater than 1, h must satisfy the expressed inequality. To illustrate, if $\tau_{0.1} = 1$ μsec and $t_o = 0.01$ μsec, then R_2' should be less than $28.9R_2$ in order that k_1 be greater than 1.

When there is *maximum transfer of energy* ($Z = R_2 = R_1$), t_o [in Eq. (10.53)] equals $t_{0.9}/1.58$ and:

$$h \le \frac{1.51}{k_1}\sqrt{\frac{\tau_{0.1}}{t_{0.9}}} \tag{10.55}$$

Suppose the rise occupies a hundredth part of the total pulse duration. Then $h \le 15.1$. When the rise occupies a tenth part of the total duration (a width-to-rise ratio of 10), $h \le 4.78$. A magnetron load, for example, is equivalent to a biased diode, and its effective resistance increases by a factor of about 10 during the off period. It is not surprising, therefore, that tail oscillations occur. In a radar transmitter each return backswing of these oscillations produces a false echo,[13] but low shunt core-loss resistance R_c (see Fig. 10.20b) will sometimes reduce or even eliminate such echoes. When the primary inductance is also nonlinear, the frequency of the tail oscillation will vary with the increase in R_2.

In some circuits the time required for the tail or backswing to die down is significant. This *recovery time* is defined as the period between the first time the pulse falls to 10 percent of its peak and the last time it decays to $\pm$ 10 percent of its peak (see Fig. 10.22c).

There are numerous constructive applications of underdamped behavior in ringing circuits. One, the TV horizontal flyback transformer, is described in Secs. 10.9e and 11.6b. Other practical applications are listed in Table 11.1.

(3) **Backswing Magnitude and Spikes.** The magnitude of the backswing when the circuit is *overdamped* may be calculated. When there is *heavy damping*, then the output voltage (normalized) is:

$$\frac{v_2}{V_2} = -\left(\frac{1}{4k_1^2} + \Delta\right) e^{-\pi x_1/k_1} \tag{10.56}$$

Note that the backswing cannot exceed the coefficient $(1/4k_1^2 + \Delta)$. This is evident from Eq. (10.56) and will be borne out by an examination of Fig. 10.21c. And if Δ is greater than 0.1 and k_1 is greater than 4, Δ is much greater than $1/4k_1^2$.

For many practical purposes, the magnitude of the backswing may be considered equal to Δ:

$$\Delta \cong \frac{R_2}{R}\frac{\tau}{T_d} = \frac{R_2}{R} D \tag{10.57}$$

Thus, backswing or Δ will not exceed R_2/R times the top droop. For example, if the top droop (R/L_P) is 10 percent, $R_1 = R_2$ during the on period, and $k_1 > 4$, then the backswing will not exceed twice the top droop, or 20 percent.

Suppose, however, that the source resistance is low enough so that $R = 0.1R_2$ and that L_P has been reduced proportionately, with the result that the top droop is still 10 percent. In this case, Δ will now be $10(0.1) = 1$, and the magnitude of the backswing will be about

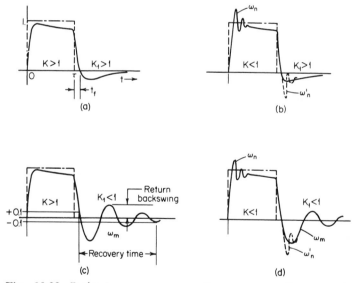

Fig. 10.22 Backswing response: nonlinear source disconnected. Linear load (Fig. 10.20a): (a) Load connected, no oscillations. (b) Load connected, rise oscillation. Nonlinear load (Fig. 10.20b): (c) Load disconnected, tail oscillation ω_m. (d) Load disconnected, rise and tail oscillations. Note: If source switch is slow to open, fall oscillation ω_n' shown dotted in (b) and (d) is possible.

100 percent! It is now clear that the reverberation of low source resistance and low primary inductance, despite the tantalizing prospect of reduced transformer size and low top droop, result in a large backswing during the decay period. A reasonable approach, therefore, would be to match source and load resistance during the off as well as the on period, when possible.

The problem of keeping Δ, and therefore backswing, small is not academic. We are already familiar with the havoc caused by large Δ when we open a switch in series with an inductor and a battery. A high potential gradient causes arcing at the switch contacts. If the backswing magnitude exceeds the rated dielectric strength of the inductor coil insulation, the insulation is also damaged. When the switch is a diode connected in a rectifier circuit with an inductive load, it will be permanently damaged when its peak inverse voltage is exceeded. Or the pulse source or load may be a transistor with a reverse breakdown voltage which must not be exceeded.[12,14]

The backswing, if excessive and brief, is called a *spike voltage* e_{pk}, whose magnitude depends upon whether the capacitance in the circuit is sizeable. Consider the typical pulse drive circuit (Fig. 10.23). Let us first assume that capacitor C is not present and that distributed capacitance is negligible. The ratio of spike to supply voltage can be computed from:

$$\frac{e_{pk}}{E_{CC}} = 1 + \left(\frac{I_1 R_2}{E_{CC}}\right) e^{-\tau/T_d} \qquad (10.58)$$

where E_{CC} is the collector supply voltage, I_1 is the peak magnetizing current at $t = \tau-$, and R_2 is the reflected load resistance. In Fig. 10.23, i_c is the instantaneous value of the collector current.

If I_1 is comparable in size to the load current, the spike ratio will be in the order of:

$$\frac{E_{pk}}{E_{CC}} \cong 1 + 1 = 2 \qquad (10.59)$$

that is, twice the magnitude of the supply voltage.

Class D switching circuits, such as the bipolar inverter and converter shown in Figs. 2.9a and 2.11, illustrate the problem. If the core is driven into saturation, then I_1 briefly exceeds the load current at $\tau-$. This condition is depicted in Fig. 10.12c: spikes greater than twice the supply voltage will occur, at $\tau+$, precisely after each discontinuity.*

However, our initial (and simplifying) assumption that capacitance

---

* The effect of the leakage inductance between each half-primary winding is discussed in Sec. 11.5c.

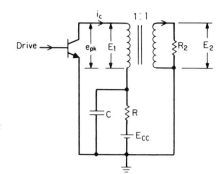

Fig. 10.23 Drive circuit for fast rise.

is negligible in the drive circuit shown in Fig. 10.23 may not be true. Indeed, the RC circuit in the figure may have been introduced in order to speed up the response during the initial rise interval, at $t = 0+$.[15] In this case, the breakdown, or spike, voltage e_{pk} is estimated with the following equation:

$$\frac{e_{pk}}{E_{CC}} \cong \frac{\pi}{2} \frac{t_{on}}{t_{off}} \qquad (10.60)$$

Compare the result of Eq. (10.60) with what happens in the flyback converter circuit shown in Fig. 2.12c. From Table 2.5, which predicts the step-up voltage E_o when the supply voltage is E_1, we derive:

$$\frac{E_o}{E_1} = \frac{t_{on}}{t_{off}} \qquad (10.61)$$

and we infer that the spike can be large: $\pi/2$ times greater than it is when capacitance is negligible. The experienced and prudent designer therefore places a suppression diode (or varistor) in shunt with the primary winding of the transformer to avoid spikes in the response.

Source Connected. If the generator is biased in such a way that the source remains connected during $t > \tau$ (i.e., R_1 remains the same and $k_1 \gg 1$, see Fig. 10.20c), then the backswing will not exceed $\Delta + 1/4k_1^2$. Since the effective load resistance is R, Eq. (10.57) can be simplified to $\Delta \cong D$. The backswing is about equal to the top droop.

The upward slope toward the time axis can be estimated from Eq. (10.56), since the exponent which determines the rate of decay is simplified to $tR_2/L_P = t/T_d'$ when the definitions of Eq. (10.48) are employed. When the source is connected, the effective load resistance is R, the time constant is T_d, and the upward slope of the backswing is the same as the top droop tR/L_P.

Some fall oscillations which occur during the backswing can be attri-

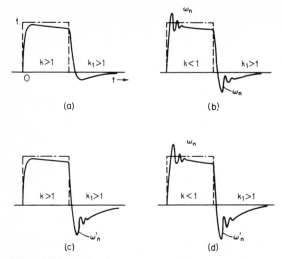

Fig. 10.24 Backswing response, linear source connected. Linear load (Figs. 10.25a and 10.20): (a) Load connected, no oscillations. (b) Load connected, rise and fall oscillations. Nonlinear load (Figs. 10.25b and 10.20d): (c) Load disconnected, fall oscillation ω'_n. (d) Load disconnected, rise ω_n and fall ω'_n.

buted to the behavior of the source while it is being disconnected. They are produced when the source is a "slow switch," e.g., a switching transistor which has an off switching time which equals or exceeds the fall time of the transformer (see Fig. 10.22b and d).

If the *load becomes disconnected* (see Fig. 10.20d), the tail remains damped but the backswing increases. If source and load are matched during the on period, the backswing (see Fig. 10.24c and d) will be the same as for the source-connected load-connected case (shown in Fig. 10.22a and b). When the resistances are not matched, the backswing may be predicted from a restatement of Eq. (10.57):

$$\Delta \cong \frac{R_1}{R} D \tag{10.62}$$

Summary of Backswing. We can now compare the elementary analysis of the backswing envelope (where $C_D = 0$) with the more general analysis (where $C_D \neq 0$) when the load is connected. In either case, the backswing magnitude does not exceed $\Delta = R_2 D/R$ when the source is disconnected and it does not exceed $\Delta = D$ when the source is connected. In either case, the backswing can exceed 100 percent.

The significant differences when $C_D \neq 0$ are:

1. There is a finite fall time, even if the switch is perfect and opens in zero time.

2. There may be backswing even when $\Delta = 0$ (see Fig. 10.21b, where $k_1 = 1$).

3. There are tail oscillations when $k_1 < 1$ (see Figs. 10.21a and 10.22c and d).

b. The Decay Period: Fall Response

Fall or decay time may be defined, analogously with t_r, as the elapsed time between 90 and 10 percent of the response at the output after the switch opens. For simplicity in the ensuing analysis, we define fall time as the time required for the response at the output to fall to zero.

(1) **Fall Response: Source Disconnected, Load Connected.** The applicable equivalent circuit is Fig. 10.20a. Fall time may be estimated from the information in Fig. 10.21.

When the underdamped equations are set to zero, x_1 becomes the normalized fall time x_f. Figure 10.11a and b depict the character of the fall when $k_l > 1$.

A study of the relevant parameters and the decay curves of Fig. 10.21 shows that fast fall time is generally obtained by increasing the top droop (larger Δ) and the backswing. However, when the top droop is held to 10 percent and the rise damping constant is $\sqrt{\frac{1}{2}}$ or 1, the fall time *does not exceed* rise time and the backswing is moderate. This result is consonant with the design principle: maximum transfer of energy results in an optimum overall pulse shape.

(2) **Fall Response: Source and Load Connected.** If the source resistance remains constant during $\tau+$, then Fig. 10.25a, which takes into account the leakage inductance, becomes the appropriate equivalent circuit. The shape of the fall response is inverse to that of the rise response and the fall time equals the rise time.

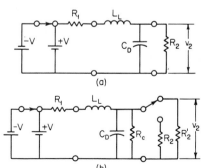

Fig. 10.25 Fall circuits, $t = \tau+$: source connected. (a) Load connected. (b) Load disconnected.

The total decay response is arrived at by superimposing the fall response onto the appropriate backswing envelope (whose slope is tR/L_P). The composite response is depicted in Fig. 10.24a and b.

(3) **Fall Response: Load Disconnected.** If the load resistance jumps to R_2' at $\tau+$ (see Fig. 10.25b), the time constant $L_L R_2'$ becomes smaller and the fall time becomes *less* than the rise time. The backswing, however, will be greater. If the rise time or front edge has an oscillation frequency ω_n [see Eq. (10.24)], the corresponding oscillation of the fall will have a different frequency ω_n' (see Fig. 10.24c and d) because of the change in load resistance. We also see why the magnitude of a spike voltage can equal or exceed the values given in Eqs. (10.59) and (10.60).

c. The Decay Period: Summary

We can summarize the variety of decay-period behavior with the aid of Figs. 10.22 and 10.24, which show the four distinct combinations of source and load switching behavior under varying degrees of damping. There are eight basic responses, depending on whether k is smaller or greater than 1 during the rise or on period, and on whether k_1 is smaller or greater than 1 during the decay or off period. The off-period response is always intimately related to and dependent on the on-period response.

In theory, these responses will depend on the behavior of the source and load impedances. Possible combinations are:

1. Nonlinear source and linear load (see Fig. 10.22a and b)
2. Nonlinear source and nonlinear load (see Fig. 10.22c and d)
3. Linear source and load (see Fig. 10.24a and b)
4. Linear source and nonlinear load (see Fig. 10.24c and d)

The response when the *source is disconnected* is shown in Fig. 10.22. The fall time may be less than, equal to, or greater than rise time. When there are no oscillations (see Fig. 10.22a), the response is similar to the elementary case (first order). Figure 10.22b shows rise oscillations when $k < 1$. If the load is also disconnected and $k_1 < 1$, tail oscillations occur (see Fig. 10.22c and d).

If the source switch is slow to open, the fall is subject to additional oscillations (shown as dotted lines in Fig. 10.22b and d). When there is a slow switch and both k and k_1 are less than 1, three different oscillations—rise, fall, and tail—may occur (see Fig. 10.22d).

The response when the *source remains connected* (shown in Fig. 10.24) is obtained by superimposing the fall response onto the damped backswing envelope. Again, Fig. 10.24a is equivalent to the symmetrical response in the elementary treatment and the fall time equals the rise time. Figure 10.24*b* shows the rise and fall oscillations when $k < 1$. Figure 10.22*c* shows the fall oscillations when $k > 1$ and the load is disconnected. Figure 10.22*d* shows the rise and fall oscillations when $k < 1$ and the load is disconnected.

In some respects the figures depict possibilities not likely to occur in practice, especially in transistor switching circuits. For example, source resistance often drops to a very low value at the completion of the rise interval, so that top oscillations simply do not occur. Or the load may be disconnected during the fall period, but become reconnected during the backswing period because a damping diode is placed across the primary. In this case, fall time can readily exceed rise time and the tail does not oscillate.

A large backswing is most likely to occur when the load is nonlinear. But then, even when the load is linear, the backswing can be large if the source is nonlinear and mismatched to the load. In such a circuit, the top droop can be very small and the backswing large. Oscillographs of pulse-transformer circuits will sometimes reveal unpredicted oscillations. These serve to remind us of the inherent limitations of our simple lumped-constant circuits and of the difficulty of accurately assessing the behavior of the source and load resistances throughout the entire cycle.

d. The Equal-Area Theorem

There is a basic question concerning the decay-period response: Is it possible, without recourse to a nonlinear component such as a suppression (or damper) diode, to design a pulse transformer and its circuit parameters so that the backswing will be zero? The answer is, No. Since a transformer cannot pass a dc voltage, the area above the time axis of the pulse output response must equal the area under the axis. One proof is begun by first expressing the output voltage as $v_o = Nd(Li_m)/dt$, and then solving for the total area of the output wave form:

$$\int_0^\infty v_o \, dt = N \int_0^\infty \frac{d(Li_m)}{dt} \, dt = NLi_m \Big|_0^\infty = 0 \qquad (10.63)$$

where Li_m is the stored energy due to the flow of magnetizing current i_m in a winding of N turns. The area is equal to zero since $i_m = 0$ at zero and infinite time.

Another proof is begun by noting that the flux density must return to B_r before a new pulse starts. Since the core ΔB will be $B_m - B_r$

during the on period and $B_r - B_m$ during the off period, the volt-second integrals before and after τ are equal but opposite in sign.

The principle of constant flux linkages reminds us that, whether we are dealing with an isolated pulse or a train, the zero axis adjusts itself so that the areas above and below the axis are equal. The area below the axis of a single pulse with an infinitesimal backswing will take infinite time after τ to achieve equality with the area above the axis.

e. The Horizontal Flyback Transformer

Ringing, or tail oscillation, is constructively exploited in some circuits. In the TV flyback circuit, for example, the fast flyback of sawtooth current at the end of a scanning cycle produces a parasitic high voltage which, stepped up by a transformer and rectified, furnishes *kilovolts* (kV) of anode voltage for the picture tube.

In the horizontal sweep of the deflection flyback circuit (Fig. 10.26a), a sawtooth of current i_Y flows in the inductance L_Y of the deflection yoke during the on (trace) interval.* The transistor and diode are represented by switches S_1 and S_2 in the equivalent circuit (Fig. 10.26b).[16] S_1 closes at t_1 and S_2 shortly after, at t_2. During the retrace interval $t_{re} = t_2 - t_1$, there is a sharp drop of current (cosinusoidal in shape) in the yoke, which produces a high voltage across the capacitor, the yoke, and the primary winding of the transformer.

The peak voltages e_c, across the capacitor, and e_Y, across the yoke, are given by:

$$e_c = e_Y = L_Y i_Y \omega \qquad (10.64)$$

These peak voltages are further magnified by the turns ratio of the transformer.

The resonant frequency ω_n is:

$$\omega_n = 2\pi f = \frac{1}{\sqrt{LC}} = \frac{\pi}{t_{re}} \qquad (10.65)$$

where L is the effective inductance (of L_Y in parallel with L_P), C is the effective capacitance (of C_Y in parallel with C_D), and t_{re} is the retrace time.

The combination of Eqs. (10.64) and (10.65) results in:

$$e_Y = \pi \frac{(L_Y i_Y)}{t_{re}} \qquad (10.66)$$

---

* There are many practical variations of the TV deflection circuit. Typical time intervals are $t_1 + t_2 = 1/15,750 = 63.5$ μsec and $t_{re} = 10$ μsec.

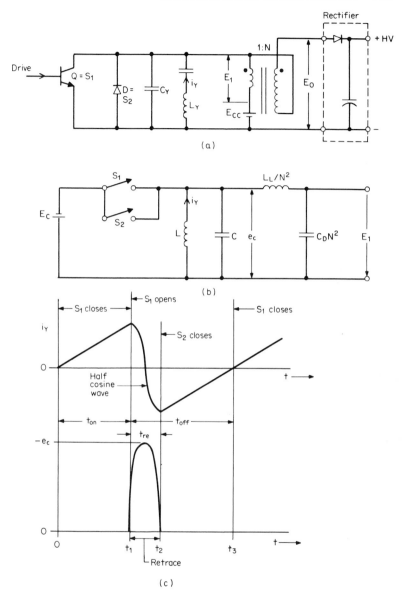

Fig. 10.26   The deflection flyback circuit.

From our previous encounters with expressions of this form, we can identify $L_Y i_Y$ with the total flux Φ^* in the yoke and the end of the on time, at $\tau-$:

---

* See Eqs. (2.29) and (7.47), as well as a discussion of the energy storage inductor in Sec. 7.6a.

$$\Phi = L_Y i_Y = 2E_c t_{on} \qquad (10.67)$$

Combining Eqs. (10.66) and (10.67) permits us to write an expression for the voltage step as the ratio of yoke voltage and supply voltage:

$$\frac{e_Y}{E_c} \cong \frac{\pi}{2} \frac{t_{on}}{t_{re}} \qquad (10.68)$$

And, as can be seen from Fig. 10.26a, a further magnification of the flyback voltage is obtained with a transformer who turns ratio is N_2/N_1. Note this equation's similarity in form to the equation for the converter flyback circuit [see Eq. (10.61)]. We can therefore infer that the capacitance of the ringing circuit increases, by $\pi/2$, the voltage step up of the nonringing circuit.

A further increment of step up (about 30 percent) is possible if the $L_L C_D$ product of the transformer is adjusted* to resonate at a harmonic of the retrace frequency of ringing.

10.10. EQUIVALENT BANDWIDTH

Earlier, in Sec. 10.4a when we introduced the linear network and the ideal filter, we stated that the bandwidth of a pulse is directly related to its rise time and inversely related to its width. We have also seen that, in a nonideal transformer, the presence of finite shunt inductance results in a droop at the top of the response.

Sometimes we know the rise and top droop response of a transformer and wish to predict the frequency response. Conversely, we may know the frequency response and wish to predict from it the rise and droop behavior.

The simplest procedure is consistent with linear wave-shaping analysis, in which the droop and rise time responses of a network to a step voltage are equated, respectively, to the 3-dB down low-frequency and the 3-dB down high-frequency responses. The top-period circuit (Fig. 10.18a) and the elementary rise-period circuit (see Fig. 8.7a) are identical to the first-order high-pass and low-pass equivalent circuits for the audio transformer. The lower frequency f_1 and upper frequency f_2 occur when:

$$2\pi f_1 L_P = R \qquad 2\pi f_2 L_L = R_S \qquad (10.69)$$

If we now rewrite Eq. (10.41) in terms of the time constant for 10 percent droop and combine it with the first expression in Eq. (10.69), we can equate the low frequency f_1 to the pulse duration $\tau_{0.1}$:

---

* The technique is called third-harmonic "tuning." The flyback transformer (described earlier in Secs. 2.10b and 5.7a) will be discussed further in Sec. 11.6b.

$$D = \tau_{0.1} \frac{R}{L_P} = 0.1$$

and

$$f_1 = \frac{R}{2\pi L_P} = \frac{D}{2\pi \tau_{0.1}} = \frac{1}{20\pi \tau_{0.1}} \qquad (10.70)$$

If the pulse width τ is part of a square wave whose PRF is $f = \frac{1}{2}\tau$, we obtain a useful expression relating the droop, the lower 3-dB down frequency, and the square-wave frequency:

$$D = \pi \frac{f_1}{f} \qquad (10.71)$$

The high frequency f_2 can be equated to the rise time t_o:

$$f_2 = \frac{R_S}{2\pi L_L} = \frac{2.30}{2\pi t_o} = \frac{0.366}{t_o} = \frac{0.35}{t_r} \qquad (10.72)$$

If f_2 is made equal to $1/\tau$, then:

$$t_r = 0.35\tau \qquad (10.73)$$

We see that the rise time consumes about 35 percent of the pulse width. Note that Eq. (10.73), our estimate of the *actual* rise time t_r, compares rather well with the estimate for the rise time in the *ideal* network:

$$t_R = 0.50\tau \qquad (10.10)$$

A pulse with so long a rise time still resembles a rectangular pulse, and this accounts for the rule of thumb: A pulse shape is preserved if the 3-dB down frequency is about equal to the reciprocal of its width (see Table 10.1).

A transformer with a bandwidth $f_2 - f_1$ might exhibit no peaking in the frequency domain and yet exhibit ringing in the time domain. The relations between the damping constant in the time domain and the phase shift, attenuation, and peaking in the frequency domain can be ascertained.

Gain and phase curves for this important second-order model are plotted in Figs. 10.27 and 10.28, in which the damping factor k [Eq. (10.26)] is used as a parameter. In both figures the abscissa is $\omega T_a = \omega/\omega_a$. The cutoff frequency ω_a is greater than the undamped resonant frequency ω_r; that is:

$$\omega_a = \omega_r \sqrt{a} \qquad (10.74)$$

The frequency ω_p, at which the response peaks, is related to the cutoff frequency:

$$\omega_p = \omega_a \sqrt{1 - 2k^2} \qquad (10.75)$$

Note that as the damping factor increases, the magnitude and frequency of the peak decrease.

The *bandwidth* ratio f_2/f_1 is the quotient of Eqs. (10.69):

$$\frac{f_2}{f_1} = \frac{R_S}{R}\frac{L_P}{L_L} = \frac{T_d}{T_o} \tag{10.76}$$

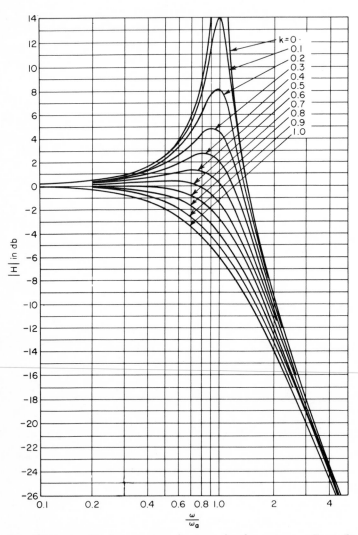

Fig. 10.27 Amplitude response of a second-order system. (*From R. Landee, D. Davis, and A. Albrecht,* Electronic Designers' Handbook, *New York: McGraw-Hill Book Company, 1957, by permission.*)

We have already derived Eq. (10.51), which relates the ratio of primary and leakage inductance to the *squareness ratio* $\tau_{0.1}/t_o$, and we can restate it as follows:

$$\frac{\tau_{0.1}}{t_o} = \frac{1}{23}\frac{R_S}{R}\frac{L_P}{L_L} = \frac{1}{23}\frac{T_d}{T_o} \tag{10.77}$$

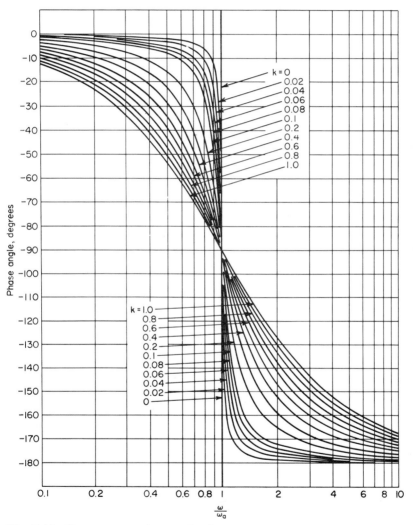

Fig. 10.28 Phase response of a second-order system. (*From R. Landee, D. Davis, and A. Albrecht,* Electronic Designers' Handbook, *New York: McGraw-Hill Book Company, 1957, by permission.*)

Note that a low ratio of source to load resistance R_1/R_2 yields both the greatest bandwidth and the greatest ratio of pulse width to rise time. We temper this conclusion, however, by noting that if the source is disconnected during the off period, this advantage will be offset by a large backswing.

A combination of the last two equations yields a simple relationship between the bandwidth and the squareness ratio:

$$\frac{f_2}{f_1} = 23 \frac{T_{0.1}}{t_o} \tag{10.78}$$

It is important to bear in mind that there are certain difficulties in using so simple a formula.[17] We know that rise time can be less or greater than predicted by Eq. (10.33). Reasonable accuracy may be expected only if the transformer constants are approximately those of the maximally flat Butterworth filter model.

The same considerations which make for a wide band span also result in good pulse fidelity. Thus Eqs. (8.52) and (10.79) can serve as guidelines for the design of the pulse as well as wide-band transformer. From Eq. (8.52) we can obtain an expression for pulse squareness, i.e., the width-to-rise ratio in terms of the circuit parameters, the material properties, and the coil geometry.

Thus, if $K_L = \frac{1}{2}$, $t_r = 2.15/\omega_2$, $\tau = D/\omega_1$, and $N \gg 1$:

$$\frac{\tau}{t_r} = \sqrt{\frac{2}{15.2^2\epsilon_0} \frac{\tau D}{R_2} \frac{AF_c}{l\lambda^2 F_L} \frac{\mu_p}{\epsilon}} \tag{10.79}*$$

Note that just as the *span ratio* decreases as the low-frequency cutoff rises, so also does the pulse-squareness ratio decrease as the pulse becomes briefer. This means that sharply delineated rectangular pulses become increasingly difficult to transmit through a pulse transformer as their duration shortens.

Typical values for the squareness ratio are 100, when pulse widths are greater than 2 μsec, and 10, when widths are less than 1 μsec. These seemingly small values become large when they are translated into the frequency domain. For example, a 400-Hz square wave whose rise time is to be 1/100th of the 1,250-μsec pulse duration requires a frequency band extending from 12.7 to 29,300 Hz. This surely confirms the impression we conveyed, at the beginning of our discussion of the pulse transformer network (Sec. 10.6), that a pulse transformer is a wide-band transformer.

---

* The parameters K_L, F_c, F_L, and so forth are defined in Secs. 8.6 and 9.4d, which deal with the bandwidth of the wide-band transformer.

REFERENCES

1. M. Van Valkenburg, *Network Analysis*, 3d ed. (Englewood Cliffs, N.J.: Prentice-Hall, 1974).
2. M. Schwartz, *Information Transmission, Modulation, and Noise*, 3d ed. (New York: McGraw-Hill Book Company, 1980).
3. F. Kuo, *Network Analysis and Synthesis*, 2d ed. (New York: John Wiley & Sons, 1966), p. 389.
4. F. Noel, J. Kolodzey, "Nomograph Shows Bandwidth for a Specified Pulse Shape," *Electronics* (April 1, 1976): 102.
5. H. Blinchikoff, A. Zverev, *Filtering in the Time and Frequency Domains* (New York: John Wiley & Sons, 1976).
6. C. Shannon, "Communications in the Presence of Noise," *Proceedings IRE*, vol. 37 (January 1949): 10–21.
7. J. Millman, H. Taub, *Pulse and Digital Circuits* (New York: McGraw-Hill Book Company, 1956), Chap. 9.
8. N. Grossner, "Pulse Transformer Circuits and Analysis: Part 1," *Electro-Technology*, Vol. 69, No. 2 (February 1962): 77–80; "Part 2," ibid., No. 3 (March 1962): 80–90.
9. H. Bostick, "Pulse Transformers," Chaps. 12–15 in *Pulse Generators*, G. Glasoe, J. Lebacqz (eds.) (New York: McGraw-Hill Book Company, 1948).
10. S. Moskowitz, J. Racker, *Pulse Techniques* (Englewood Cliffs, N.J.: Prentice-Hall, 1951), pp. 91–93, 95–98.
11. K. Macfadyen, *Small Transformers and Inductors* (London: Chapman & Hall, 1953), pp. 168–75.
12. J. Watson, *Applications of Magnetism* (New York: John Wiley & Sons, 1980), pp. 289–305.
13. Reuben Lee, *Electronic Transformers and Circuits*, 2d ed. (New York: John Wiley & Sons, 1955), pp. 333–38.
14. J. Millman, H. Taub, *Pulse, Digital, and Switching Waveforms* (New York: McGraw-Hill Book Company, 1965), pp. 290–92.
15. W. Stover (ed.), *Circuit Design for Audio, AM/FM, and TV* (New York: McGraw-Hill Book Company, 1967), Chap. 21.
16. J. Tucker, "Computer Aided Design of Horizontal High-Voltage Transformers for Solid State Deflection Circuits," *IEEE Transactions on Broadcast and TV Receivers*, Vol. BTR-16, No. 2 (May 1970): 112–18.
17. C. Wilds, "Pulse Transformers, Frequency Response, and Wide Band Transformers," *Electronic Engineering*, Vol. 34 (September 1962): 608–13.

CHAPTER **11**

The Pulse Transformer: Synthesis

As we have already observed, the task of the transformer designer is to synthesize, from a set of specifications, a satisfactory design. The behavior of the linear pulse transformer was described in Chap. 10, and we know that many parameters must be taken into account if we are to achieve the desired performance. Because the pulse transformer is a wide-band transformer, our approach to its synthesis parallels that described, in Chap. 9, for synthesizing the wide-band transformer.

We start with the general applications of the pulse transformer. Next we discuss the feasibility of some specifications and define fidelity and distortion in the time domain. We then examine certain major objectives of transformer design. This leads to a discussion of the design of the core. Then the design of the coil is discussed briefly. We conclude with a summary of the various considerations entailed in the design of a pulse transformer. Throughout, we shall try to clarify the limitations as well as the capabilities of the pulse transformer.

420

11.1. CIRCUIT APPLICATIONS

Since our principal concern is with the design process, we have not attempted to present the details of the large variety of circuits in which the pulse transformer is used. Additional information about the broad range of its applications is available in the works cited throughout this section. Here we distinguish two major classes: *linear* and *nonlinear.*

a. Linear Pulse Transformers[1,2]

Linear pulse transformers perform the classic functions of isolation, inversion, and impedance transformation. In many transistor and active circuits the transformer is used for some simple purpose, for example, to isolate direct current between input and output, to invert the polarity of a pulse, or to couple a single-coded (unipolar) device to a double-ended (bipolar) circuit. In driver and modulation circuits, the transformer may be required to handle high levels of pulse power. If faithful monitoring of the shapes of current pulses is required, a current transformer is used.

Linear pulse transformers are used in a large variety of circuits. The important categories are:

1. The familiar class A, B, and C amplifiers, in which the transformer performs a coupling function. These were listed in Table 8.2.

2. The newer class S, D, E, and F coupling transformers, also listed in Table 8.2, which are used under specified conditions. (It should be noted that the designations S and D are sometimes interchanged in the literature.)

3. Low-power and high-power linear pulse transformers used in power electronic circuits. Some were mentioned earlier, in Secs. 2.10 and 5.7. They include the inverter and converter transformers which process square-wave and rectangular pulses over a wide range of the duty cycle.

4. Inductive energy storage components which make constructive use of the behavior of the pulse transformer during the off period between pulses. (Some of these important components are listed in Table 11.1.)

5. Pulse transformers used in radar circuits for the efficient transfer, in brief bursts, of megawatts (MW) of peak power to a microwave oscillator such as the magnetron.[3]

6. Components used in pulsed power technology. This technology, which may be viewed as an outgrowth of radar, involves "the storage

TABLE 11.1   Applications of Inductive Energy Storage
Transformers

Type of transformer	Basic features	See
Ringing choke*	High step-up voltage	Eq. (10.47)
Blocking oscillator	Fast fall time; precise pulse width	Ref. 8
TV flyback transformer	High step-up voltage	Secs. 10.9*e*, 11.6*b*
Ignition coil	High step-up voltage	Sec. 2.9*b*
Pulse-forming network	Pulse sharpening	Ref. 6
Flyback converter	Step up of dc voltage	Secs. 2.10*b*, 5.7

* In the photoflash charging circuit, e.g., a capacitor stores energy in spurts over
an interval of time which is the reciprocal of the pulse repetition frequency.

of energy and its subsequent release to a load in the required form."[4]
Applications include the linear accelerator, controlled thermonuclear
reaction (CTR), the resonant charging inductor, the blocking oscillator
transformer, the pulse-forming network (PFN), the magnetic modulator, the pulse power transformer (PT), and the pulse current transformer.

b. Nonlinear Pulse Transformers

When nonlinear functions are to be performed, two basic characteristics of the transformer, remanence and saturation, are exploited.
(These characteristics were described in Sec. 1.5.)

Remanence in the core of the transformer is the salient characteristic
involved in digital storage, shift register, counting, switching, and logic
circuits.

Saturation is exploited to perform functions such as wave shaping
and the control of power.[5]   Important examples of the use of saturation
are:

1. The dc inverter, used to produce symmetrical pulses of any width[6]
2. The blocking oscillator, which generates narrow pulses over a
range of pulse repetition rates, at low power[7]
3. The pulse-sharpening reactor (thyractor) and the saturating transformer, used in the magnetic pulse modulator (pulser)
4. The pulse-forming network (PFN), which generates narrow steep
pulses of high power[8]

5. The ferrite coaxial transmission line, used to produce fast-rise, high-voltage pulses[9]

6. The ferroresonant or *constant-voltage transformer* (CVT), which produces a regulated square-wave output, described in Sec. 5.5.

7. The pulse magnetic amplifier, comprising either a pair of saturable reactors or a saturable transformer behaving as a variable impedance, which provides continuous control of pulses of power[10,11,12]

8. The trigger or differentiating transformer, used to time, trigger, or synchronize various devices such as the thyristor[13]

9. The magnetic frequency multiplier and divider, which provide power at a multiple or submultiple of the input line frequency[14]

The distinctly nonlinear transformers, inductors, and magnetic amplifiers are now viewed as part of the growing field of nonlinear magnetics and comprise too large a group to be treated adequately in this volume. Nevertheless, during some part of their performance cycle these transformers behave like quasi-linear transformers. Their designer can therefore benefit from some of the basic concepts discussed in this and other chapters of this book.

11.2. SPECIFICATIONS AND FEASIBILITY

A formal set of specifications for a pulse transformer is an explicit statement of the electrical, mechanical, and environmental requirements. Quality and reliability are assured by requiring compliance with appropriate industry specifications, e.g., IEEE 272, 390, and 391 for computer-type, low-power, and high-power pulse transformers and appropriate military specifications, e.g., MIL-T-55,631, MIL-T-21,038, or MIL-T-39,026 for low-power pulse transformers and MIL-T-27 for high-power transformers.

We can specify a transformer at any stage of design or manufacture. When constraints are placed on size or cost, an effort is made, as early as possible, to verify the feasibility of the electrical specifications.

In the quest for a feasible design, transformer designers, like other engineers who must juggle many variables, are not infrequently faced with mutually contradictory requirements. Consequently, we need explicit expressions of the relationships among the basic parameters. This helps us, especially when feasibility is a problem, to see which tradeoffs are indicated and which factors determine limit-of-the-art performance. The most important determination to make is the major objective of the design, that is, the salient aspect of the transformer's performance in the circuit in which it is to be used. In considering the various, and sometimes mutually exclusive, design objectives, we discuss in some detail the design concepts with which the designer must deal, and sometimes wrestle.

11.3. DESIGN OBJECTIVES

There are several approaches to the design of a pulse transformer. Each approach is influenced by some specific objective for the performance of the circuit. The most important are:

1. The maximum transfer of energy
2. Fidelity of transmission of the applied wave shape
3. Accuracy in the replication of a coded train of pulses
4. Fast rise time
5. A limitation on the magnitude of *radio-frequency interference* (RFI)

a. The Maximum Transfer of Energy

Many times, our prime objective is the transfer of energy of the pulse, fidelity being of secondary importance. The transformer, as a black box interposed between source and sink, must transfer this energy with the least possible disturbance to the circuit. To fulfill this requirement, two criteria must be met: minimum storage of energy and minimum reflection of power.

(1) **Minimum Storage of Energy.** Bostick's approach to the problem of minimizing the energy stored in the core and coil during the transmission of a pulse is now considered classic.[15] It is especially suited to the design of high-power transformer circuits such as the magnetron pulse modulators used in radar systems. The analysis demonstrates the desirability of meeting the following set of conditions for a given pulse width and load resistance R_2:

$$Z = \sqrt{\frac{L_L}{C_D}} = R_2 \tag{11.1}$$

$$2L_P C_D = \tau_{\text{opt}} = \tau \tag{11.2}$$

$$\frac{2L_L}{L_P} = \text{a minimum} \tag{11.3}$$

where L_L (leakage inductance), C_D (distributed capacitance), and L_P (primary inductance) are the major variables of the transformers.

In Eq. (11.1) the characteristic impedance Z equals the load resistance R_2. In Eq. (11.2) τ_{opt} is defined as that pulse width τ which will produce the optimum output pulse shape, described by Bostick as "a skillful compromise among high rate of rise, low overshoot, small droop, high rate of fall, and low backswing." When a transformer is to pass a range of pulse widths from τ_{min} to τ_{max}, then τ_{opt} should equal $\sqrt{\tau_{min}\, \tau_{max}}$. An important implication of these equations is that, for any given pulse width and load, there is a distributed capacitance C_D which is optimum.

(2) **Reflections.**[16] Reflections of power between the source and the load must also be minimal. The extent of mismatch between generator and load can be defined in terms of a voltage reflection coefficient (the input reflection factor) ρ introduced earlier [in Eq. (8.12)] in our discussion of the wide-band transformer. An alternative expression is:

$$\rho = \frac{R_L - Z_o}{R_L + Z_o} = \sqrt{1 - \frac{1}{H^2}} \tag{11.4}$$

where R_L is the resistance of the load and Z_o is the characteristic impedance of the transformer. The transmission factor H, defined in Secs. 8.4a and 8.5b, is related to the high-frequency attenuation A_H (stated in decibels, dB):

$$A_H = 10 \log H^2 \tag{8.28}$$

The *voltage standing-wave ratio* Γ, or VSWR, is defined by:

$$\Gamma = \text{VSWR} = \frac{1+\rho}{1-\rho} \tag{11.5}$$

If $\rho = 0$, then VSWR $= 1$ and there are no reflections.

The return loss A_r, defined in Sec. 8.3 in terms of ρ, can also be expressed in terms of the standing-wave ratio. Thus:

$$A_r = 20 \log \left| \frac{1}{\rho} \right| \tag{8.14}$$

$$A_r = 20 \log \Gamma \tag{11.6}$$

In practice, it is difficult to obtain low values of VSWR at the extreme operating frequencies. To illustrate, if attenuation A_H is 1 dB for a first-order model, then $H^2 = 1 + 1/(2)^2 = 5/4$ and VSWR $= 2.6$. It can be shown, however, that VSWR approaches unity if we satisfy the criterion:

$$Z_o = \sqrt{\frac{L_L}{C_D}} = R_1 \tag{11.7}$$

In the analysis of the behavior of reflected pulses in a transmission line or delay line, we encounter the same criterion, $Z_0 = R_1$, the resistance of the source.   It is met by the Butterworth filter models (discussed in Sec. 9.2a) and by the transmission-line transformer (discussed in Sec. 9.5).

Designers of digital pulse circuits are likely to specify the limitation on reflections in terms of VSWR rather than return loss.   This comes about naturally, since the linear dimensions of the circuit and the coil winding are comparable in magnitude to the wavelength of the pulse. One can literally visualize the consequences of a mismatch, simply by measuring the peaks of radiated power at discrete distances between crests or valleys of a standing wave.

The connection between the transfer of energy and the geometry of the coil is discussed later, in Sec. 11.5b.

b. Fidelity of the Wave Shape

The second objective of design is fidelity of wave shape.   For many purposes, fidelity is but slightly compromised by small overshoot and a moderate departure from linearity of phase.   Such a compromise is obtained with the Butterworth maximally flat amplitude model.

There are occasions, however, when the specification is more stringent.   Closely limited overshoot and fairly uniform time delay can become important when it is necessary to transmit trains or groups of pulses, as in video or coded-pulse circuits.   The problem is most acute when a substantial number of pulse circuits, including their transformers, are connected in tandem.

When a signal is nonsinusoidal and discontinuous (but not necessarily rectangular), it is desirable to have a guideline for fidelity.   The principle of linear phase or flat delay is such a guideline.*

If all the frequencies in the Fourier spectrum which represent arbitrary, complex, and discontinuous signals are equally delayed, there is no distortion of the waveform; the delayed signal retains its identity. Perfect response is not possible, however; thus it is convenient to define an MFED response—a *m*aximally *f*lat *e*nvelope (or group) *d*elay which is analogous to the *m*aximally *f*lat *a*mplitude (MFA).

According to the theory of network synthesis, a maximally flat delay may be represented by a Bessel polynomial and may be realized physically by the Bessel filter model comprising an LC ladder network. Our analysis of this response begins with the delay t_0, called the dc

---

* The desirability of such a guideline has been mentioned at several junctures: in our discussion of distortion in the wide-band transformer (Sec. 8.5d) and of group delay in the linear network (Sec. 10.4b).   It is discussed again here in the context of the design synthesis of a pulse transformer.

or zero-frequency delay, which serves to define the corresponding zero frequency w_o:

$$t_{(\omega=0)} = t_o = \frac{1}{\omega_o} \tag{11.8}$$

However, some deviation ϵ_d from a specified delay at zero frequency is unavoidable. First we define a normalized frequency U, or span ratio:

$$U = \frac{\omega_d}{\omega_o} = \omega_d t_o \tag{11.9}$$

where ω_d is the frequency corresponding to the group delay t_g.

The deviation from the zero-frequency delay is not to exceed some specified value of ϵ_d (typically 5 to 50 percent):[17]

$$\left| \frac{t_g - t_o}{t_o} \right| \leqq \epsilon_d \tag{11.10}$$

Setting ω_o equal to unity also normalizes the zero-frequency delay t_o; thus U becomes a measure of the relative bandwidth necessary to achieve a desired uniformity, or flatness, of delay.* Some idea of the bandwidth needed for reasonably flat delay can be obtained if we first use the three-pole MFA model and consider its behavior as a delay line. With an integral number n_d of sections, it can be shown that, inside the pass band:

$$\theta = 2 n_d \sin^{-1} \Omega_a \tag{11.11}$$

$$t_g = \frac{d\theta}{d\omega} = \frac{2 n_d / \omega_2}{\sqrt{1 - \Omega_a{}^2}} \qquad t_o = \frac{2 n_d}{\omega_2} \tag{11.12}$$

where $\Omega_a = \omega / \omega_2$, ω_2 is the cutoff frequency, and θ is the phase shift.

At low frequencies, t_g approaches t_o, the delay increasing with the number of sections. Equation (11.10) can be defined in terms of U and Ω_a:

$$U - 1 = \frac{1}{\sqrt{1 - \Omega_a{}^2}} - 1 \cong \tfrac{1}{2} \Omega_a{}^2$$

We can now specify the criteria as follows:

$$\omega_d = \omega_2 \sqrt{2\epsilon_d} \tag{11.13}$$

$$\Omega_a = \sqrt{1 - \frac{1}{U^2}} \cong 1 - \frac{1}{2U} \tag{11.14}$$

---

* Zero frequency in the band-pass filter corresponds to the mid-band frequency in the wide-band transformer.

If we make ϵ_d equal 5 percent, then $\omega_2 = 3.16\ \omega_d$. That is, the cutoff frequency ω_2 must be three times the upper operating frequency ω_d, if group delay is to be reasonably flat.* Now suppose U is fixed at some arbitrary value in order to prescribe the maximum permissible deviation from a fixed delay. Then, according to Eq. (11.9), the product of delay and bandwidth is fixed.

Bode has considered the relationship from a more general point of view.[18] In a network whose average dissipation in the reactive elements is $1/Q$ in the vicinity of cutoff, the relations between attenuation and phase can be expressed by:

$$A - A_o \cong \frac{\omega}{Q}\frac{dB}{d\omega} \qquad (11.15)$$

$$B - B_o \cong -\frac{\omega}{Q}\frac{dA}{d\omega} \qquad (11.16)$$

where $A - A_o$ and $B - B_o$ represent, respectively, the variation of attenuation and phase† with frequency.

Note that if attenuation is to be constant, then $dB/d\omega$, the group delay, should be constant. The phase shift should therefore be linear. Since the form of these equations is identical, we also infer that a small variation of phase in the pass band implies a constant and low rate of attenuation in the stop band. (This was noted in the discussion of feedback in Sec. 9.1b).

Small delay entails large bandwidth; conversely, long delay implies a small bandwidth. This principle is analogous to that of the constant gain-bandwidth concept referred to in Sec. 10.5. It is also consistent with a conclusion arrived at in our discussion of pulse code modulation in Sec. 10.5c: the product of transmission time and bandwidth is a constant.

More specifically, provided $\omega \ll \omega_2$, it can be shown that:

$$B = \frac{2}{\pi}\ n\ \frac{\omega}{\omega_2} \qquad \frac{dB}{d\omega} = \frac{2}{\pi}\ \frac{n}{\omega_2} \qquad (11.17)$$

Here n is an integer, 1, 2, or 3, corresponding to a slope of 6, 12, and 18 dB per octave, respectively.

In the three-pole model, $t_g \cong 6/\pi\ \omega_2 = 1.91/\omega_2$. This compares favorably with $2/\omega_2$ obtained from Eq. (11.12) when $\omega \ll \omega_2$. These considerations have an important bearing on our choice of a suitable

---

* A very wide bandwidth is also indicated for the Butterworth filter, as can be inferred from Table 10.2.

† We supplant our previous symbol θ, for phase, with B [in radians (rad)] in order to retain the clarity afforded by Bode's equations.

network model: We should favor a low-order model because of its more gradual slope at cutoff. We can now conclude that increasing the number of sections (i.e., the poles) of the Butterworth model will increase the delay and the flatness of amplitude but will not increase the flatness of delay.

There are several approaches to the objective of good linearity of phase, or flatness of delay. One alternative to the Butterworth filter is the Bessel filter, which has a good transient response—unusually small overshoot and reasonably quick rise. In the frequency domain, the Bessel filter is characterized by an amplitude response with a slow rate of cutoff, more gradual than that of the Butterworth filter. (This gradual rolloff is characteristically associated with linear phase response and excellent transient response.) To obtain a desired Bessel response it is customary to specify the permissible attenuation, in decibels, at a frequency ω. With the delay t_o specified at ω_o, U is established, and the percentage of deviation from t_o is now strictly dependent on the order of the filter model.

Another approach is to exploit the basic virtues of the transmission line. The result is the so-called transmission-line transformer described in Sec. 9.5.

c. Fidelity and Accuracy

In coded pulse transmission, as in the *pulse code modulation* (PCM) scheme described in Sec. 10.5, the accuracy of the coded pattern is of paramount importance. Shape is less important than maintaining the separation between successive pulses in a train whose *pulse repetition frequency* (PRF) is high. Fast rise and quick recovery after the fall are vitally important; droop and overshoot can be large. In other words, we may tolerate distortion if good impulse response (defined as the rate of rise) is provided.

In these circuits, the presence or absence of a pulse (a *bit* of information) is the central concept. Distortion is defined as the loss of information from an ensemble or train of pulses (bits of the code), and its constraint is embodied in the following inequality:

$$\text{Noise} + \text{intersymbol interference} < \tfrac{1}{2} \text{ pulse height}$$

It is easy to see that distortion of the wave shape is permissible, provided that the tail amplitude (interpreted as an intersymbol) is less than half the pulse height. Phase response should be linear. The identity of the pulses can be assured with reshaping circuits, but resolution at a high PRF requires that rise time be fast and recovery quick. A wave shape other than rectangular (e.g., the raised cosine and Gaussian waveforms mentioned in Sec. 10.2c) may provide better resolution.

Schwartz has shown that zero intersymbol interference is possible when the skirt of the pulse is sinusoidal.[19] If ω_x denotes a frequency on the skirt beyond ω_c, the cutoff frequency, the *roll-off factor r* is defined as:

$$r = \frac{\omega_x}{\omega_c} \tag{11.18}$$

which falls between zero (in an ideal low-pass filter) and unity (in the raised-cosine spectrum).

The practical problems associated with realizing an ideal filter, one whose bandwidth $B = 1/2T$, (T is the period between pulses), are then resolved by increasing the bandwidth to $(1 + r)/2T$. The increment can be moderate (e.g., 19 percent).

d. Fast Rise Time

There are many circuits in which one portion of the pulse response, the rise time, is of major importance and in which distortions of other portions of the envelope (e.g., droop or backswing) are of little consequence. This is because fast rise times result in the maximal rate of transmission of information [see Eq. (10.15)]. Fast rise is also desirable in trigger circuits to assure precise timing and synchronization of pulse trains.

Selective distortion occurs in certain wave shaping and digital applications. In class D switching circuits, the nonlinear source resistance (e.g., saturating or switching transistor) falls abruptly to a small value at the conclusion of the rise time. As a consequence, the overshoot which occurs during the on period when the source resistance is linear is not encountered. The adroit circuit designer reshapes the distorted pulse by ensuring that its top becomes clipped and by eliminating the backswing spike with a suppressor diode. A large top droop (which favors fast fall time) can then be tolerated and attention can be shifted to the problem of achieving a minimum rise time.

It is instructive to examine the relationship of rise time to each of the following:

1. The damping constant k
2. The filter model
3. The volt-second product

In Sec. 10.7b we observed that fast rise, accompanied by large overshoot, results from choosing very low damping constants ($k \ll 1$). The filter model which satisfies the criterion for maximum transfer of energy has a k of 0.707. This is the Butterworth (MFA) filter, when source and load resistances are matched. Since this model requires that the characteristic impedance be matched to the load, there are, theoretically, no reflections and the standing-wave ratio is unity. Both the two-pole and three-pole models have moderate overshoot (4.3 and 8.15 percent, respectively). Thus they qualify as optimum models when fast rise, minimum reflection, and good fidelity are required simultaneously.

Figure 11.1 depicts the step response and impulse response of increasingly complex Butterworth models. Note that the magnitude of the impulse response provides a direct measure of the rate of rise in the time domain. As the number of poles increases, the rate of rise decreases and the delay increases. The fastest rise time t_r is obtained with the two-pole model which was discussed at length in Sec. 10.7b.

As we have noted several times, a small transformer yields a fast rise. Since size is intimately related to the volt-second product [see Eqs. (11.21 and 11.23)], rise time can be expressed as a function of

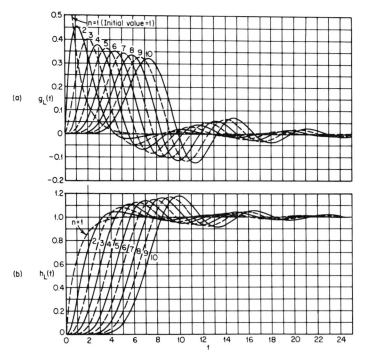

Fig. 11.1 Transient response of the low-pass Butterworth filter. (*a*) Impulse response. (*b*) Step response. (*By permission, IEEE.*)

this product.   Such a relationship is particularly useful in the development of high-power pulse transformers:[20]

$$t_r \geqslant 0.04 \sqrt{\frac{R_L(V_2\tau)}{\mu_p D}} \qquad (11.19)$$

where load voltage V_2 is in kilovolts (kV), t_r and τ are in microseconds (μsec), droop D is in percent, load resistance is in kilohms (kΩ), and μ_p is the pulse permeability of the core.   The constant is based on certain typical design values: ϵ, the dielectric constant of the insulation, equals 3; the winding configuration has a low $L_L C_D$; the flux density is 5 kilogauss (kG); and the insulation is rated at 150 volts per milli-inch (V/mil).

Equation (11.19) confirms again the advantages of small load impedance and brief pulses.   It also makes clear that rise time is lengthened when high voltages are used.

e. Limits on RFI

Pulses such as the rectangle and triangle are discontinuous.   Since a rectangle is the derivative of a triangle, the triangular pulse has a spectrum of frequencies f whose amplitude falls off as $1/f^2$ while the rectangular pulse falls off at the rate of $1/f$.   Hence the RF energy associated with a rectangular pulse is greater than that of the triangular pulse.   This reasoning can be used to justify choosing a trapezoidal rather than a rectangular pulse shape.

The RF energy associated with discontinuous waves creates circuit problems in the form of *radio-frequency interference* (RFI) conducted or radiated to other circuits in the neighborhood of the switch.   To some extent, a low standing-wave ratio and matched impedances help to reduce such radiated interference.   When there is *line-conducted interference* (LCI), it may become necessary to add a filter network to the input terminals of the power supply.[21]

In a balanced configuration, common mode chokes are combined with capacitors to obtain an effective power-line filter (Fig. 11.2). The filter can be designed so as to create a deliberate mismatch of

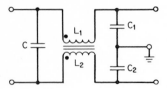

Fig. 11.2   Common mode choke.

its impedance to the load at frequencies in the RFI band.* In low-power circuits, ferrite beads are strung onto wires which conduct digital signals. These reduce noise currents above 1 megahertz (MHz).

11.4. THE DESIGN OF THE CORE

We have noted, in the previous two chapters, that a high cutoff frequency and fast rise time are obtained with a small $L_L C_D$ product. We have also observed that this makes a physically small transformer an actual design objective. However, a low cutoff frequency is necessary to the transmission of a wide pulse, and thus a lower bound on size is established. This is a consequence of certain constraints on the transformer's minimum shunt inductance L, maximum practicable permeability μ, and the maximum flux density B in the core. As we shall see in the course of discussion, these constraints can sometimes be lessened.

a. Core Parameters

It is desirable to connect the variables L and B with the stored magnetic energy and the size of the core. Let us start with an expression arrived at in our analysis of the energy storage inductor:

$$\frac{LI^2}{V_{Fe}} = \frac{10^{-8}\,\beta B_o}{0.4\pi} \qquad (7.20)$$

Here LI^2 is proportional to stored energy W, and V_{Fe} is the volume of the core. B_o is the operating flux density in a core whose gap ratio β equals the reciprocal of μ_e, the effective permeability.

An alternative basic equation expresses energy density (stored magnetic energy per unit volume) in terms of flux density and permeability:

$$\frac{W}{Al} = \frac{1}{2}\frac{B^2}{4\pi\mu}10^{-7} \qquad (11.20)$$

where the energy W equals $LI^2/2$, and the volume of the core V_{Fe} equals the product of the core area A and the length of the magnetic path l.

We can extend the usefulness of this expression by relating it to the four circuit parameters—pulse width τ, load voltage V_2, load resistance R_2, and load power $P_L = V_2{}^2/R_2$—and to two transformer variables—L and B (or, more strictly, ΔB, the incremental flux density).

---

* Limits on RFI are listed in standards set forth by the FCC (Federal Communications Commission) and the VDE (Verband Deutscher Elektrotechniker).

Pulse energy delivered to the load is $W_L = P_L \tau$. The relation between input voltage V_1, flux ΔBA, and τ is:

$$V_1 = \frac{N \, \Delta BA}{\tau} 10^{-8} \tag{10.45}$$

If we now eliminate the turns variable N from Eqs. (6.7) and (10.45) and solve for size Al (which equals volume), we obtain:

$$\text{vol} = \frac{(10^7/2)P_L \tau^2 \, (R_2/L)}{B^2/8\pi\mu} = \text{const} \frac{\mu}{L} \left(\frac{V_1 \tau}{\Delta B} \right)^2 \tag{11.21}$$

The denominator on the left will be recognized as the energy density [in ergs per cubic centimeter (ergs/cu cm)] given by Eq. (11.20).

Equation (11.21) is a most pregnant expression, since it describes size as a function of seven major parameters. We see that volume is inversely proportional to energy density but directly proportional to load power, pulse duration, and load resistance.

Volume is also inversely proportional to incremental flux density, a conclusion consistent with Eq. (4.12) and our earlier analysis of the conventional power transformer. High power usually entails operation at high flux density, as in the high-power wide-band transformer. If miniaturization is our objective, however, it will be easier to achieve if we operate with short pulses at low power into a low value of load resistance.

Equation (11.21) may also be recast into two other forms which let us see the relationship of the matching factor $m = R_2/R_1$ to droop $D \cong \tau/T_d$ (where T_d is the time constant of the top period) and to the volt-second product $V_1\tau$. One such expression is:

$$\text{vol} = \frac{(10^7/2) \, W_L(\, R_2/R) \, (\tau/T_d)}{\text{energy density}} = \frac{(10^7/2) \, W_L(1 + m)D}{\text{energy density}} \tag{11.22}$$

The other expression is derived from Eq. (11.22) by noting that $W_L = V^2\tau/R_2$ and $T_d = L/R$:

$$\text{vol} = \frac{(10^7/2)(V_1\tau) \, [(V_1/L)\tau]}{\text{energy density}} = \frac{(10^7/2)(V_1\tau)I_m}{\text{energy density}} \tag{11.23}$$

Here I_m is the magnitude of magnetizing current at the end of the pulse duration τ.

In each of the last three equations, the numerator depicts some combination of circuit parameters necessary to the transfer of pulse energy, as well as a finite top droop and a final magnetizing current at the termination of the pulse. In every case, core volume is proportional to the pulse energy to be transferred and inversely proportional to the density of the magnetic energy.

The foregoing discussion leads to the conclusion that the transformer designer should utilize a core material which can be operated at a high incremental flux density ΔB. We must note, however, that ΔB and μ are interdependent. It is not desirable to obtain a high ΔB by sacrificing μ, for this will affect shunt inductance and droop. The relations between ΔB, μ, and losses are of fundamental importance and are discussed further in the sections which follow.

b. Pulse Permeability

It has already been made clear (in Sec. 10.8) that the permeability of the core depends on the type of waveform that is applied—its shape, its symmetry, the duration of the pulse, and the frequency of its repetition. Hence, it is desirable to distinguish between sine-wave permeability (discussed in Sec. 6.1) and pulse permeability μ_p, the permeability of the core under excitation by pulses.

The choice of a suitable core material is also intimately connected with the power losses in the core (discussed more fully later, in Sec. 11.4d). Because of its pertinence here, we shall state the customary guideline. In high-power applications, when maximum power per unit volume is desired, grain-oriented silicon is favored. When power is low and PRF (expressed in *p*ulses *p*er *s*econd, pps) high, we use ferrites and powdered iron (see Table 11.2).

(1) The Square-Loop Core. Square-loop core material poses special problems. Suppose the remanent flux density B_r is only slightly less than the saturation density B_s and that the parameters N and A in Eq. (10.46) are such that B_s is reached at time $\tau_1 < \tau$. The slope $\Delta B/\Delta H$, and therefore L_P become almost zero at τ_1 and the output voltage will drop abruptly to zero at τ_1+.*

A square-loop core is disadvantageous for faithful transmission through a single-ended transformer because its μ is inherently low. But suppose that the pulse train contains symmetrical positive and negative pulses and that push-pull drive is used. Now the flux density within the core will traverse a range from $+B_r$ to $-B_r$. This results in high μ, but high B_r creates a memory characteristic. For example, if the last pulse in a train is positive and leaves the core in a $+B_r$ state, and if the first pulse in a subsequent train is also positive, the core response, $B_s - B_r$, will be small and the droop will be large. This dependence of the response on the prior history of the core is exploited in nonlinear digital magnetic devices which perform storage,

---

* This phenomenon is exploited in the dc inverter, which uses transistors connected to a transformer whose constants are deliberately selected so that B_s will occur at τ_1. Such circuits transform zero frequency, or dc, of a battery into a periodic square wave.

TABLE 11.2 Core Materials for Pulse Transformers

Material	Typical shapes	Typical circuit conditions	Pulse width, μsec	PRF, pps
Supermendur (cobalt iron)	Wound toroid (2- and 4-mil)	Rest bias desirable; low voltage, medium power	1–1,000	To 1,000
Nickel-iron alloys	U-I, D-U, L-L laminations (6-mil)	Reset bias desirable; high voltage, high power	2–1,000	To 1,000
	Wound toroid (1-, 2-, and 4-mil)	Medium voltage, medium power	1–10	To 2,000
Metallic glass*	Wound toroid (1-mil)	Reset bias desirable	0.05–10	To 2,000
Silicon, grain-oriented	C core (1- and 2-mil)	High voltage	1/4–2	To 1,000
	D-U, U-I, L-L (4–6-mil)	High power and low power	2–1,000	To 1,000
Ferrites, high μ	Toroid, cup-core	Low power	0.1–10	10K–10M
	U-I core	Medium power	0.1–10	10K–10M
Ferrites, low μ; powdered iron	Toroid, cup-core	Low power	0.01–0.1	10M–100M

* METGLAS (Allied Corporation) ribbon is used as the core material of high voltage magnetic switches, and in pulse-forming networks for lasers and accelerators.[22] Also see Table 6.2.

counting, and logic functions. And when a discrete value of saturation density is the central principle on which the nonlinear performance is predicated, square-loop material is used on a toroidal geometry. In linear circuits, however, square-loop material is usually contraindicated.

(2) **Round-Loop Cores.** When strip material such as silicon, Supermendur, or nickel-iron is used, its thickness will depend on the pulse width. When there is voltage excitation, pulse permeability μ_p is proportional to incremental flux density ΔB but inversely proportional to the square of the thickness:

$$\mu_p = \frac{3.19\rho(NA\ \Delta B)}{\delta^2 V_1} \qquad (11.24)$$

where ρ is the strip restivity in ohm centimeters ($\Omega \cdot$ cm) and δ is the strip thickness in inches (in).[23]

Since $NA\ \Delta B$ can be equated with the volt-second product, μ_p should be directly proportional to the pulse width (provided the flux can penetrate the strip). Table 11.3 is Reuben Lee's summary of values μ_p obtained with 2-mil Hipersil grain-oriented core over a wide range

TABLE 11.3 Pulse Permeability,
2-mil Hipersil

Pulse width, μsec	ΔB, G	μ_p
0.10	1,000	100
0.25	1,500	200
0.50	3,000	350
1.0	6,000	600
2.0	8,000	930
5.0	10,000	1,200
10	11,000	1,420
20	12,000	1,600

SOURCE: Reuben Lee, "Pulse Transformer Design Chart," *Electronic Equipment*, Vol. 7, No. 9 (September, 1957): 36. Copyright 1957, Sutton Publishing Co.

of pulse widths. Manufacturers of silicon strip C cores guarantee minimums of pulse permeability under typical conditions by performing standard tests.*

The fact that some ferrites have high pulse permeability (in the order of 1,000), even when the PRF is high, makes them particularly suited for low-power applications.

c. Optimum Permeability

Because pulse permeability tends to be low, considerable effort is expended to increase it. The type of drive is usually tackled first. *Balanced biopolar drive* (with a center-tapped primary) substantially reduces polarization and permits the core to be driven over the entire $\pm B_m$ range of the major hysteresis curve (see Fig. 10.7b). A core usually has a lower μ (typically 300 to 600) when it is subjected to rectangular pulses rather than to sine-wave excitation.

It is important to try to increase the slope of the minor *B-H* loop, shown in Fig. 10.19c. This is usually done in one of the following ways:

1. The selection of an *optimum air gap l_a* by a procedure analogous to that used to establish maximum incremental permeability in audio transformers and chokes (see Sec. 7.2 and Fig. 10.19d).

---

* Typical guarantees are: $\mu_p = 600$ for 2-mil Silectron, grain-oriented C core at 0.25 μsec, 2.5 kG, and 1,000 pps; $\mu_p = 300$ for 1-mil Silectron at 2 μsec, 10 kG, and 400 pps.

2. The establishment of a reset bias by means of a permanent magnet in the magnetic circuit.[24]

3. The introduction of a *reset pulse* of opposite polarity during the off time between pulses or, alternatively, the introduction of a *reset bias* by introducing a constant reverse current in series with the primary or in a tertiary winding (discussed in Sec. 7.8c).

(1) Optimum Air Gap.   When the transformer is excited by an asymmetrical train of pulses, the core becomes dc polarized. Consequently, as we have noted previously (Chap. 7 and Sec. 10.8), permeability is low. Just as there is an optimum gap for maximum incremental permeability when the excitation is a sine wave, so there is also an optimum air gap which will result in a maximum effective pulse permeability.

A proper air gap can reduce the magnetic circuit reluctance by reducing the remanent flux density B_r, thereby increasing ΔB more than ΔH. While it is easier to determine μ_p empirically than analytically, it is instructive to determine the gap ratio l_a/l from theoretical considerations. A high average dc incremental permeability is desirable. In Sec. 7.2 we noted that the introduction of an air gap tilts the hysteresis loop, thereby modifying the incremental permeability. A similar analysis of the pulse circuit shows that the air gap has been adjusted to produce maximum permeability when the following expressions have been satisfied:

$$\frac{l_a}{l} = \frac{H_c}{B_r} \qquad \left(\frac{l_a}{l}\right)_{\text{opt}} = \frac{4H_c}{B_s} - \frac{1}{\mu_{\text{dc}}} \qquad (11.25)$$

where μ_{dc} represents the average dc permeability, H_c is coercive force, and B_r/H_c is the slope of the side of the hysteresis loop.

If F represents the squareness ratio B_r/B_s of the core, then $B_r = FB_s$ and Eq. (11.25) becomes:

$$\left(\frac{l_a}{l}\right)_{\text{opt}} = \frac{1}{\mu_{\text{dc}}}(4F - 1) = \frac{H_c}{B_s}\left(4 - \frac{1}{F}\right) \qquad (11.26)$$

When Hipersil, for which $F \cong 0.75$ and $\mu_{\text{dc}} \cong 10^4$, is used:

$$\left(\frac{l_a}{l}\right)_{\text{opt}} \cong 0.2(10)^{-3} = 0.02 \text{ percent}$$

Since the typical Hipersil C core has an effective total gap length of 0.001 in, an optimum gap ratio can be obtained only when the mean core length l exceeds 5 in. In miniature cores, thus, we do not obtain sufficiently small gaps. In general, a production run of pulse transformers will exhibit large variations in μ_p because of the difficulty of building them with uniformly small air gaps.

When permeability must remain very stable over a range of temperatures, the ferrite core must be built with a larger gap than the silicon core (see Sec. 6.4d). As a consequence, silicon or nickel-iron strip is preferred when the operating temperature range is extreme. Since saturation also varies with extremes of temperature, adequate circuit performance can depend on both a judicious choice of core material and a suitable air gap.

The need for a definite or precise air gap, either to maximize or to stabilize the temperature dependency of μ_p, can radically alter our choice of the core geometry. Since this is especially so for miniature transformers, the feasibility of precision grinding of ferrite surfaces has made the cup-core and U-I assemblies especially attractive even when other considerations might lead us to prefer the wound strip C core or the U-I lamination.

(2) Reset Bias. When flux density must necessarily be high, as in high-voltage high-power pulse transformers, then fast rise time can be obtained only if the coil is small. However, when pulses are unidirectional, permeability is low.

A large reduction in coil and core size does become feasible if a large increase in permeability is feasible. The designer must try to overcome the effects of remanence or residual bias. One way is to employ a reverse or reset bias, so that the bias becomes steady direct current. An alternative technique is to use pulses during the off period and introduce, via a tertiary (clamp) winding, reset pulses of the proper polarity and magnitude. Then one can, by designing appropriate circuitry, use a toroidal geometry which operates with very large swings of flux.[25]

Figure 11.3 shows how a negative reset bias, $|H_p| > H_r$, can produce a much larger flux swing ΔB than the swing ΔB_g produced by a conven-

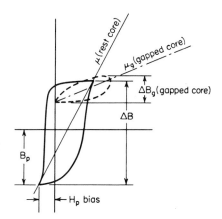

Fig. 11.3 Reset bias. Reset core (solid line) $\Delta B < 2B_p$. Gapped core (dashed curve).

tional core gap. Reset techniques effectively change the remanent flux density to a $-B_r$ value. Consequently the effective flux change, now $B_m - (-B_r)$, produces a much higher μ of $\Delta(B_m +B_r)/\Delta H$. This results in higher pulse permeability.

Melville, in his thoroughgoing analysis, sets forth three basic reset conditions:[26]

1. $H_r = 0$ $\Delta B < 2B_p$
2. $H_r = 0$ $\Delta B > 2B_p$
3. $H_r > 0$ $\Delta B > 2B_p$

In the last of these cases, the residual positive bias occurs when the off time between pulses is comparable to the time constant of the load. The other two cases result in physically lower values of ΔB than the desirable maximum $2B_p$. This is because the lower tip of the minor loop rises with each successive pulse. The first case is the preferred operating condition, and there is an optimum negative bias for this case.

The reset techniques make it possible to obtain pulse permeabilities exceeding 2,000 at 2 μsec and 5,000 at 6 μsec when 2-mil uncut Hipersil and Supermendur are used. Flux excursions exceeding 22 kG become practical. Melville has shown that it is feasible to obtain a pulse permeability of 5,000 at 1 μsec ($\Delta B = 7,500$) with reasonably low core loss per cycle by using 1-mil 80 percent nickel (Mumetal) strip.

Lord describes the use of grain-oriented nickel-iron to obtain a μ_p of 2,500 at 1 μsec ($\Delta B = 20,000$) and a μ_p of 14,000 at 10 μsec ($\Delta B = 25,000$).[4] The consequences are impressive. The core's cross section can be reduced to a size which results in a 200-kW power rating with a rise time of 0.2 μsec and a pulse width of 10 μsec.*

d. Power Losses and Resolution

Efficiency is most conveniently studied in terms of power losses, which fall into three groups: copper, dielectric, and core. In the pulse transformer, the core losses will ordinarily predominate, because most pulse transformers are operated with pulses shorter than 100 μsec. Since the turns are comparatively few, the copper resistance is markedly smaller than that of the audio, or power, transformer, which operates below a few kilohertz. Hence, winding resistance can often be ignored in the analysis of the pulse transformer. Dielectric losses, which are proportional to the loss tangent of the insulation and the

---

* High pulse permeability can be achieved in the pulse-forming network in which core flux is reset, between pulses, by the recharging current.

energy stored in the coil $C_D V^2/2$, are also small. They become appreciable, however, when kilovolts of voltage are involved.

Power losses in a core vary directly with the duty cycle. ($\tau \times$ PRF) and inversely with the pulse duration τ. They are also proportional to the core volume and inversely proportional to pulse permeability (which is low). Core loss may be calculated with the aid of manufacturers' charts which plot ergs per cubic centimeter per pulse vs. τ.

The measurement of the power loss in the core is not a simple matter. It requires an accurate simulation of the excitation: either a constant rate of flux change (equivalent to a triangular current drive) or a variable rate of flux change (equivalent to a sine flux drive.) It is of practical, as well as theoretical, interest to compare losses under sine-wave excitation with losses due to excitation by square and rectangular pulses. The analysis of oscillographs shows that sine-wave excitation produces greater losses of power in eddy currents than does square-wave drive, while losses due to hysteresis are independent of the wave shape.[27] Measurements show that total power losses in the core are smaller when square-wave excitation is used than when sine-wave excitation is used.

If we invoke the hypothesis that power losses due to eddy currents are proportional to the root-mean-square (rms) magnitude of the frequency components, then:

$$\frac{\text{Sine-wave eddy current loss}}{\text{Triangular-wave eddy current loss}} = \frac{1/\sqrt{2}}{1/\sqrt{3}} = 1.225$$

In sheet magnetic materials, one can expect power losses to be about 20 percent less under square-wave excitation. However, power losses due to hysteresis predominate over eddy current losses in ferrites, and total power losses in the core are only 7 to 12 percent less when they are subjected to square waves.[28] In the design of pulse transformers it is therefore possible to use sine-wave charts (with a suitable correction factor) as a source of data about power losses in the core.

Analyses of toroidal transformers show that pulse permeability is inversely proportional to the lamination time constant T_k and an eddy current function F:[29]

$$\mu_p = \frac{\mu\Delta}{1 + (T_k/\tau)F} \tag{11.27}$$

$$T_k = (\mu_o \mu\Delta\sigma\delta^2)/12 \tag{11.28}$$

Here, $\mu_o = 4\pi(10)^{-7}$, $\mu\Delta$ is the low-frequency incremental permeability, and τ is the pulse width. F is an eddy current function (related to τ/T_k) which is determined by tests on specimens of strip. T_k is a function of the conductivity σ of the material (the reciprocal of resistivity) and the thickness δ of the tape. The use of medium nickel-iron strip 50-μm (2-mil) thick results in a time constant of 1.7 μsec and values of permeability on the order of 2,200 for a 5-μsec pulse and 3,000 for a 20-μsec pulse.

(1) Power Losses in Core Materials. In Table 11.2 we listed the most popular core materials, the most common shapes employed, and typical operating conditions. Several materials are of overlapping suitability—silicon-iron strip and ferrite, for example.

The shortest pulse for which a strip material should be used can be estimated from the maximum depth of flux penetration or from estimates of the magnitude of a lamination time constant. (See also the discussion of the magnetic skin effect of Sec. 6.1e.)

One-mil silicon strip is not generally used when pulses are briefer than $1/4$ μsec, and the PRF is usually limited to a few thousand pps. To use it when the repetition rate is higher would be to risk overheating the core, although this can be dealt with, when necessary, by special methods of heat transfer such as forced circulation (see Sec. 4.5). Thus, when the PRF is to be high, ferrite cores are used because of their very high resistivity. On the other hand, when high power is required at the peak of the pulse, silicon cores are indicated because they can be operated at high flux density; and high flux density makes it possible to build a smaller transformer, which is advantageous in itself. We have already noted that in the power pulse transformer the power losses in the core predominate over those in the copper. This is in contrast to the low-frequency power transformer operated at about 60 Hz in which power losses in the copper predominate. The coil of the typical power pulse transformer has a small temperature rise, and it is possible to achieve a much higher ratio of power to weight than can be achieved in the conventional low-frequency power transformer.

When the power level and PRF are low and the pulses are longer than 0.2 μsec, silicon and ferrite may perform equally well. Under these conditions, designers incline toward the lower-cost ferrite. A useful qualitative guide is to choose silicon when flux density is high and PRF low, and to choose ferrite when flux density is low and PRF high.

(2) Switching Time and Resolution. Resolution is sometimes affected by our choice of a particular grade of ferrite. This is likely to occur when the duty cycle is high and the PRF is very high, in the order of *mega*pulses *per* *second* (Mpps). If the core is large, we choose a ferrite with a low loss tangent because of its low losses (see Sec.

6.4d). Such a material has a low permeability and a high μQ product. Since the loss is proportional to $fB^2/\mu Q$, this is a good choice.

Good resolution implies that there is sufficient time between pulses for B_m to fall to B_r. If the time is too small, the top droop can blur the resolution. Since ferrites of low μ have a faster recovery time than those of high μ, we reverse our customary practice and choose a low-μ ferrite when the PRF is very high. The upper bound of the PRF appears to depend ultimately on the switching time of the core, an observation that has spurred the development of core materials to be used in high-speed timing circuits and high-density digital storage schemes such as the magnetic bubble memory system.

The time required to switch between discrete levels of flux (e.g., B_r to B_s) is a measure of the resolution (and therefore the accuracy) of the individual pulses in a train. The *switching time* t_s, therefore, sets a limit on the maximum rate of transmission and storage of digits. Tests performed on miniature square-loop (bobbin) cores show a simple linear relation between the reciprocal of switching time and the drive current or coercive force H_c:[30]

$$\frac{1}{t_s} \propto (\text{const})\, H_c \tag{11.29}$$

A large slope [the constant in Eq. (11.29)], as is shown in Fig. 11.4, means that a small increment in the drive current produces switching. Note that a minimum (threshold) level of current is needed to switch the core.

When the PRF is around 0.1 Mpps (megapulses per second), *thin metallic tapes* and *metallic glass* (½ to 2 mil) are used;[31] when the PRF is around 0.1 Mpps, *ultrathin metallic tapes* (⅛ to ¼ mil) and

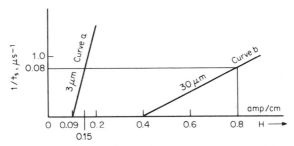

Fig. 11.4 Reciprocal switching time vs. magnetizing force. Square-loop nickel alloy. Curve a, 80 percent nickel alloy. Curve b, 50 percent nickel alloy. (*Based on R. Boll, ed., Soft Magnetic Materials, London: Heyden & Son, 1979, p. 257.*)

ferrites are used; and when it is between 10 and 100 Mpps, deposited metallic films (about 4 μin) are used.

11.5. THE DESIGN OF THE COIL

In the pulse and wide-band transformer, the geometry of the core and the geometry of the coil are mathematically, as well as physically, intertwined. When our objective is to effect a maximum transfer of energy to the load (see Sec. 11.3), we arrive at several conclusions about the geometry of the coil.

a. Transfer of Energy

If Eq. 11.22 is solved for load energy per unit volume, we obtain:

$$\frac{W_L}{Al} = 2\left(\frac{10^{-7}(\Delta B)^2}{8\pi\mu}\right)\frac{R}{R_2}\frac{T_d}{\tau} = 10^{-7}\frac{(\Delta B)^2}{4\pi\mu}\frac{L}{R_2}\frac{1}{\tau} \qquad (11.30)$$

We can see from this formula that a large transfer of pulse energy is facilitated by very brief pulses, small load resistance, high inductance, and high incremental flux density. But further analysis is of value.

One can approach the problem of designing a transformer for maximum transfer of pulse energy without preconceived notions about the relative duration of rise, droop, or fall times. The approach is basic, since all conclusions arise from energy considerations.[15]

When characteristic impedance is expressed as a function of coil geometry and the density of energy in the coil is made proportional to that in the core at the end of the pulse, an expression for the optimum number of turns is:

$$N_{\text{opt}} = K_N \left(\frac{\tau}{\sqrt{\mu\epsilon}\,d_2}\right)^{1/2} \qquad (11.31)$$

where ϵ is the dielectric constant of the insulation and d_2 is the diameter of the wire. The coefficient K_N combines three constants:

F_1, the voltage distribution factor arising from a particular winding configuration and turns ratio

F_3, the degree of equality between core and coil energy densities

F_4, the linear space factor associated with the spacing of turns on the secondary winding

Lord has shown that the concept of a turns index is a convenient starting point in the design of high-power pulse transformers.[32] When the turns ratio $N > 3$, for example, then:

$$N_2 = \left(\frac{10 \,\Delta B \sqrt{L_L' C_D'}}{I_m' \mu \sqrt{\epsilon}}\right)^{1/2} \tag{11.32}$$

where the primes mean that the parameters are referred to the turns of the secondary winding N_2. L_L', C_D', and I_m' are dictated by the required circuit performance; ΔB, μ, and ϵ, which are associated with the core and coil materials, are chosen by the transformer designer.

From the geometric formula for Z_0 [in Eq. (9.28)] and the required spacing between windings, an expression for core area A as a function of turns N_{opt}, load R_2, dielectric constant ϵ, and wire diameter d_2 are obtained. When A is substituted into Eq. (10.45), we obtain the following expression for optimum incremental flux density:

$$\Delta B_{opt} = \frac{V_2 \tau 10^8}{N_{opt} A} = K_B \frac{V_2}{R_2} \frac{\sqrt{\mu}}{d_2} \tag{11.33}$$

Here V_2 is the secondary pulse voltage appearing across the load R_2. The coefficient K_B combines the constants F_3 and F_4, as well as a factor F_2 which relates L_L to the winding configuration. A significant conclusion is that ΔB varies with the magnitude of the pulse voltage. A high pulse voltage and, therefore, high pulse power, implies the desirability of a large swing in flux density; low pulse power implies the desirability of a small change.

When Eq. (11.31) is combined with an expression for core area A, the following expression for core volume results:

$$\text{vol} = K_{vol} R_2^{3/2} \tau^{3/4} \left(\frac{d_2^{9/4} \epsilon^{3/8}}{\mu^{3/8}}\right) \tag{11.34}$$

Here K_{vol} combines the constants F_1, F_2, F_3, and F_4.

Equation (11.34) is the end result of optimization. It is gratifying to note that, when it is compared with the three previous equations for core size, the desirability of low load resistance, very brief pulses, high core permeability, and a low dielectric constant for the insulation is confirmed.

In Chaps. 8, 9, and 10, we noted that a low $L_L C_D$ product is advantageous because it results in a high cutoff frequency and fast rise time. We also noted (in Sec. 9.4d) the mutual interdependence of L_L and C_D. L_L is directly proportional to d/W_L (the quotient of the spacing between windings and the winding length), while C_D is proportional to W_L/d. A decrease in L_L produces an increase in C_D, and vice versa. Since the product $L_L C_D$ is proportional to $N_2 \lambda$, the length of wire in the winding, fast rise and high cutoff frequency are obtained with a physically small coil.

Characteristic impedance, we recall, can be expressed in terms of coil geometry. [see Eq. (9.24)]. In the single-layer construction:

$$Z = \sqrt{\frac{L_L}{C_D}} = 377 \sqrt{\frac{F_2}{F_1}} \frac{dN_2}{W_L \sqrt{\epsilon}} \qquad (11.35)$$

Just as in the design of the wide-band coil, we can strive for a specified impedance by adjusting the parameters d, W_L, and N_2, but this is attempted only when we are synthesizing an optimum filter model.

b. The Geometry of the Coil

In designing the coil of the pulse transformer, we follow essentially the same procedures we use for designing the wide-band transformer (Sec. 9.4). In the bipolar pulse circuit (the counterpart of the push-pull sine-wave circuit), however, good resolution requires that we take care to reduce the magnitude of a spike voltage on the leading edge of a square waveform. Hence we need to provide close coupling between each half of the primary winding. This is frequently accomplished with the bifilar winding technique, which results in a very small leakage inductance between the primary windings.

In the design of small, high-speed pulse transformers, the design objective of a flat group delay requires a minimum of reflections as well as a high degree of resolution. A logical choice of coil geometry is the transmission line, described in Sec. 9.5.

11.6. THE DESIGN OF A PULSE TRANSFORMER

Procedures for designing the pulse transformer, in the main, parallel those for designing the wide-band transformer. Since the volume of the core is proportional to $(\epsilon/\mu)^{3/8}$, we can use this relationship as a rough gauge of the extent to which the pulse transformer's performance can be enhanced by basic improvements in magnetic and dielectric materials.

In practice, the important circuit objectives are: (1) minimum size; (2) minimum rise time; (3) maximum bandwidth and span ratio (or pulse width-to-rise ratio); and (4) the resolution of consecutive pulses, if the PRF (or the clock rate of the computer) is high.

The first two objectives reinforce each other, since minimum size favors fast rise. And the size of a transformer is, of course, related to its geometry. For a given filter model and a specified impedance and voltage level, there is an appropriate coil geometry.[33] The coil design, including the window aspect ratio, is affected by essentially the same considerations which go into the design of the wide-band transformer (see Secs. 9.4 and 11.5).

The last two objectives require that we choose a suitable material and a suitable shape for the core.

On occasion, the coil is small enough to justify the more novel technique of mechanically interchanging the roles of coil and core. The core, in the form of ribbon, is then wound spirally through the coil. Another interesting departure from standard practice is the transmission-line transformer. In this transformer, a bifilar coil winding produces a response which has a minimum of reflections. The physical dimensions of the connecting wires in the circuit are comparable to the wavelength of the pulse. The technique is most suitable for pulses which last only nanoseconds. Figure 11.5 illustrates the characteristic smallness of the nanosecond-pulse transformer, about the size of the point of a lead pencil.

If the impedance ratio is small, the transmission-line geometry becomes an attractive alternative to the standard coil. Values of L_L and C_D in a coaxial cable are uniform (and consistent) and the characteristic impedance is accurately known; hence, designing the coil is a simple matter.[34] (See the example given in Sec. 9.5c.)

a. The Design Procedure

In the design of the pulse transformer, as in the design of other types of transformers, the first step is to *review the specifications*. It is important, particularly when one is to design a nonlinear switching circuit, to make an accurate assessment of the behavior of the source and load resistances. If errors of judgment have been made in these specifications, our prediction of fall time and tail amplitude may be affected. Once satisfied of the internal consistency of the specifications, the designer can turn to the task of establishing the *feasibility* of the requirements. Then the design can begin.

Fig. 11.5 Nanosecond pulse transformers. (*Courtesy Sprague Electric Company.*)

There is more than one way to design a pulse transformer. The most elegant, perhaps, is to attempt to synthesize the design directly (see Sec. 11.5a). Another is to use similitude theory and scaling techniques to derive the design from one of a group of proven designs. In this sensible approach, used here, tabulated information provides a springboard. Our guideline is to strive to progressively reduce the size of the transformer (since we know that small size facilitates good overall transient performance) until we encounter one or more of the basic *magnetic, thermal,* or *dielectric* limitations: (1) the maximum practical operating flux density, (2) the maximum permissible temperature rise (which affects our choice of core material), and (3) the maximum voltage gradient which is safe for the insulation system to be used (which affects our choice of coil geometry).

We begin by establishing the maximum ratings of a standard set of cores. Such a focus on standardization has the advantage of producing an economical set of designs which will result in transformers that perform consistently. Since many pulse transformers must handle substantial levels of power, the initial procedure is similar to that used in evolving a table of ratings for the conventional power transformer (see Sec. 5.3).

Each core and coil assembly is capable of dissipating a specific maximum amount of heat and supporting a maximum voltage, depending, respectively, on the coil's surface area and on the nature of its insulation and margins. Accordingly, each core is assigned a maximum power rating for a pulse of standard duration. The rating is based on the maximum permissible temperature rise and winding voltage. A standard class of insulation and type of coil construction are assumed.

The next task is to determine the number of turns in the high-voltage winding necessary to provide a volt-second rating and inductance adequate to yield a specified droop for the given pulse duration. We want, if possible, to arrive at a single-layer winding. In addition, we need to assign an incremental flux density which is suitable to the material and pulse duration. A set of values for oriented silicon has been listed in Table 11.3. If we insert these into Eq. (10.45), we can determine N_2, the number of turns in the high-voltage winding. Now we can assign a maximum volt-microsecond product to each core. It should be noted that ΔB is not linearly proportional to τ. The volt-microsecond product of each core is therefore not strictly invariant, but is constant over only a narrow range of pulse durations.

Once a standard number of turns is associated with a core, it becomes possible to compute the minimum shunt winding inductance and the minimum leakage inductance. From the coil geometry and the standard pad thickness (the interwinding spacing), we can compute C_D.

Since C_D is a complex function of the turns ratio and polarity, it is convenient to compute it for the arbitrary condition of $n = -1$ (unity turns ratio, with polarity inverted).   Finally, it is possible to assign to each core a minimum $L_L C_D$ product based on N_2 and a standard coil geometry.

From a knowledge of L_P and $L_L C_D$, we can determine the maximum L_P/L_L ratio and the pulse-squareness ratio $\tau_{0.1}/t_r$ for a standard circuit impedance.*

Thus it is possible to assign to each core (given a standard coil geometry and turns ratio) the following major ratings:

1. Maximum average pulse power
2. Maximum winding voltage
3. Maximum volt-microsecond products for a range of standard pulse durations
4. Maximum ratio of shunt inductance to leakage inductance
5. Minimum $L_L C_D$ product

It is our intention to choose the smallest core which appears to satisfy, simultaneously, all the requirements.

The number of design steps and calculations can be considerable. Thus, when a large number of stereotyped designs are to be developed, the programmed digital computer can be of help.

b. Design Example: The Horizontal Flyback Transformer

We have chosen to discuss the TV flyback transformer again because its design calls into play virtually the entire spectrum of considerations in the design of a wide-band or pulse transformer.

We described this circuit in Sec. 10.9 and noted that its basic function is to provide a sawtooth of current to the yoke, which results in the horizontal deflection of the electron beam on the raster (screen) of the cathode-ray tube (CRT).   The ramp-shape current during the flyback (retrace) interval produces an output voltage (Fig. 10.26) which can be estimated from Eq. (10.68).

A perusal of the literature on the TV flyback transformer reveals the diverse approaches to its design.[35,36]   The problems we describe here have been chosen for their basic relevance to the design of a pulse transformer as well as to the design of a high-frequency, high-voltage transformer.   The significant problems and tasks which face the designer can be enumerated as follows:

---

* $\tau_{0.1}$ is the pulse width corresponding to a top droop of 10 percent.

1. Limitations on the voltage step-up ratio (see Sec. 9.2b). In the horizontal flyback circuit, a voltage in the order of 30 kV is needed, and a voltage tripler is chosen to reduce the need for a very large step-up ratio of turns and also to ease the task of dealing with the spectre of corona.

2. Dielectric reliability. The high-voltage coil will be sliced into pie sections (tires) in order to reduce the voltage gradient per layer of the winding (see Sec. 3.14) and its distributed capacitance C_D.

3. The need to eliminate ringing and to increase bandwidth. An effort to optimize the $L_L C_D$ product will require close attention to the geometry of the high-voltage coil (see Sec. 9.4d).

4. Improved regulation and timing. Much attention has been devoted in the literature to the concept of harmonic tuning and other ways to improve regulation of the deflection circuit when the current of the CRT beam varies widely with the magnitude of the video signal.[37]

5. Control of ringing with sectional windings. When high ac voltage is to be converted to dc voltage, one solution to the problem of ringing is to use individual diodes in series with each section (ideally a single layer) and divide the load branch into a serial string of capacitors (see Fig. 11.6a). This technique is used in high-voltage dc converter circuits.[38]

6. Stray capacitance. There are several schemes which enable the designer to reduce the undesirable effects of stray capacitance in a

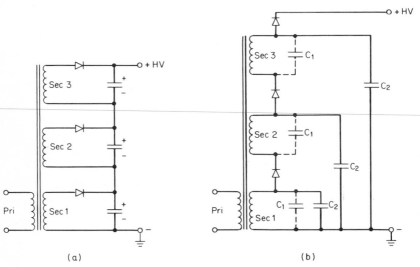

Fig. 11.6 Rectifier connection with sector windings. (*a*) Additive sectors, preferably on single layers. (*b*) Tripler circuit; each secondary winding is divided into four to six sections.

voltage multiplier circuit.[39]   One scheme, shown in Fig. 11.6b, has been used in a TV flyback transformer.   In this design, self-capacitances C_1 and stray capacitances C_2 are integrated into the topology of the voltage tripler circuit.   For clarity, the figure is shown with but three windings; actually, each winding represents a group of four to six sections which, in this example, results in a total of sixteen sections.[40]

c. The Limits of Design

Our guideline in design is to strive for progressive reduction in the size of the transformer (since we know that small size facilitates good overall transient performance) until we encounter one or more of the basic magnetic, thermal, or dielectric limitations: (1) the maximum practical operating flux density, (2) the maximum permissible temperature rise (which affects our choice of core material), and (3) the maximum voltage gradient which is safe for the insulation system to be used (which affects our choice of coil geometry).

When limit-of-the-art performance is desired, we reexamine our choice of core material and reevaluate our coil and core geometry. Next we reexamine our basic concepts and review the optimum techniques of network theory.   And still, when the component will be complex or costly, we often seek compromises in the interest of economy, the all-pervasive desideratum.

The suitability of a specific design is confirmed by testing it in the circuit in which it is to be used.   However, it is usually advantageous to test pulse transformers under standard, readily reproducible conditions which have been agreed on by both the customer and manufacturer.

REFERENCES

1. L. Giacoletto, *Electronics Designers' Handbook* (New York: McGraw-Hill Book Company, 1977), Sec. 23.4g, pp. 94–109.
2. D. Fink, *Electronics Engineers' Handbook* (New York: McGraw-Hill Book Company, 1975), Sec. 14-33 and 15-33.
3. G. Simcox, "The Evolution of Pulsed Power," *IEEE International Pulsed Power Conference Proceedings* (Lubbock, Tex., 1976), Vol. 76–CH1147–8, Region 5.
4. H. Lord, "Pulse Transformers," *IEEE Transactions on Magnetics*, Vol. MAG-7, No. 1 (March 1971): 17–28.
5. W. Geyger, *Nonlinear-Magnetic Control Devices* (New York: John Wiley & Sons, 1964).
6. J. Watson, *Applications of Magnetism* (New York: John Wiley & Sons, 1980), pp. 293–98.

7. J. Millman, H. Taub, *Pulse Digital and Switching Waveforms* (New York: McGraw-Hill Book Company, 1965), Chap. 16.

8. K. Busch, A. Hasley, C. Neitzert, "Magnetic Pulse Modulators," *Bell System Technical Journal*, Vol. 34 (September 1955): 943–93.

9. M. Weiner, L. Silber, "Pulse Sharpening Effects in Ferrites," *IEEE Transaction on Magnetics*, Vol. MAG-17, No. 4 (July 1981): 1,472–77.

10. H. Hart, R. Kakalec, "A New Feedback Controlled Ferroresonant Regulator Employing a Unique Magnetic Component," *IEEE Transactions on Magnetics* (September 1971): 571–74.

11. P. Hunter, "Thyristor Controlled Ferroresonant Regulator Utilizing a Double Shunt Magnetic Structure," *IEEE Proceedings: Workshop on Magnetics*, Vol. 72CH-629–6-MAG (May 1972).

12. D. Leppert, "A High Frequency Ferroresonant Transformer," ibid.

13. W. Dull, A. Kusko, T. Knutrud, "Pulse and Trigger Transformers," *EDN* (August 20, 1976): 57–62.

14. L. Gyugyi, B. Pelley, *Static Power Frequency Changers,* (New York: John Wiley & Sons, 1976).

15. W. Bostick, "Pulse Transformers," in *Pulse Generators,* R. Glasoe, G. Lebacqz (eds.) (New York: McGraw-Hill Book Company, 1948), Chap. 13.

16. R. Matick, *Transmission Lines for Digital and Communication Networks* (New York: McGraw-Hill Book Company, 1969).

17. J. Storer, *Passive Network Synthesis* (New York: McGraw-Hill Book Company, 1957), pp. 116–18.

18. H. Bode, *Network Analysis and Feedback Amplifier Design* (Princeton, N.J.: D. Van Nostrand Company, 1945), pp. 220–22, 286–90, 310, 312–18, 322–25, and 343.

19. M. Schwartz, *Information Transmission, Modulation, and Noise,* 3d ed. (New York: McGraw-Hill Book Company, 1980), pp. 183–88 and 229–30.

20. R. de Buda, J. Vilcans, "Limitations of the Output Pulse Shape of High Power Pulse Transformers," *IRE National Convention Record*, Part 8 (1958), pp. 87–93.

21. E. Hnatek, *Design of Solid-State Power Supplies,* 2d ed (New York: Van Nostrand Reinhold Company, 1981), Chap. 8.

22. J. VanDevender, R. Reber, "High Voltage Magnetically Switched Power Systems," *IEEE International PULSED Power Conference* (Albuquerque, New Mexico, June 1981), pp. 256–261.

23. A. Ganz, "Applications of Thin Permalloy in Wide-Band Telephone and Pulse Transformers." *Transactions AIEE*, Vol. 65 (April 1946): 177–83.

24. F. Assmus, "Mixed Cores with Displaced Hysteresis Loops and High Pulse-Induction Ranges," *IEEE Transactions on Magnetics*, Vol. MAG-14, No. 5 (September 1978): 987–89.

25. Reuben Lee, "Properties of Reset Cores in Radar Pulse Transformers," *Journal of Applied Physics*, Vol. 33, No. 3, Suppl. (March 1962): 1,261–62.

26. W. Melville, "The Measurement and Calculation of Pulse Magnetization

Characterization of Nickel Irons from 0.1 to 5 Microseconds," *Proceedings IRE,* Vol. 97, Part 2 (London, 1950): 165–98 and 229–34.

27. J. Triner, "Analyze Magnetic Loss Characteristics Easily Using a High Power Wideband Operational Amplifier," *Power-conversion International,* Vol. 7, No. 3, (March 1981): 55–63.

28. D. Chen, "Comparisons of High Frequency Magnetic Core Losses under Two Conditions: A Sinusoidal Voltage and a Square-Wave Voltage," *IEEE PESC 1978 Record,* Vol. 78CH 1337-15 AES: 237–41.

29. P. Hooke, A. Piercy, "The Prediction of Pulse Permeability and Loss in Pulse Current Transformers," *Journal of Physics: Applied Physics* (Great Britain), Vol. 11 (April 1978): 937–42.

30. R. Boll (ed.), *Soft Magnetic Materials* (London: Heyden & Son, 1979), pp. 251–58.

31. C. H. Smith, M. Rosen, "Amorphous Metal Reactor Cores For Switching Applications," *Proceedings, International PCI Conference,* (Munich, September 1981): 13–28.

32. H. Lord, "A Turns Index for Pulse Transformer Design," *Transactions AIEE,* Vol. 71 (1952): 165–68.

33. V. Merchant, H. Seguin, J. Dow, "Novel Transformer Designs for High-power High-repetition-rate Applications," *Review of Scientific Instruments,* Vol. 50, No. 9 (September 1979): 1,151–53.

34. R. Dollinger, "Design of Coaxial-cable Pulse Transformers," *IEEE Transactions on Electron Devices,* Vol. ED-26, No. 10 (October 1979): 1,549–51.

35. E. Cherry, "Third Harmonic Tuning of E.H.T. Transformers," *Proceedings Institution of Electrical Engineers,* Part B (London, March 1961), pp. 227–36.

36. J. Tucker, "Computer Aided Design of Horizontal High Voltage Transformers," *IEEE Transactions on Broadcast and TV Receivers,* Vol. BTR-16, No. 2 (May 1970): 112–18.

37. R. Woodhead, "The Influence on EHT Regulation of the Harmonic Content in the Retrace Voltage of a TV Horizontal Output Stage," *IEEE Transactions on Consumer Electronics,* Vol. CE-21, No. 1, (February 1975): 20–31.

38. L. Jansson, "A Survey of Converter Circuits for Switched-Mode Power Supplies," *Mullard Technical Communications,* Vol. 12, No. 119 (July 1973).

39. H. Roth, "Applying Switching Technology to Low Current, Regulated HV Supplies," *Powercon I Conference Proceedings,* 1975, pp. 170–74.

40. R. Takeuchi et al., "Multi-Stage Singular Flyback Transformer," *IEEE Transactions on Consumer Electronics* (February 1977): 107–12.

Index